Applied Nanophotonics

With full color throughout, this unique text provides an accessible yet rigorous introduction to the basic principles, technology, and applications of nanophotonics. It explains key physical concepts such as quantum confinement in semiconductors, light confinement in metal and dielectric nanostructures, and wave coupling in nanostructures, and describes how they can be applied in lighting sources, lasers, photonic circuitry, and photovoltaic systems. Readers will gain an intuitive insight into the commercial implementation of nanophotonic components, in both current and potential future devices, as well as challenges facing the field. The fundamentals of semiconductor optics, optical material properties, and light propagation are included, and new and emerging fields such as colloidal photonics, Si-based photonics, nanoplasmonics, and bioinspired photonics are all discussed. This is the "go-to" guide for graduate students and researchers in electrical engineering and physics interested in nanophotonics, and students taking nanophotonics courses.

Sergey V. Gaponenko is a professor and Head of the Laboratory of Nano-optics at the National Academy of Sciences of Belarus. He is also the author of *Optical Properties of Semiconductor Nanocrystals* and *Introduction to Nanophotonics* (Cambridge University Press, 1998, 2010).

Hilmi Volkan Demir is a professor at the Nanyang Technological University, Singapore, and a founder and director of the University's Luminous! Centre of Excellence for Semiconductor Lighting and Displays. He is also a professor at Bilkent University UNAM of Turkey, his alma mater.

Applied Nanophotonics

SERGEY V. GAPONENKO

National Academy of Sciences of Belarus

HILMI VOLKAN DEMIR

Nanyang Technological University, Singapore

Bilkent University UNAM, Turkey

CAMBRIDGE
UNIVERSITY PRESS

CAMBRIDGE
UNIVERSITY PRESS

University Printing House, Cambridge CB2 8BS, United Kingdom

One Liberty Plaza, 20th Floor, New York, NY 10006, USA

477 Williamstown Road, Port Melbourne, VIC 3207, Australia

314–321, 3rd Floor, Plot 3, Splendor Forum, Jasola District Centre, New Delhi – 110025, India

79 Anson Road, #06–04/06, Singapore 079906

Cambridge University Press is part of the University of Cambridge.

It furthers the University's mission by disseminating knowledge in the pursuit of education, learning, and research at the highest international levels of excellence.

www.cambridge.org
Information on this title: www.cambridge.org/9781107145504
DOI: 10.1017/9781316535868

© Cambridge University Press 2019

First published 2019

Printed and bound in Great Britain by Clays Ltd, Elcograf S.p.A.

A catalogue record for this publication is available from the British Library.

Library of Congress Cataloging-in-Publication Data
Names: Demir, Hilmi Volkan, author. | Gaponenko, S. V. (Sergey V.), 1958– author.
Title: Applied nanophotonics / Hilmi Volkan Demir (Nanyang Technological University, Singapore), Sergey V. Gaponenko (National Academy of Sciences, Belarus).
Description: Cambridge, United Kingdom ; New York, NY : Cambridge University Press, 2018. | Includes bibliographical references and index.
Identifiers: LCCN 2018016271| ISBN 9781107145504 | ISBN 1107145503
Subjects: LCSH: Nanophotonics.
Classification: LCC TA1530 .D46 2018 | DDC 621.36/5–dc23
LC record available at https://lccn.loc.gov/2018016271

ISBN 978-1-107-14550-4 Hardback

To our wives and parents:
To Olga, Vasily, and Alina Gaponenko,
To Çiğdem Gündüz, Rahşan, and Salih Demir

CONTENTS

PREFACE

Nanophotonics looks at light–matter interactions at the nanoscale – covering all of the processes of light propagation, emission, absorption, and scattering in complex nanostructures. We found that looking at nanophotonics starting from the very basics and taking it all the way to the applications, which would be important and very useful for practitioners of nanophotonics, has been missing from the literature. The idea for this book, *Applied Nanophotonics*, was born at NTU Singapore as a result of our long discussions of how academic education and technical training in the field of nanophotonics should be. This book is therefore intended to be a self-contained textbook that can be used for both graduate and undergraduate students as well as engineers, scientists, and R&D experts who would like to have a complete treatment of nanophotonics.

This book was made possible as a result of the research work carried out by the authors over the period 2000–2018 at Stanford University, Bilkent University, NTU Singapore, and the Belarussian National Academy of Sciences. For that we are grateful to all of our colleagues, collaborators, and students, with whom we have explored the world of nanophotonics and learned a great deal in this joyful and fun adventure. To this end, special thanks go to Prof. D. A. B. Miller and Prof. J. Harris of Stanford University. At the final stage of this book project the critical reading of the selected chapters by Dr. A. Baldycheva, Dr. P. L. Hernandez-Martinez, Dr. S. Golmakaniyoon, and Dr. R. Thomas was of great help, as was the assistance of K. Güngör, who helped to produce the cover design. S. V. G. gratefully acknowledges the creative atmosphere and promotional support from NTU in 2014–2016.

Also, we would like to thank the publishing house of Cambridge University Press, for the excellent and fruitful cooperation, without which this book would never have been accomplished. In particular, we bestow our thanks to Heather Brolly, Anastasia Toynbee, Gary Smith, and Jane Shaw from Cambridge. We are also indebted to the referees and colleagues for their encouraging comments and helpful advice in the early stages of this book project.

Last but not least, we thank our wives, families, and friends for their never-ending encouragement and support.

H. V. Demir and S. V. Gaponenko
Singapore, Ankara, Minsk
May 2018

NOTATION

A	spontaneous emission probability (rate), the Einstein coefficient
A	size of a quantum well; length; period in space
$\mathbf{a}$	acceleration
a_{B}^{*}	exciton Bohr radius
a_{B}	$= 5.2917... \cdot 10^{-2}$ nm, electron Bohr radius
a, b, c	periods of a three-dimensional lattice
a_{L}	crystal lattice period
$\mathbf{B}$	magnetic induction vector
B	stimulated emission factor (the Einstein coefficient)
C	concentration
c	$= 299{,}792{,}458...$ m/s, speed of light in vacuum
$\mathbf{D}$	electric displacement vector
D	density of modes, density of states
D	optical density ($-\lg(\text{transmission})$)
$\mathbf{d}$	dipole moment; unit vector along dipole moment
d	dimensionality of space; thickness
e	$= 1.6021892 ... \cdot 10^{-19}$ C, elementary electric charge
$\mathbf{E}$	electric field vector
E	kinetic energy
E_{F}	Fermi level (energy)
E_{g}	band gap energy
$\mathbf{F}$	force
f	volume-filling factor; fraction
f_{BE}	Bose–Einstein distribution function
f_{FD}	Fermi–Dirac distribution function
G	Green's function
h	$= 6.626069... \cdot 10^{-34}$ J·s, Planck constant
$\hbar$	$\equiv h / 2\pi$
$\mathbf{H}$	Hamiltonian
$\mathbf{H}$	magnetic field vector
I	intensity
i	imaginary unit
$\mathbf{J}$	electric current density
$\mathbf{k}, k$	wave vector, wave number

k_B	$= 1.380662 \ldots \cdot 10^{-23}$ J/K, Boltzmann constant
l	orbital quantum number
$\mathbf{L}, L$	angular momentum
L, l	thickness
ℓ	mean free path
$\mathbf{M}$	magnetic polarizations
M	exciton mass
M	mass
m_0	$= 9.109534 \cdot 10^{-31}$ kg, the rest mass of an electron
m^*	effective mass
$\mathbf{n}$	unit vector
N, n	concentration; integer number
$n\mathrm{r}$	refractive index; real part of complex refractive index for absorbing materials
$\mathbf{P}$	electric polarization
P	hole concentration in a semiconductor
$\mathbf{p}, p$	momentum, quasi-momentum
Q	quantum efficiency; quantum yield
R	reflection coefficient for intensity
r	reflection coefficient for amplitude
$\mathbf{r}$	radius vector
R, r	radius, distance
r, ϑ, φ	spherical coordinates
Ry	$= 13.605 \ldots$ eV, Rydberg energy
Ry^*	exciton Rydberg energy
$\mathbf{S}$	pointing vector
$\mathbf{T}$	translation vector
T	time period; temperature; transmission coefficient
t	time; transmission coefficient for amplitude
U	potential energy; energy
u	spectral energy density per unit volume
V	volume
$\mathbf{v}, v$	velocity
$\mathbf{v}_g, v_g$	group velocity
W	emission rate
x, y, z	coordinates
α	absorption coefficient
Γ	dephasing rate
γ	decay rate
$\gamma_{\mathrm{rad}}^{\mathrm{vacuum}} \equiv \gamma_0$	radiative (spontaneous) decay rate in vacuum
ε	relative dielectric permittivity; molar absorption coefficient

κ	imaginary part of the complex refractive index; evanescence parameter in tunneling
λ	wavelength
μ	reduced mass; chemical potential; relative magnetic permeability
μ_0	permeability of a vacuum
ν	frequency
ξ	set of all coordinates of the particles in a quantum system
ρ	electric charge density
σ	absorption cross-section
τ	time constant in various processes (decay, transfer, scattering)
$r\vartheta, \varphi$	spherical coordinates
χ	dielectric susceptibility
χ_{nl}	roots of the spherical Bessel functions
Ψ	wave function, time-dependent
ψ	wave function, time-independent
ω	circular frequency
ω_{p}	plasma circular frequency

ACRONYMS

Terms

2DPC	two-dimensional photonic crystal
3DPC	three-dimensional photonic crystal
AFM	atomic force microscope
CCD	charge-coupled device
CCT	correlated color temperature
CD	compact disk
CD-ROM	compact disk read-only memory
CFLs	compact fluorescent lamps
CIS	copper indium sulfide
CMOS	complementary metal-oxide-semiconductor (technology)
CQD	colloidal quantum dot
CQS	color quality scale
CRI	color rendering index
CVD	chemical vapor deposition
CW	continuous wave
DBR	distributed Bragg reflector
DFB	distributed feedback
DOM	density of modes
DOS	density of states
DVD	digital versatile disk
DWDM	dense wavelength division/multiplexing
EBL	electron blocking layer
EQE	external quantum efficiency
ESU	electrostatic unit
ETL	electron injection layer
FCC	face-centered cubic
FMN	flavin mononucleotide
FRET	Förster resonance energy transfer
FTTH	fiber to the home
HOMO	highest occupied molecular orbital
HTL	hole injection layer
ICP	inductively coupled plasma

IJE	injection efficiency
IQE	internal quantum efficiency
IR	infrared
ITO	indium tin oxide
LAN	local area network
LCD	liquid crystal display
LDOS	local density of states
LED	light-emitting diode
LEE	light extraction efficiency
LER	luminance efficacy of optical radiation
LUMO	lowest unoccupied molecular orbital
MBE	molecular beam epitaxy
MDM	mode division multiplexing
MEG	multiple exciton generation
MIXSEL	mode-locked integrated external-cavity surface-emitting laser
MOCVD	metal–organic chemical vapor deposition
MOVPE	metal–organic vapor-phase epitaxy
NP	nanoparticle
NW	nanowire
OLED	organic light-emitting diode
PC	personal computer
PC	photonic crystal
PECVD	plasma-enhanced chemical vapor deposition
PL	photoluminescence
PON	passive optical network
PSS	patterned sapphire substrate
RDE	radiative efficiency
RET	resonance energy transfer
RIE	reactive ion etching
RIU	refractive index unit
ROM	read-only memory
SAM	saturable absorber mirror
SDL	semiconductor disk laser
SEM	scanning electron microscope
SERS	surface enhanced Raman scattering
SESAM	semiconductor saturable absorber mirror
SOI	silicon-on-insulator
TAC	time-to-amplitude converter
TCO	transparent conducting oxide
TEM	transmission electron microscope
TNT	trinitrotoluene

UV	ultraviolet
VCSEL	vertical cavity surface-emitting laser
VECSEL	vertical external-cavity surface-emitting laser
VTE	voltage efficiency
WDM	wavelength division/multiplexing
WPE	wall-plug efficiency
XRD	x-ray diffraction
YAG	yttrium aluminum garnet

Companies and Organizations

AAAS	American Association for the Advancement of Science
ACS	American Chemical Society
AIP	American Institute of Physics
APS	American Physical Society
CIE	Commission Internationale de l'Éclairage
EPFL	École Polytechnique Fédérale de Lausanne
ETHZ	Swiss Federal Institute of Technology at Zurich
IBM	International Business Machines
MIT	Massachusetts Institute of Technologies
NREL	National Renewable Energy Laboratory
NTSC	National Television System Committee
OSA	Optical Society of America
RCA	Radio Corporation of America
RSC	Royal Society of Chemistry

1 Introduction

1.1 FROM PASSIVE VISION TO ACTIVE HARNESSING OF LIGHT

Light plays a major role in our perceptual cognition through vision, which occurs in the very narrow spectral range of electromagnetic radiation from approximately 400 nm (violet) to approximately 700 nm (red), whereas the whole wealth of electromagnetic wavelengths extends from picometers (gamma rays) to kilometers (long radiowaves) and beyond. In the early stages of technology, humans strived for enhancement of visual perception. Major highlights are: the invention of eye glasses (Salvino D'Armate, Italy, at the end of the thirteenth century); the invention of the microscope c.1590 by two Dutch spectacle makers, Zacharias Jansen and his father Hans; and the invention of the telescope in 1608 by Hans Lippershey, also a Dutch eyeglass maker (Figure 1.1). These devices only enhanced our passive vision by means of light coming from the sun or another source, e.g., a candle or an oven, and scattered toward our eyes. At that time no attempt had been made to generate light other than through the use of fire, or to store images in any fashion. Much later, in 1839, Louis Daguerre introduced the first photocamera (daguerreotype) in which a light-sensitive AgI-coated plate was used for image recording. This was the first manmade optical processing device. In 1873–1875, Willoughby Smith in the USA and Ernst Werner Siemens (Siemens 1875) in Germany described the remarkable sensitivity of a selenium film conductivity to light illumination, thus creating a basis for photoelectric detectors. Very soon after, in 1880, using a selenium plate as a photodetector, Alexander Bell in the USA ingeniously demonstrated wireless optical communication over 200 m using a sunbeam modulated by a microphone membrane (Bell 1880). The modulated signal was received by a photodetector, converted into an electric signal and eventually into acoustic vibrations by a telephone. Later, during World War I, this idea was implemented in a communication across a few kilometers for military purposes, using modulated incandescent lamp radiation. This is the approach to wireless optical communication that has more recently gained the notation "Li-Fi" (light fidelity). For a practical, far-reaching implementation of optical communication, reliable light sources with fast modulation were needed. Notably, at this time, Charles Fritts (1883) in New York developed a solar cell with 1% efficiency using selenium on a thin layer of gold, thus introducing the possibility of industrial-scale photovoltaics.

Photonics milestones:
From passive vision to active harnessing of light

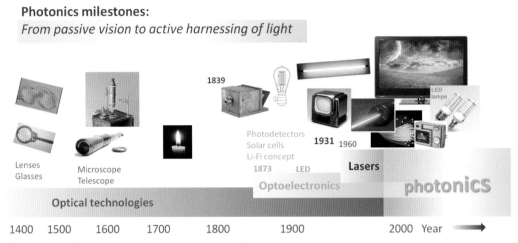

Figure 1.1 Milestones of photonics technologies.

Systematic experiments with electric gas discharge resulted in the first electric gas lamps being introduced in Europe and the USA around the early 1900s. These were the first devices directly converting electricity into light without heating. The emergence of television (TV) in the USA in 1923–1930 was allowed by the iconoscope, invented by Vladimir Zworykin (1988–1982), a brilliant Russian mind, at the time working at Westinghouse and RCA. His TV transmitter and receiver were both based on cathode ray tubes. The advent of the TV marked the beginning of the optoelectronic era in which humans attempted to transmit, receive, store, and reproduce optical data using electrical processes.

Notably, in the same period the light-emitting diode was invented (see Zheludev 2007). In 1907, in a brief 22-line letter to the *Electric World* editor, Henry J. Round reported on curious observations of visible light starting from 10 V applied to a crystal piece of SiC, at that time referred to as carborundum (Round 1907). This preliminary report was not noticed at the time, and in the period 1928–1933 Oleg Losev (Losev 1928) from Nizhniy Novgorod (Russia) performed a systematic study of electroluminescent properties of metal–semiconductor junctions with a carborundum crystal, thus marking the emergence of semiconductor light-emitting diodes. His works were published in *Philosophical Magazine* and *Physikalische Zeitschrift*. Sadly, Oleg Losev died in 1942 at the age of only 39 in Leningrad, suffering from severe hunger during the time of the fascists' siege. His works were forgotten for decades. Only much later, in 1961, were the modern generation of III–V LEDs reported by J. R. Biard and G. Pittman from Texas Instruments (USA), and the first visible LED was made by N. Holonyak at General Electrics in 1962.

The first lasers appeared in 1960 in the USA (solid-state lasers with optical pumping, T. H. Maiman; P. Sorokin and M. Stevenson; a gas discharge laser, A. Javan, 1960). The term LASER (light amplification by stimulated emission of radiation) was introduced by G. Gould, the author of the first patent of an optically pumped laser (1957). In 1962–1963 the first semiconductor laser diodes was reported (USA: R. Hall et al.; M. Nathan et al.;

T. Quist et al.; USSR: V. S. Bagaev et al.). Semiconductor lasers enabled modern fiber optical communication, which now reaches almost every building and home. These lasers have also made it possible to develop laser disk (CD-ROMs, and also rewritable CDs) systems for data processing and audio/video recording/reproduction. Owing to double heterostructures with quantum wells, modern semiconductor diodes have achieved unparalleled wall-plug efficiency levels exceeding 50%, the exceptionally high direct conversion of electric current into light.

Thanks to quantum well InGaN semiconductor structures, blue LEDs have made a revolutionary step toward all-solid-state lighting. The detector components have also made great progress, with outstanding performance of familiar charge-coupled device (CCD) arrays in modern digital photo- and video-cameras. Progress in semiconductor optoelectronics was marked by the three Nobel prizes awarded to Zh. Alferov and R. Kremer (2000, double heterostructures), W. S. Boyle and G. E. Smith (2009, CCD array), and I. Akasaki, H. Amano, and S. Nakamura (2014, blue LEDs).

To emphasize the merging of optics, optoelectronics, and laser technologies, the notion of *photonics* was coined a couple of decades ago. Photonics embraces all technologies and devices in which electromagnetic radiation of the visible, infrared, and ultraviolet ranges is harnessed for our needs.

1.2   WHAT IS NANOPHOTONICS?

Nanophotonics makes use of a wealth of size-dependent phenomena arising when space and matter feature confinement at a nanoscale. First, since the wavelength of electromagnetic radiation used in photonics is of the order of 1 μm in air and reduces to a few hundreds of nanometers in dielectric and semiconductor media, fractioning of matter and space on the scale of the order of 100 nm changes conditions for electromagnetic radiation propagation and the related lightwave confinement phenomena appear. Light–matter interaction in photonics mainly reduces to the interaction of electromagnetic radiation with *electrons* in atoms, molecules, and solids. Therefore, optical absorption and emission rates and spectra are defined to a large extent by *electron properties of the matter*. Owing to the wave properties of electrons featuring de Broglie wavelength of the order of 10 nm in semiconductors, the confinement of matter on this scale gives rise to a number of size-dependent phenomena referred to as *quantum confinement effects*. The term "quantum" here highlights that these phenomena result from quantum mechanical consideration.

Nanophotonics studies light–matter interactions at the nano scale and makes use of the aforementioned lightwave confinement and electron confinement phenomena in various structures and devices. Each of the two types of the confinement effects suggest a number of ways to improve and/or modify the parameters of existing optoelectronic devices, including light-emitting diodes, laser diodes, solid-state lasers, solar cells, and optical communication circuitry. Because of the different scales of lightwave and electron confinements, both groups of phenomena can be applied simultaneously in the same device, as often occurs, e.g., in modern semiconductor lasers.

This book represents the authors' attempt to introduce engineers and researchers to the realm of nanophotonics. It is broken down into two parts: Part I, Basics; and Part II, Advances and Challenges.

Part I presents the minimal tutorial description of electron and lightwave properties in nanostructures and the basic physical principles of nanophotonics, namely: properties of electrons in various potential wells (Chapter 2); electron properties of semiconductor nanostructures (Chapter 3); properties of electromagnetic waves under size confinement (Chapter 4); modification of spontaneous emission rates in nanostructures (Chapter 5); principles of optical gain and lasing (Chapter 6); and energy transfer phenomena (Chapter 7). The content of this part implies knowledge of solid-state theory and optics at the undergraduate and introductory graduate levels. The interested reader can find more specialized works (e.g., Miller [2008], Gaponenko [2010], Joannopoulos et al. [2011], Klingshirn [2012], Novotny and Hecht [2012], and Klimov [2014]) if they wish to dive deeper into the basics of nanophotonics. In Part I, references to the plentiful original publications are reduced to the reasonable minimum while focusing at the physical essence of the phenomena involved and experimental prerequisites for their implementation.

Part II describes current advances of applications of the principal electron confinement and light confinement phenomena in various practical devices. This part of the book covers applications of nanostructures in lighting (Chapter 8); lasers (Chapter 9); optical communication circuitry (Chapter 10); and photovoltaics (Chapter 11). Each of these chapters ends with a list of challenges the designers face, and a tentative forecast for the near-future progress. The very last chapter, Chapter 12, explores emerging nanophotonics. Here, new trends in nanophotonics are traced that offer promise of new photonic components (densely integrated optical chips, subwavelength lasers, various sensors, bioinspired and biocompatible devices) and technological platforms (colloidal photonics, silicon photonics). In Part II, we attempt to overview the principal achievements in nanophotonic devices, which makes it necessary to extend the list of references to the original publications.

The style and materials covered meet the level of graduate students. Advanced readers can skip Part I and consider only Part II. Since students in engineering may lack an optical background, all necessary introductory data from semiconductor optics, material optical properties, light propagation in inhomogeneous media and nanostructures are included. Challenges discussed in the textbook show modern trends and emphasize that nanophotonics is an open and active field bordering optical engineering, materials science, electronics engineering, and sometime colloidal chemistry and even biophysics.

This book highlights the basic principles of nanophotonics and their successful implementation in photonic devices, and will be a good companion book for a number of specialist publications devoted to specific types of optoelectronic devices that have been published lately: Chow and Koch (2013), Liu (2009), Cornet et al. (2016), Schubert (2006) with the summarized and emphasized role of nanostructures in existing and emerging photonics.

References

Bell, A. G. (1880). On the production and reproduction of sound by light. *Amer J Sci*, **118**, 305–324.

Chow, W. W., and Koch, S. W. (2013). *Semiconductor-Laser Fundamentals: Physics of the Gain Materials*. Springer Science & Business Media.

Cornet, C., Léger, Y., and Robert, C. (2016). *Integrated Lasers on Silicon*. Elsevier.

Fritts, C. E. (1883). On a new form of selenium cell, and some electrical discoveries made by its use. *Amer J Sci*, **156**, 465–472.

Gaponenko, S. V. (2010). *Introduction to Nanophotonics*. Cambridge University Press.

Joannopoulos, J. D., Johnson, S. G., Winn, J. N., and Meade, R. D. (2011). *Photonic Crystals: Molding the Flow of Light*. Princeton University Press.

Klimov, V. (2014). *Nanoplasmonics*. CRC Press.

Klingshirn, C. F. (2012). *Semiconductor Optics*. Springer Science & Business Media.

Liu, J. M. (2009). *Photonic Devices*. Cambridge University Press.

Losev, O. V (1928). Luminous carborundum detector and detection effect and oscillations with crystals. *Phil Mag*, **6**, 1024–1044.

Miller, D. A. B. (2008). *Quantum Mechanics for Scientists and Engineers*. Cambridge University Press.

Novotny, L., and Hecht, B. (2012). *Principles of Nano-Optics*. Cambridge University Press.

Round, H. J. (1907). A note on carborundum. *Electrical World*, **49**, 308.

Schubert, E. F. (2006). *Light-Emitting Diodes*. Cambridge University Press.

Siemens, W. (1875). On the influence of light upon the conductivity of crystalline selenium. *Phil Mag*, **50**, 416.

Zheludev, N. (2007). The life and times of the LED: a 100-year history. *Nat Photonics*, **1**, 189–192.

Part I

Basics

2 Electrons in potential wells and in solids

Absorption and emission of light by atoms, molecules, and solids arise from electron transitions. When in a potential well, an electron has discrete energy spectra; when in a periodic potential it has energy bands separated by gaps. This chapter presents an introductory overview of electron confinement phenomena and electron properties of crystalline solids. A summary of the data for real semiconductor materials used in photonics is provided.

2.1 ONE-DIMENSIONAL WELLS

2.1.1 Wave properties of electrons

In the macroscopic world, a mass point particle features mass, m; velocity, v; momentum, $p = mv$ (boldface type corresponds to vectors); and kinetic energy, E, which reads

$$E = \frac{mv^2}{2} = \frac{p^2}{2m}.$$ (2.1)

A classical wave is described by wavelength λ, which is the distance a wave travels in a single period of oscillations, T, i.e.

$$\lambda = vT,$$ (2.2)

where v is the speed. Wavelength is related with wave vector $\mathbf{k}$ as

$$k = |\mathbf{k}| = \frac{2\pi}{\lambda},$$ (2.3)

where k is referred to as the wave number. Wave vector direction coincides with the direction of the phase motion, i.e., it is normal to the wave front propagation direction.

In the microscopic world – on the atomic and molecular scale measuring lengths of the order of nanometers – quantum particles, e.g., electrons, exhibit wave properties with wave vector and wavelength related to momentum $\mathbf{p}$ in accordance with the relations first proposed by Louis de Broglie in 1923:

$$\mathbf{p} = \hbar\mathbf{k}, \qquad \lambda = \frac{h}{p}.$$ (2.4)

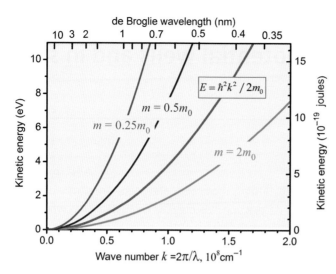

Figure 2.1 Kinetic energy versus wave number and de Broglie wavelength for a particle with various mass, where $m_0 = 9.109534 \cdot 10^{-31}$ kg is the free electron mass.

Here, $h = 6.626069 \cdot 10^{-34}\,\mathrm{J \cdot s}$ is the Planck constant and $\hbar \equiv h/2\pi$. Now kinetic energy E related to momentum (Eq. (2.1)) acquires dependence on the particle wave number as

$$E = \frac{p^2}{2m} = \frac{\hbar^2 k^2}{2m}. \tag{2.5}$$

In terms of kinetic energy E, the particle de Broglie wavelength reads, using Eqs. (2.4),

$$\lambda = \frac{h}{\sqrt{2mE}}. \tag{2.6}$$

The last relation is rather instructive for an immediately intuitive idea about the typical wavelength scale inherent in the electron world. For example, if an electron with the rest mass $m_0 = 9.109534 \cdot 10^{-31}$ kg has gained kinetic energy $E = 1$ eV while being accelerated in a vacuum between a couple of electrodes with potential difference 1 V, its de Broglie wavelength equals 1.23 nm.

We arrive at an important conclusion. The nanometer scale corresponds to wavelengths that electrons possess in real solid-state electronics at voltages available from a primitive galvanic cell or a solar cell. The relation $E = \hbar^2 k^2 / 2m_0$ is presented in Figure 2.1 within the reasonable energy range 0–10 eV.

2.1.2 Quantum particle in a rectangular well

Waves obey steady-state standing oscillations when confined between solid walls. These are shown in Figure 2.2. Amplitude profiles for a few bigger wavelengths are plotted. The notions "amplitude" and "solid walls" carry different meanings for different waves – e.g., in acoustics, amplitude means air pressure and "solid walls" mean real physical walls in

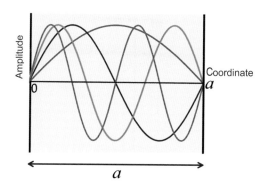

Amplitude

Coordinate

0

a

a

Figure 2.2 Standing waves between two hard walls spaced by distance *a*. "Amplitude" means air pressure for acoustic waves in air, electric field amplitude for electromagnetic waves, and wave function for a quantum particle like an electron.

which sound does not propagate; in optics, amplitude refers to the electric field, and "solid walls" means perfect mirrors; in quantum mechanics, amplitude refers to the particle wave function and "solid walls" means infinite potential barriers.

Presentation of an electron (or any other quantum particle) as a wave together with Figure 2.2 allows us to arrive at the principal quantum mechanical relation for a particle confined in a box with infinite potential barriers. The standing waves in Figure 2.2 obey a set

$$\lambda_n = 2a, \frac{2a}{2}, \frac{2a}{3}, ..., \frac{2a}{n}, \qquad n = 1, 2, 3, ... \tag{2.7}$$

which corresponds to the set of wave numbers

$$k_n = \frac{\pi}{a} n, \qquad n = 1, 2, 3, ... \tag{2.8}$$

This set, in turn, gives rise to the discrete set of momenta (see Eq. (2.4)) and energy (Eq. (2.5)), with the latter obeying the law

$$E_n = \frac{\pi^2 \hbar^2}{2ma^2} n^2, \quad n = 1, 2, 3, ... \tag{2.9}$$

The energy spectrum of a particle in a box with infinite potential barriers therefore obeys a discrete set with infinite number of energy values E_n, typically called "energy levels." The first three levels are shown in Figure 2.3(a). Notably, spacing between neighboring levels expands with growing level numbers. In the case of a three-dimensional rectangular box, confinement in every dimension gives rise to the three sets of energy states similar to Eq. (2.9).

2.1.3 Wave function and Schrödinger equation

Further analysis requires introduction of the wave function and Schrödinger equation. In quantum mechanics, a particle state is described by a wave function Ψ which depends on the number of variables in accordance with the degrees of freedom available for a particle.

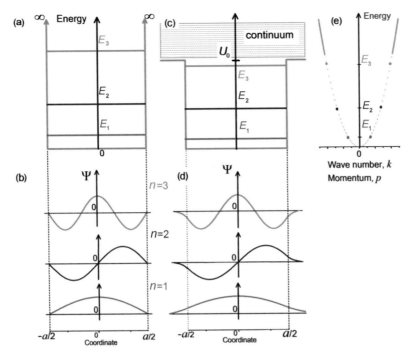

Figure 2.3 One-dimensional potential well with (a,b) infinite and (c,d) finite potential walls, and (e) energy versus wave number dependence for the case of the finite potential walls. In the case of infinite walls, the energy states obey a series $E_n \sim n^2$ and the wave functions vanish at the walls. The total number of states is infinite. The probability of finding a particle inside the well is exactly equal to unity. In the case of finite walls, the states with energy higher than U_0 correspond to infinite motion and form a continuum. At least one state always exists within the well.

For a single particle, Ψ depends on time and three coordinates (x, y, z), whereas, e.g., for a couple of particles Ψ depends on time and six coordinates $(x_1, y_1, z_1, x_2, y_2, z_2)$. A wave function is considered to give the probability of a system to be found in certain positions in space. The value

$$|\Psi(\xi)|^2 d\xi = \Psi^*(\xi)\Psi(\xi)d\xi \qquad (2.10)$$

is proportional to the probability of finding in the measurements coordinates of the particle or the system of particles in the range $[\xi, \xi + d\xi]$, ξ being the set of all coordinates of the particles in the system, i.e., for a single particle

$$d\xi = dxdydz$$

holds, whereas for a system of two particles it reads

$$d\xi = dx_1 dy_1 dz_1 dx_2 dy_2 dz_2, \qquad (2.11)$$

and so on. The probabilistic interpretation of a wave function dates back to 1926, when it was proposed by Max Born. It constitutes the major postulate of wave mechanics.

The wave function can be normalized to have

$$\int |\Psi(\xi)|^2 \, d\xi = 1. \tag{2.12}$$

Then, $|\Psi|^2 d\xi$ equals probability $dW(\xi)$ that a particle is located within the interval $[\xi, \xi + d\xi]$.

Note that a normalized wave function is always determined with the accuracy of the factor $e^{i\alpha}$, where α is an arbitrary real number. This ambiguity has no effect on physical results since all physical values are determined by the product $\Psi\Psi^*$.

To know the wave function of a particle or a system, one needs to solve the Schrödinger equation proposed in 1926 by Erwin Schrödinger. This equation constitutes another postulate of quantum mechanics. It reads:

$$\mathbf{H}\Psi = i\hbar \frac{\partial \Psi}{\partial t}, \tag{2.13}$$

where $\mathbf{H}$ is the system *Hamiltonian*. The time-independent Hamiltonian coincides with the energy operator. For a single particle it reads

$$\mathbf{H} = -\frac{\hbar^2}{2m}\nabla^2 + U(\mathbf{r}), \tag{2.14}$$

where $-\dfrac{\hbar^2}{2m}\nabla^2$ is the kinetic energy operator and $U(\mathbf{r})$ is the potential energy of a particle.

If the Hamiltonian does not depend on time, the time and space variables can be separated, i.e.,

$$\Psi(\xi,t) = \Psi(\xi)\phi(t). \tag{2.15}$$

In this case, the time-dependent Eq. (2.13) reduces to the *steady-state equation*

$$\mathbf{H}\Psi(\xi) = E\Psi(\xi), \tag{2.16}$$

where E is a constant value. The steady-state Schrödinger equation is a problem for eigenfunction and eigenvalues of the Hamiltonian operator $\mathbf{H}$. E values are the energy values of a system in a state $\Psi(\xi)$. States with the defined energy E are called steady states.

Consider a particle in a one-dimensional space with a constant potential energy U_0. Eq. (2.16) takes the form

$$\boxed{\frac{d^2\psi(x)}{dx^2} + \frac{2m}{\hbar^2}(E - U_0)\psi(x) = 0.} \tag{2.17}$$

Introducing a notation of a wave number

$$k^2 = \frac{2m(E - U_0)}{\hbar^2}, \tag{2.18}$$

we arrive at a compact form

$$\frac{d^2\psi(x)}{dx^2} + k^2\psi(x) = 0. \tag{2.19}$$

The latter equation resembles the one for a pendulum, for harmonic motion, and for an LC circuit. It has a general solution

$$\psi(x) = A\exp(ikx) + B\exp(-ikx), \tag{2.20}$$

where A and B are constants to be derived based on the problem conditions. Eq. (2.20) can be written in the other form inherent to a plane harmonic wave:

$$\Psi = A'\sin kx + B'\cos kx. \tag{2.21}$$

Eq. (2.18) gives

$$E - U_0 = \frac{\hbar^2 k^2}{2m} = \frac{p^2}{2m}. \tag{2.22}$$

The difference $E - U_0$ is the particle kinetic energy, whereas E is the particle full energy. The de Broglie wavelength of a particle reads

$$\lambda = \frac{2\pi}{k} = \frac{2\pi\hbar}{\sqrt{2m(E-U_0)}}. \tag{2.23}$$

We can now apply the Schrödinger equation to find wave functions of an electron in a rectangular one-dimensional box, as shown in Figure 2.3(a). We consider Eq. (2.17), with U_0 replaced by $U(x)$, which reads

$$U(x) = \begin{cases} 0 \ \ for \ \ |x| < a/2 \\ \infty \ \ for \ \ |x| > a/2 \end{cases}. \tag{2.24}$$

The symmetry of the potential $U(x) = U(-x)$ results in a symmetry of the probability density,

$$|\Psi(x)|^2 = |\Psi(-x)|^2,$$

whence

$$\Psi(x) = \pm\Psi(-x),$$

and we arrive at two independent solutions with different parity. The odd and even types of solutions are

$$\psi^- = \frac{\sqrt{2}}{a}\cos\frac{\pi n}{a}x \qquad (n=1,2,5,...), \tag{2.25}$$

$$\psi^+ = \frac{\sqrt{2}}{a}\sin\frac{\pi n}{a}x \qquad (n=2,4,6,...). \tag{2.26}$$

The energy spectrum consists of a set of discrete levels expressed by Eq. (2.9), as we have intuitively predicted above.

Electron wave functions are shown in Figure 2.3(b) for the first three energy values. Levels for different states (along with the corresponding wave functions) are biased vertically in correlation with energy positions for clarity. The total number of states is infinite. The probability of finding a particle inside the well is exactly equal to unity.

It is useful to check a few reference numbers for an electron in a real well. For $m = m_0$ and $a = 2$ nm, we get $E_1 = 0.094$ eV, $E_2 = 0.376$ eV, ..., and the minimal energy spacing $E_2 - E_1 = 0.282$ eV. The latter is one order of magnitude higher than the $k_B T = 0.027$ eV value at room temperature. If a transition from the state with E_1 to the state with E_2 is promoted by means of photon absorption, the corresponding wavelength of electromagnetic radiation $\lambda = 4394$ nm belongs to the middle infrared range.

The above results can be extended to the three-dimensional case. If a well has sizes a, b, c in three directions, then particle energy will be defined by a set of three quantum numbers, n_1, n_2, n_3, namely:

$$E_{n_1 n_2 n_3} = \frac{\pi^2 \hbar^2}{2m} \left(\frac{n_1^2}{a^2} + \frac{n_2^2}{b^2} + \frac{n_3^2}{c^2} \right), \qquad n_1, n_2, n_3 = 1, 2, 3, \ldots \tag{2.27}$$

One can see that different sets of quantum numbers (and, accordingly, different wave functions) may give the same value of energy. Such states are called *degenerate states*.

2.1.4 A rectangular well with finite walls

In the case of finite walls, U_0 (Figure 2.3(c)), the states with energy higher than U_0 correspond to infinite motion and form a continuum. At least one state always exists within the well. The total number of discrete states is determined by the well width a and height U_0. The parameters in Figure 2.3(c) correspond to the three states inside the well. Unlike the case of infinite walls, the wave functions now extend to the classically forbidden regions $|x| > a/2$ (Figure 2.3(d)). Wave functions are no longer equal to zero at the walls, but extend outside the walls exponentially. The Schrödinger equation should be solved based on the assumption of wave function and its first derivative continuity. Extension of wave function outside the borders means that a nonzero probability arises to find a particle outside the well. Accordingly, the probability to find a particle inside the well is always less than unity and decreases with increasing E_n. Extension of wave functions outside the well grows with n value. This is because lower potential walls enable deeper penetration of wave function outside.

A particle wavelength in the case of finite walls gets longer as compared to the well with the same width but with infinite walls. Accordingly, the energy levels are lower than in the case of infinite walls. The total number of states inside the well is controlled by the relation

$$a\sqrt{2mU_0} > \pi\hbar(n-1). \tag{2.28}$$

This holds for $n = 1$ for any combination of a, m, U_0 – i.e., at least one state always exists inside the well. The maximal number of levels is equal to maximal n for which Eq. (2.28) still holds. For deeper levels (smaller n), Eq. (2.9) can be used as a reasonable approximation.

Notably, though at least one localized state always exists in a rectangular one-dimensional well, this may not be the case for more complicated situations – e.g., a one-dimensional asymmetric well (left and right potential walls have different heights) may not contain a state inside. This means a quantum particle may not be captured by such a well. The same is true for two- and three-dimensional wells. Shallow and narrow wells may have no state inside.

A relation between E and k in the case of a free particle has the form $E = \hbar^2 k^2 / 2m$ (dashed curve in Figure 2.3(e)). In the case of the finite potential well, a part of the $E(k)$ function relevant to confined states is replaced by discrete points.

2.1.5 Potential shape and energy spectrum

Every time an electron or any other quantum particle experiences localization in space owing to the potential well, its energy spectrum becomes discrete. There are a number of important potential well profiles in quantum mechanics that are useful in practical problems and can provide intuitive insight on electron properties in various wells. From Figure 2.3(a) and Eq. (2.9) one can see that a rectangular potential well gives rise to an infinite set of expanding energy levels.

Notably, wave function strictly vanishes to zero only at the border of an infinite potential well. In all other cases wave function penetrates outside the potential barrier, i.e., a

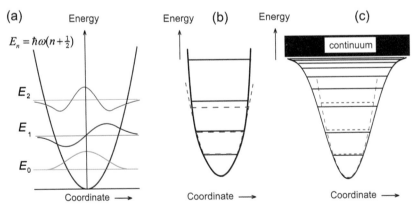

Figure 2.4 (a) Quantum harmonic oscillator, the first three wave functions and energy levels; (b) U-like and (c) V-like potential wells. The quantum harmonic oscillator represents a unique potential well that gives rise to equidistant energy levels of a quantum particle. U-like and V-like wells near the bottom can be approximated by parabolas (red dashes) and therefore a few lower states have energies close to those in a harmonic oscillator. The higher states in U-like potential diverge, similar to a rectangular well, whereas in V-like potential higher states converge to form a continuum, similar to a Coulomb well.

quantum particle "spreads" beyond a well. The above example of a finite potential well (Figure 2.3(c,d)) presents this important property of quantum particles resembling the general property inherent in all waves. In a case of an asymmetric well with one infinite and one finite potential wall, the wave function will always vanish at the infinite wall and will extend outside the well at the finite wall.

A potential well with vertical walls gives rise to expanding energy levels, i.e., the spacing between neighbor states $\Delta E_n = E_{n+1} - E_n$ rises with the state number n. There is an important case of the potential well, *quantum harmonic oscillator*, that features the unique property of equidistant energy levels, i.e., $\Delta E_n = $ const always holds for every n.

The quantum harmonic oscillator (Figure 2.4) has potential energy

$$U(x) = \tfrac{1}{2}m\omega^2 x^2. \tag{2.29}$$

The steady-state Schrödinger equation has the form

$$\nabla^2 \psi(x) + (k^2 - \lambda^2 x^2)\psi(x) = 0, \tag{2.30}$$

with $k^2 = 2mE/\hbar^2$ and $\lambda = m\omega/\hbar$. Like in a rectangular well with infinite walls, the energy spectrum of a particle resembles an infinite number of discrete states with energy values E_n. The symmetry of potential gives rise to odd and even solutions. The general solution reads

$$\psi_n(x) = u_n(x)\exp(-\lambda x^2/2), \tag{2.31}$$

where $u_n(x)$ stands for complex polynomials. The first three solutions have the form

$$\psi_0(x) = \exp(-\lambda x^2/2),$$

$$\psi_1(x) = \sqrt{2\lambda}\cdot x \exp\left(-\frac{1}{2}\lambda x^2\right), \tag{2.32}$$

$$\psi_2(x) = \frac{1}{\sqrt{2}}(1 - 2\lambda x^2)\exp\left(-\frac{1}{2}\lambda x^2\right).$$

These are shown in Figure 2.4. The number of nodes in the wave function equals n. The energy values are

$$\boxed{E_n = \hbar\omega(n + \tfrac{1}{2}),\ n = 0,\ 1,\ 2,\ \dots} \tag{2.33}$$

Every U-like or V-like potential well can be approximated by a parabolic potential near the minimum in accordance with the Taylor series (Figure 2.4(b,c)). Therefore, the lowest energy levels in these wells can be approximated by harmonic oscillators. In the case that potential walls are steeper than the harmonic law, the higher energy levels will expand, as is shown for the U-like potential well in Figure 2.4(b), whereas in the case in which the potential is more plain than the parabolic one, the higher levels converge to eventually form a continuum (Figure 2.4(c)).

2.1.6 Double-well and multi-well potentials

Consider two identical wells of width a separated by a barrier with finite height and thickness (Figure 2.5). We can trace evolution of a particle energy spectrum and its wave function when barrier height and width tend to zero.

If the distance R between wells is big compared to the well width ($R \ll a$), the particle wave function is close to zero between the wells. Solution of the Schrödinger equation in this case nearly coincides with wave function for an isolated well Ψ_1 with the only difference that $|\Psi_1|^2$ becomes two times smaller because of the normalization (a particle with equal probability can be found in either of the two wells). This is shown in the upper panel of Figure 2.5(a). Notably, another solution of the Schrödinger equation exists for which Ψ has a different sign in one of the two wells (lower panel in Figure 2.5(a)). Energy values for the two wave functions coincide. The $\Psi^{(+)}$ function is symmetric, whereas the $\Psi^{(-)}$ function is referred to as an asymmetric wave function. When inter-well distance shrinks, $\Psi^{(+)}$ and $\Psi^{(-)}$ functions change (Figure 2.5(b)). At close distances, the $\Psi^{(+)}$ state has lower energy as compared to the $\Psi^{(-)}$ state. In the limit, functions $\Psi^{(+)}$ and $\Psi^{(-)}$ convert into functions Ψ_1 and Ψ_2 of the ground and the first excited states, respectively, of a particle in a well whose width equals $2a$ (Figure 2.5(c)). If potential walls are high enough, a particle's lower

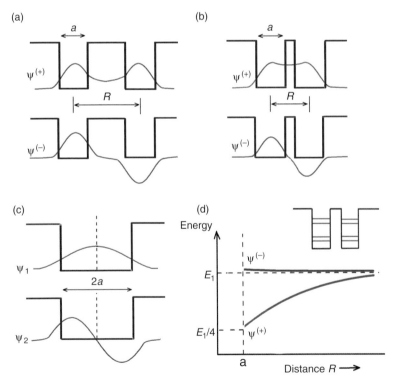

Figure 2.5 Double-well potential. (a,b) Potential and wave functions for different inter-well spacing R; (c) extreme case of $R = 0$; (d) the two lowest energy levels versus inter-well distance.

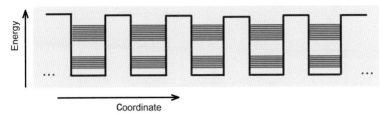

Figure 2.6 Splitting of energy levels into minibands in a potential consisting of periodically arranged identical wells/barriers.

states can be approximated by the formula derived for a well with infinite walls (Eq. (2.9)), i.e., $E_n = E_1 n^2$, $n = 1, 2, 3, ...$, $E_1 = \pi^2 \hbar^2 / (2ma^2)$. Comparing energy set for a well width a (E_1, $4E_1$, $9E_1$,...) with that for the well width $2a$ ($E_1 / 4$, E_1, $9E_1 / 4$, $4E_1$, $25E_1 / 4$, $9E_1$,...), one can see that the double-width well has the same set of energies as the original well, along with additional levels lying lower than original ones. Exact solutions show that every level (twice degenerate) in individual wells splits into two non-degenerate levels continuously upon $R \to a$. The lowest states are shown in Figure 2.5(d). Notably, such splitting occurs in any double-well potential, not only in rectangular wells.

In a set of N identical wells separated by finite barriers, N-fold splitting of every energy level occurs (Figure 2.6). For very large N, one can speak about evolution of a discrete set of energy levels into energy bands. This consideration gives an instructive hint on evolution of electron states from atoms to crystal. It is also important in understanding minibands in multiple semiconductor quantum wells and *superlattices*. In all cases of multiple wells, electron wave functions spread over the whole set of wells that means the electronic states become delocalized.

2.2 TUNNELING

A number of examples of quantum particles in various potentials show that the wave function of a quantum particle always penetrates through a potential wall of finite height (see Figures 2.3 and 2.5). This property in the case of a finite potential barrier gives rise to *tunneling*, an important phenomenon for many applications.

When a particle travels in space with the finite potential barrier U_0, one should search for the properties of transmitted and reflected waves in the forms satisfying the Schrödinger equation (Figure 2.7). In the case of a rectangular potential barrier and a particle energy $E < U_0$, the solution of the steady-state Schrödinger equation takes the form

$$\psi_1(x) = A \exp(ikx) + B \exp(-ikx) \text{ for } x < 0, \ k = \sqrt{2mE} / \hbar$$
$$\psi_2(x) = C_1 \exp(-\kappa x) + C_2 \exp(\kappa x) \text{ for } 0 < x < a, \ \kappa = \sqrt{2m(U_0 - E)} / \hbar \qquad (2.34)$$
$$\psi_3(x) = D \exp(ikx) \qquad\qquad\qquad \text{for } x > a.$$

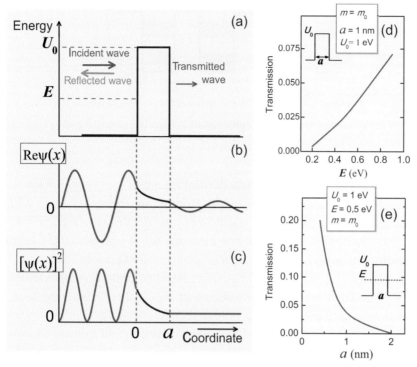

Figure 2.7 Tunneling in quantum mechanics. The left panel shows a rectangular potential barrier (a), a particle wave function (b), and probability density (c) for the case when the electron energy E is lower than the potential height U_0. The right panel presents probabilities of an electron tunneling through a rectangular barrier with height $U_0 = 1$ eV as a function of (d) electron energy E and (e) barrier width a.

Here, A is defined from normalization and can be put as $A = 1$, whereas coefficients B, C, D are to be found from continuity of the wave function and its first derivative in the points $x = 0$, $x = a$. The amplitudes of reflected and transmitted waves are defined by the B and C coefficients, respectively. The probability density oscillates in front of the barrier because of interference of incident and reflected waves, and takes the constant value outside the barrier. The final formulas for the transmittance T and reflection R read

$$T_{QM} = \left|\frac{D}{A}\right|^2 = \left[1 + \frac{U_0^2}{4E(U_0 - E)} \sinh^2(\kappa a)\right]^{-1} > 0, \tag{2.35}$$

$$R_{QM} = 1 - T_{QM} = \left[1 + \frac{4E(U_0 - E)}{U_0^2} \sinh^{-2}(\kappa a)\right]^{-1} < 1, \tag{2.36}$$

where $\sinh(x) = (e^x + e^{-x})/2$ is the hyperbolic sine function. Subscript QM means the quantum mechanical phenomena are considered.

To give the reference absolute values of tunneling probability in real situations, the transmittance function for an electron versus its energy E and barrier width a is plotted for the barrier height $U_0 = 1\,\mathrm{eV}$ to give an idea about the absolute transmittance values (Figure 2.7(d, e)). One can see that transmittance almost linearly grows versus E and rapidly, almost exponentially, drops versus a.

For $\kappa a \gg 1$, $\sinh^{-1}(x)$ reduces to $e^x/2$ and Eqs. (2.35) and (2.36) can be written in a simpler form as

$$T_{\mathrm{QM}} \approx \frac{U_0^2}{4E(U_0 - E)} \exp\left[-\frac{a}{\hbar} \sqrt{2m(U_0 - E)} \right] \text{ for } \kappa a \gg 1. \qquad (2.37)$$

For example, when $U_0 - E = 1\,\mathrm{eV}$ and $a = 10\,\mathrm{nm}$, the value of $\kappa a \approx 10$. Note that formally $\kappa = \sqrt{2m(U_0 - E)}/\hbar$ resembles wave number of a particle with kinetic energy $U_0 - E$, or in other words, $\kappa = 2\pi/\lambda$ where λ is the de Broglie wavelength of a particle with kinetic energy $U_0 - E$, i.e., the energy deficiency for a particle to move over the barrier in question without tunneling. This presentation helps to estimate the absolute value of κa as a ratio of the barrier width and a fictitious particle wavelength with the same mass but with the kinetic energy $U_0 - E$. Figure 2.1 can be used to estimate de Broglie wavelength of electrons. Whenever $\kappa a \gg 1$ holds, tunneling probability is always much less than 1.

When a quantum particle moves through a couple of potential barriers there is a non-trivial effect referred to as *resonant tunneling*. For certain energies a particle gains a unit probability to pass through two barriers without reflection. Figure 2.8 shows an example of the accurate solution of the steady-state Schrödinger equation for an electron in a double-barrier potential. One can see resonant transmission corresponds to the condition of the integer number of de Broglie half-wavelengths in the well between barriers. Resonant tunneling can be treated as a direct consequence of interference of waves passing through barriers and reflecting back. Resonant conditions multiply wave function amplitude in the well, enabling high overall transmission under conditions of low transparency of every barrier (compare wave functions inside and outside barriers). The ratio of wave function amplitudes inside and outside the well rises with barrier height, as is shown in Figure 2.8. The sharpness and finesse of transmission resonance are both enhanced (right panel of Figure 2.8). In many instances, resonant tunneling resembles phenomena inherent in an optical cavity comprising a couple of plane mirrors.

Tunneling phenomena play an important role in many nanophotonic and nanoelectronic components. Notably, tunneling itself is a direct consequence of wave properties of electrons, holes, and atoms in solids. In classical physics, i.e., in the macroscopic world, tunneling is inherent in waves like electromagnetic waves and acoustic waves. Optical analogs of tunneling phenomena are, e.g., transparency of thin metal films and frustrated total reflection. Tunneling of light is considered in detail in Chapter 4.

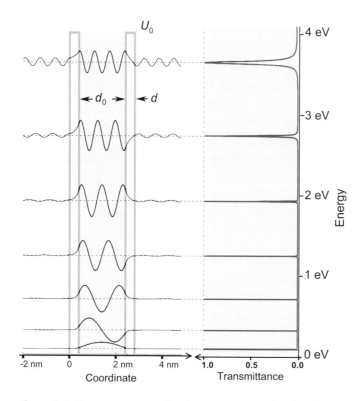

Figure 2.8 Resonance tunneling in quantum mechanics. Accurate numerical solution for an electron when its full energy outside the barriers E is lower than the height of the two potential barriers, U_0 with thickness d and spacing d_0. Parameter values are $d_0 = 2$ nm, $d = 0.4$ nm, $U_0 = 4$ eV. The left panel shows the particle wave function corresponding to reflectionless propagation. The right panel shows the transmittance on the same energy axis.

2.3   CENTRALLY SYMMETRIC POTENTIALS

2.3.1   Centrally symmetric wells

A spherical quantum well with infinite potential is the first representative example of the three-dimensional centrally symmetric potentials (Figure 2.9). It plays an important role in modeling of semiconductor quantum dots.

In the case of a spherically symmetric potential $U(r)$, we deal with a Hamiltonian

$$H = -\frac{\hbar^2}{2m}\nabla^2 + U(r),\tag{2.38}$$

where $r = \sqrt{x^2 + y^2 + z^2}$. The problem is considered in spherical coordinates, r, ϑ, and ϕ:

$$x = r\sin\vartheta\cos\phi,\, y = r\sin\vartheta\sin\phi,\, z = r\cos\vartheta.\tag{2.39}$$

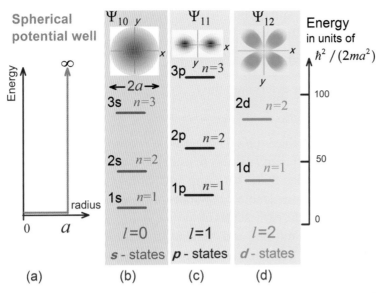

Figure 2.9 Spherical quantum well. (a) Potential shape; (b,c,d) energy levels for *s*-, *p*-, and *d*-states, respectively. Wave functions for the lowest *s*-, *p*-, and *d*-states are shown in the upper part of each panel.

Then, the Hamiltonian (Eq. (2.38)) reads

$$H = -\frac{\hbar^2}{2mr^2}\frac{\partial}{\partial r}\left(r^2\frac{\partial}{\partial r}\right) - \frac{\hbar^2\Lambda}{2mr^2} + U(r), \tag{2.40}$$

where the Λ operator is

$$\Lambda = \frac{1}{\sin\vartheta}\left[\frac{\partial}{\partial\vartheta}\left(\sin\vartheta\frac{\partial}{\partial\vartheta}\right) + \frac{1}{\sin\vartheta}\frac{\partial^2}{\partial\phi^2}\right]. \tag{2.41}$$

The wave function can be separated into functions of r, ϑ, and φ:

$$\psi = R(r)\Theta(\vartheta)\Phi(\phi), \tag{2.42}$$

and can be expressed in the form

$$\Psi_{n,l,m}(r,\vartheta,\phi) = \frac{u_{n,l}(r)}{r}Y_{lm}(\vartheta,\phi), \tag{2.43}$$

where Y_{lm} are the *spherical Bessel functions*, and $u(r)$ satisfies a one-dimensional equation,

$$-\frac{\hbar^2}{2m}\frac{d^2u}{dr^2} + \left[U(r) + \frac{\hbar^2}{2mr^2}l(l+1)\right]u = Eu. \tag{2.44}$$

To obtain the energy spectrum, one must deal with the one-dimensional Eq. (2.44) instead of the three-dimensional equation with Hamiltonian (Eq. (2.40)). The state of the system is characterized by the three quantum numbers, namely the principal quantum number n, the orbital number l, and the magnetic number m. The orbital quantum number determines the angular momentum value $\mathbf{L}$:

$$\mathbf{L}^2 = \hbar^2 l(l+1), \quad l = 0, \ 1, \ 2, \ 3, \ ... \tag{2.45}$$

The magnetic quantum number determines the L component parallel to the z axis:

$$L_z = \hbar m, \quad m = 0, \ \pm 1, \ \pm 2, \ ... \ \pm l. \tag{2.46}$$

Every state with a certain l value is $(2l + 1)$-degenerate accordingly to $2l + 1$ values of m. The states corresponding to different l values are usually denoted as s-, p-, d-, and f-states, and further in alphabetical order. States with zero angular momentum ($l = 0$) are referred to as s-states; states with $l = 1$ are denoted as p-states; and so on. The origin of s-, p-, d-, and f-notations dates back to the early stages of atomic spectroscopy and is related to the obsolete "sharp," "principal," "diffuse," and "fundamental" line notations, respectively.

The above properties of the Schrödinger equation are inherent in every centrally symmetric potential. The specific values of energy are determined by $U(r)$ function. In the case of the rectangular spherical potential with infinite barrier, the energy values read

$$E_{nl} = \frac{\hbar^2 \chi_{nl}^2}{2ma^2}, \tag{2.47}$$

where χ_{nl} are roots of the spherical Bessel functions, with n being the number of the root and l being the order of the function. χ_{nl} values for several n, l values are listed in Table 2.1. Note that for $l = 0$ these values are equal to πn ($n = 1, 2, 3, ...$) and Eq. (2.47) converges with the relevant expression in the case of a one-dimensional box (Eq. (2.9)). This results from the fact that for $l = 0$, Eq. (2.40) for the radial function $u(r)$ reduces to Eq. (2.17). To summarize, a particle in a spherical well possesses the set of energy levels $1s$, $2s$, $3s$, ..., coinciding with the energies of a particle in a rectangular one-dimensional well, and additional levels $1p$, $1d$, $1f$, ..., $2p$, $2d$, $2f$, ..., that arise due to spherical symmetry of the well (Figure 2.9).

Figure 2.9 presents also wave functions for $1s$, $1p$, and $1d$ states. Every wave function equals zero outside the well. The quantum number n determines the number of nodes of the wave function inside the well (not at the borders). $n = 1$ corresponds to having no nodes,

Table 2.1 **Roots of the Bessel functions χ_{nl} for small n and l values**

l	$n = 1$	$n = 2$	$n = 3$
0 (s-states)	3.142 (π)	6.283 (2π)	9.425 (3π)
1 (p-states)	4.493	7.725	10.904
2 (d-states)	5.764	9.095	12.323

$n = 2$ means there is a single node, $n = 3$ gives rise to two nodes, etc. The l quantum number is equal to the number of nodes in the wave function as the polar angle ϑ varies between 0 and π. Note that, for the case of an infinite potential well, the only restrictions on the n and l values are that n should be a positive integer and l must be a non-negative integer.

In the case of the spherical well with the finite potential, U_0, Eq. (2.47) can be considered a good approximation only if U_0 is large enough, namely if $U_0 \gg \hbar^2/8ma^2$ holds. The right side of this inequality is a consequence of the *uncertainty relation*. In the case when

$$U_0 = U_{0\,min} = \frac{\pi^2\hbar^2}{8ma^2},$$

just a single state exists within the well. For $U_0 < U_{0\,min}$, no state exists in the well at all.

The *Coulomb potential* well is the most important centrally symmetric potential since it is used to explain properties of electrons in atoms (Figure 2.10). An electron with charge e is supposed to interact with another very heavy particle with the same charge,[1] i.e., the potential reads

$$U(r) = -\frac{e^2}{r}. \tag{2.48}$$

For this problem, introduction of the atomic length unit a^0 and atomic energy unit E^0 is useful. In the SI system these read

$$a^0 = 4\pi\varepsilon_0 \frac{\hbar^2}{m_0 e^2} \approx 5.292 \cdot 10^{-2}\,\text{nm}, \tag{2.49}$$

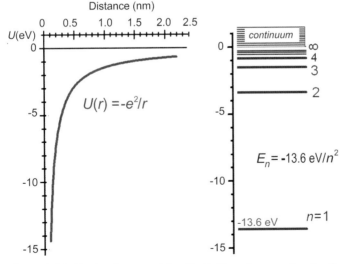

Figure 2.10 Coulomb potential $-e^2/r$ (left) and energy levels of an electron (right).

---

[1] It is assumed that the particle interacting with the electron is fixed in space, forming the origin of coordinates, or that it is so heavy compared to an electron that an electron cannot disturb this particle by its charge.

$$E^0 = \frac{1}{4\pi\varepsilon_0}\frac{e^2}{2a^0} \approx 13.60 \text{ eV.} \qquad (2.50)$$

Then, for dimensionless length and energy

$$\rho = \frac{r}{a^0}, \quad \varepsilon = \frac{E}{E^0}$$

the following equation for the radial part of the wave function can be written:

$$\left[\frac{d^2}{d\rho} + \varepsilon + \frac{2}{\rho} - \frac{l(l+1)}{\rho^2}\right]u(\rho) = 0. \qquad (2.51)$$

The solution of Eq. (2.47) leads to the following results. Energy levels obey a series,

$$\varepsilon = -\frac{1}{(n_r+l+1)^2} \equiv -\frac{1}{n^2}, \qquad n = 1,2,3,..., \qquad (2.52)$$

which is shown in the right-hand panel of Figure 2.10. The *principal quantum number* is $n = n_r + l + 1$. It takes positive integer values beginning with 1. The energy is unambiguously defined by a given n value. The *radial quantum number* n_r determines the quantity of nodes of the corresponding wave function. For every n value, exactly n states exist, differing in l which runs from 0 to $(n-1)$. Additionally, for every given l value, $(2l+1)$-degeneracy occurs with respect to $m = 0, \pm1, \pm2, ...$ Therefore, the total degeneracy is

$$\sum_{l=0}^{n-1}(2l+1) = n^2.$$

For $n = 1$, $l = 0$ (1s-state), the wave function obeys a spherical symmetry with a^0 corresponding to the most probable distance at which an electron can be found. The relevant value in a hydrogen atom is called the "Bohr radius," a_B. It differs from a^0 as in a problem including a proton and an electron, the reduced mass $\mu = m_p m_0 / (m_p + m_0)$ should be used, i.e.,

$$a_B = 4\pi\varepsilon_0 \frac{\hbar^2}{\mu e^2} = 5.2917 \cdot 10^{-2} \text{ nm.} \qquad (2.53)$$

The corresponding energy value,

$$Ry = \frac{1}{4\pi\varepsilon_0}\frac{e^2}{2a_B} = 13.605... \text{ eV} \qquad (2.54)$$

is called *Rydberg energy*, and gives the value of a hydrogen atom ionization energy from the ground state.

2.4 PERIODIC POTENTIAL

2.4.1 Bloch waves

The basic electronic and optical properties of crystalline solids arise from the properties an electron acquires in a periodic potential formed by ions arranged in a crystal lattice. Therefore, it is important to recall the basic properties of a quantum particle in a periodic potential.

Consider a one-dimensional Schrödinger equation for a particle with mass m:

$$-\frac{\hbar^2}{2m}\frac{d^2}{dx^2}\psi(x)+U(x)\psi(x)=E\psi(x), \tag{2.55}$$

with $U(x)$ having the period a, i.e.,

$$U(x)=U(x+a). \tag{2.56}$$

Eq. (2.56) leads to a condition,

$$U(x)=U(x+na), \tag{2.57}$$

where n is an integer number. Thus we have translational symmetry of potential, i.e., invariance of U with respect to coordinate shift over integer number of periods. To give a primary insight into the properties of wave functions, consider replacement $x \to x+a$. Then we arrive at the equation

$$-\frac{\hbar^2}{2m}\frac{d^2}{dx^2}\psi(x+a)+U(x)\psi(x+a)=E\psi(x+a), \tag{2.58}$$

where periodicity of the potential (Eq. (2.56)) has been implied. It is clear that $\psi(x)$ and $\psi(x+a)$ satisfy the same second-order differential equation. This means that either $\psi(x)$ is periodic or it differs from a periodic function by a complex factor whose square of modulus equals 1. In fact, the *Floquet* theorem known since 1883 states that the solution of Eq. (2.55) with a periodic potential reads

$$\psi(x)=e^{ikx}u_k(x), \quad u_k(x)=u_k(x+a), \tag{2.59}$$

i.e., the wave function is a plane wave (e^{ikx}) with amplitude modulated in accordance with periodicity of the potential $U(x)$. The k subscript for u_k means the function u_k is different for different wave numbers.

An important property of the wave function (Eq. (2.59)) is its periodicity with respect to the wave number. One can see that substitutions $x \to x+a$ and $k \to k+2\pi/a$ do not change $\psi(x)$.

In three dimensions, a periodic potential reads

$$U(\mathbf{r})=U(\mathbf{r}+\mathbf{T}), \tag{2.60}$$

where

$$\mathbf{T} = n_1\mathbf{a}_1 + n_2\mathbf{a}_2 + n_3\mathbf{a}_3 \tag{2.61}$$

is a translation vector, and $\mathbf{a}_i$ are elementary translation vectors defined as

$$\mathbf{a}_1 = a_1\mathbf{i} \, , \, \mathbf{a}_2 = a_2\mathbf{j} \, , \, \mathbf{a}_3 = a_3\mathbf{l} \tag{2.62}$$

with a_1, a_2, a_3 being the periods and $\mathbf{i}, \mathbf{j}, \mathbf{l}$ the unit vectors in x, y, z directions, respectively. The solution of a three-dimensional Schrödinger equation

$$-\frac{\hbar^2}{2m}\nabla^2\psi(\mathbf{r}) + U(\mathbf{r})\psi(\mathbf{r}) = E\psi(\mathbf{r}) \tag{2.63}$$

with periodic potential has the form

$$\psi(\mathbf{r}) = e^{i\mathbf{k}\cdot\mathbf{r}}u_k(\mathbf{r}), \quad u_k(\mathbf{r}) = u_k(\mathbf{r}+\mathbf{T}). \tag{2.64}$$

This statement is the essence of the *Bloch theorem*, derived by Felix Bloch in 1928. Functions represented by Eq. (2.64) are referred to as *Bloch waves.*

2.4.2   Brillouin zones and quasi-momentum

In what follows we consider a one-dimensional case for brevity and simplicity. Bloch waves describing a quantum particle in a periodic potential feature a number of important properties which to a large extent are common for all waves in periodic media.

First, there are standing waves for every wave number like π/a, $2\pi/a$, $3\pi/a$, etc. The standing waves appear from interference of waves scattered from periodically arranged scatterers – i.e., ions in a crystal lattice. At every point the dependence of a particle energy E on its wave number k (and, accordingly, on its momentum $p = \hbar k$) breaks to meet the standing wave condition,

$$\frac{dE}{dk} = \frac{1}{\hbar}\frac{dE}{dp} = 0, \tag{2.65}$$

which is equivalent to the conditions $p = 0$, $v = 0$. This is shown in Figure 2.11(a). Therefore, the continuous energy spectrum of a formerly free quantum particle now breaks into branches separated by gaps. These branches are called *energy bands*, and gaps between them are referred to as *band gaps*. For every k_n satisfying the condition

$$k_n = \frac{\pi}{a}n; \qquad n = \pm 1, \, \pm 2, \, \pm 3, \, \dots \tag{2.66}$$

two standing waves exist with different potential energies. Different energy values for the same k (or p) value can be understood if standing waves are imagined with the same k but with different displacement along the coordinate. In one case, wave function concentrates mainly in the portions of space with high potential, whereas in another case wave function locates mainly in portions of space with low potential.

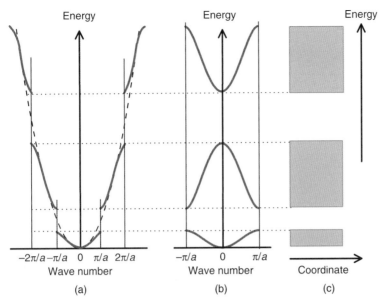

Energy · Energy · Energy

−2π/a −π/a 0 π/a 2π/a
Wave number
(a)

−π/a 0 π/a
Wave number
(b)

Coordinate
(c)

Figure 2.11 Development of energy bands in crystals. (a) Standing waves at $k = n\pi/a$ give rise to gaps in the $E(k)$ law. (b) The first Brillouin zone. Quasi-momentum conservation allows for the reduced presentation with every curve in (a) section shifted along the k axis by the integer number of $2\pi/a$. Thus all portions of the dispersion curve appear within the first Brillouin zone. (c) Energy bands in space.

Second, periodicity of $u_k(x)$ in the Bloch function along with the periodicity of the phase coefficient e^{ikx} with respect to kx with period 2π gives rise to the important property of a particle in periodic potential. Any pair of wave numbers k_1 and k_2 differing in an integer number of $2\pi/a$, i.e.,

$$k_1 - k_2 = \frac{2\pi}{a}n, \qquad n = \pm 1, \ \pm 2, \ \pm 3, \ \ldots,$$

become equivalent. This is a direct consequence of the translational symmetry of space in the problem under consideration. Therefore, the whole multitude of the k values consists of the equivalent intervals with the width of $2\pi/a$ each. Every such interval contains the full set of the non-equivalent k values. These intervals are called the *Brillouin zones* to acknowledge the principal contribution to this notion by Leon Brillouin. It is convenient to choose the first Brillouin zone around $k = 0$, i.e.,

$$-\frac{\pi}{a} < k < \frac{\pi}{a}. \tag{2.67}$$

Then the second zone will consist of the two symmetrical equal intervals,

$$-\frac{2\pi}{a} < k < -\frac{\pi}{a}, \quad \frac{\pi}{a} < k < \frac{2\pi}{a} \tag{2.68}$$

and so on. Because of the equivalence of wave numbers differing in integer number of $2\pi/a$, it is possible to move all branches of the dispersion curve toward the first Brillouin zone by means of a shift along the k axis by the integer number of $2\pi/a$. Therefore, the original dispersion curve (Figure 2.11(a)) can be modified to yield the reduced zone scheme (Figure 2.11(b)). The whole realm of a particle dynamics in a periodic potential can be thus treated in terms of events within the first Brillouin zone, the energy being a multivalued function of the wave number. Presentation of the dispersion law in Figure 2.11(b) is referred to as the *band structure*.

The value $p = \hbar k$ is called "quasi-momentum." It differs from the momentum by a specific *conservation law*. It conserves with an accuracy of $2\pi\hbar/a$, which is, again, a direct consequence of the translational symmetry of space. The quasi-momentum conservation law is to be considered in line with the other general conservation laws, namely momentum conservation (resulting from space homogeneity), energy conservation (resulting from time homogeneity), and circular momentum conservation (results from the space isotropy).

The relation $p = \hbar k$ gives rise to Brillouin zones for the momentum. The first Brillouin zone for momentum is the interval

$$-\hbar\frac{\pi}{a} < p < \hbar\frac{\pi}{a}, \tag{2.69}$$

and other zones consist of the two symmetric intervals with width equal to $\hbar\pi/a$.

2.4.3 Effective mass of electron

In the center and at the edges of the first Brillouin zone, the expansion of the $E(k)$ function in the Taylor series,

$$E(k) = E_0 + k\frac{dE}{dk}\bigg|_{k=k_0} + \frac{1}{2}k^2\frac{d^2E}{dk^2}\bigg|_{k=k_0} + \dots \tag{2.70}$$

can be reduced to a parabolic $E(k)$ law

$$E(k) = \frac{1}{2}k^2\frac{d^2E}{dk^2}\bigg|_{k=0} \tag{2.71}$$

by putting $E_0 = 0$, i.e., counting energy from the extremum point, and omitting the higher-order derivatives in the series. This becomes possible when recalling that extrema correspond to standing waves, i.e., $\dfrac{dE}{dk}\bigg|_{k=0} = 0$. In turn, the parabolic $E(k)$ law means the *effective mass* of the particle under consideration can formally be introduced in the vicinity of every extremum of the $E(k)$ function as

$$\frac{1}{m^*} = \frac{1}{\hbar^2}\frac{d^2E}{dk^2} \equiv \frac{d^2E}{dp^2} = \text{const.} \tag{2.72}$$

Note that for a free particle from the relation $E = p^2 / 2m = \hbar^2 k^2 / 2m$, we have everywhere

$$\frac{1}{\hbar^2} \frac{d^2 E}{dk^2} = \frac{d^2 E}{dp^2} \equiv m^{-1}.$$

The effective mass (Eq. (2.72)) determines the reaction of a particle to the external force, **F**, via a relation

$$m^* \mathbf{a} = \mathbf{F} \tag{2.73}$$

where **a** is the acceleration. Eq. (2.71) coincides formally with Newton's second law.

Figure 2.11(a,b) shows that, e.g., in the vicinity of $k = 0$ for the uppermost branch (band), the effective mass is noticeably smaller than the intrinsic inertial mass of a particle. Thus, a particle in a periodic potential can sometimes be "lighter" than in free space. Sometimes, however, it can be "heavier." This occurs, e.g., far from $k = 0$ when $E(k)$ becomes essentially nonparabolic. This means that in a periodic potential, a particle first accelerates as a lighter one, then with growing wave number and momentum it becomes heavier, and after crossing the inflection point its effective mass becomes negative. It is also always negative in the top of the bands near $k = 0$ because of the positive curvature of the $E(k)$ dependence in the vicinity of the maximum. The negative effective mass is an important property of a quantum particle interacting simultaneously with a background periodic potential and an additional perturbative potential. The negative mass means momentum of a particle decreases in the presence of an extra potential. This results from reflection from the periodic potential barriers/wells. The difference in momentum does not vanish, but is transferred to the material system responsible for the periodic potential – e.g., to the ion lattice of the crystal.

In the vicinity of every extremum, the Schrödinger equation with periodic potential reduces to the equation

$$-\frac{\hbar^2}{2m^*} \frac{d^2 \psi(x)}{dx^2} = E \psi(x), \tag{2.74}$$

which describes the *free motion* of a particle with the effective rather than the original mass.

2.5 ENERGY BANDS IN SEMICONDUCTORS AND DIELECTRICS

The $E(k)$ relation in Figure 2.11 represents the band structure for an electron in the simplest one-dimensional periodic potential. In real crystalline solids, periodic potential for an electron arises from the periodic displacement of ions, which in turn can be understood in terms of crystal lattices. The most simple lattices are presented in Figure 2.12. A *simple cubic lattice* can be found in a few elemental solids, e.g., Po and Te. A *body-centered cubic (BCC) lattice* occurs in certain binary ionic compounds, e.g., CsCl. For common semiconductor materials, *face-centered cubic* (FCC) and *hexagonal* lattices are important. Many semiconductors have cubic lattices based on the FCC arrangement. The typical lattices are

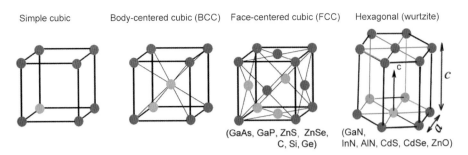

Figure 2.12 Examples of simple crystal lattices. Semiconductors are listed whose lattices are based on cubic, FCC, and hexagonal arrangements.

referred to as *zinc blende* (or *sphalerite*), representing two interpenetrating FCC lattices of two elements (ZnS, ZnSe, CdTe, GaP, GaAs, InP, InSb, CuCl), *diamond* lattice, which reproduces the zinc blende but with only one element (diamond, Ge, and Si), and *rocksalt*, which is the FCC lattice with an additional atom at the very center of the cube (NaCl, PbS, PbSe). Many compounds that are important in photonics have different versions of the hexagonal package, referred to as *wurtzite*.[2] Among those are GaN, InN, CdS, CdSe. The hexagonal lattice forms intrinsically anisotropic displacement of atoms along one axis (labeled the *c* axis), which results in anisotropy of the main physical properties, including optical properties.

The electronic properties[3] of solids are determined by occupation of the bands and by the absolute values of the forbidden gap between the uppermost completely occupied and lowermost unoccupied band whenever this is the case. If a crystal has a partly occupied band, it exhibits *metal* properties because electrons in this band provide electrical conductivity. If all the bands at $T = 0$ are either occupied or completely free, material will show *dielectric* properties. Electrons within the occupied band cannot provide any conductivity because of *Pauli's exclusion principle*: Only one electron may occupy any given state. Therefore, under an external field an electron in the completely occupied band cannot accelerate, i.e., cannot change its energy, because all neighboring states are already filled. The highest occupied band is usually referred to as the "valence band" (*v-band*) and the lowest unoccupied or partially occupied band is called "conduction band" (*c-band*).

---

[2] Interestingly, the origins of the notations *sphalerite*, *zinc blende*, and *wurtzite* are all related to the ZnS compound. *Sphalerite* originates from the Greek "sphaleros," i.e., treacherous; zinc blende originates from the German "Zinkblend," with "blenden" meaning illusive, delusive. Both names mean the cubic form of the ZnS mineral, which can be easily misidentified as other minerals. The *wurtzite* hexagonal form of ZnS is named after Charles-Adolph Wurtz (1817–1884), a French chemist.

[3] While using "electronic" we imply everything related to or resulting from electrons, not necessarily related directly to electronics. In this sense, all optical phenomena are electronic processes.

Spacing between the top of the valence band, E_v, and the bottom of the conduction band, E_c, is called the *band gap energy*, E_g:

$$E_g = E_c - E_v. \tag{2.75}$$

Depending on the absolute E_g value, solids that show dielectric properties (i.e., zero conductivity) at cryogenic temperatures close to $T = 0$, are classified into *dielectrics* and *semiconductors*. If E_g is less than 3–4 eV, the conduction band has a non-negligible population at room or slightly above room temperatures in accordance with the Boltzmann factor $-\exp(E_g/k_B T)$, with k_B being the Boltzmann constant, and these types of crystals are called "semiconductors." If E_g is significantly higher than 5 eV, then reasonable heating does not result in noticeable conductivity, and crystals are therefore referred to as *dielectrics*.

The relation between energy and wave number (i.e., momentum) of an electron for different bands, typically presented in graphic form, is called the "band structure." For real crystals it is a rather complicated multivalued function containing many bands and branches, very often differing in different directions. The electron effective mass cannot be considered as a constant, it appears to be energy-dependent and differs for various branches (so-called "valleys," i.e., local minima in the c-band). In a number of cases, the electron effective mass has to be described as a second-rank tensor. This means acceleration gained by an electron differs in direction from the applied force.

Notably, in optics, events within the close vicinity of v-band and c-band extrema are most important. The band structures of a number of "photonically important" semiconductors are sketched in Figure 2.13. One can see that GaN, GaAs, and CdSe v-band and c-band extrema have the same wave number. These are called "direct-gap semiconductors." Contrary to these, for Si, Ge, and GaP crystals, the top of the v-band does not coincide in

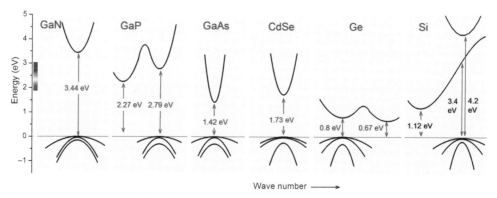

Figure 2.13 Sketch of the band structures of a number of representative semiconductors. GaN, GaAs, and CdSe are direct-gap and GaP, Si, and Ge are indirect-gap semiconductors. The top of the valence band is chosen as the zero-energy level for convenience only. With respect to a free electron in a vacuum, crystals have different energy scales.

k value with the lowest minimum in the c-band. These types of crystals are usually referred to as "indirect-gap semiconductors."

The most important semiconductor materials used in modern photonic components and devices are presented in Table 2.2. The right-hand column shows that nanophotonics can be helpful in extension and/or improvement of their applications.

Table 2.2 **Strategic semiconductor materials in photonics, their application fields, and options for nanophotonic implementations**

Material	Current status	R&D: trends and forecast	Nanophotonics
GaN, GaInN	High-power blue, UV, and white LEDs, lasers for high-density DVD and Blu-ray disks	Progress in current applications	+
GaAs, GaAlAs, GaInAs, GaAlInP, GaInAsP	Lasers in CD, DVD, laser printers, IR LEDs, near-IR-Vis photodetectors, fiber-optics communication, high-cost efficient solar cells for critical applications (space)	Progress in current applications	+
Si	Detectors, incl. CCD matrices in video- and photo-cameras, solar cells	Light-emitting devices, electro-optical modulators	+
GaP and GaInAlP	Low-power green (yellow) LEDs		+
Ge and SiGe	IR detectors (**obsolete**)	Far-IR and THz lasers (quantum cascade)	+
CdTe and CdHgTe	Far-IR detectors incl. night-vision cameras	Solar cells	+
CdSe, CdS	photoresistors	Colloidal nanophotonics incl. luminophores, LEDs, lasers	+
ZnO and ZnMgO	UV detectors	UV LEDs and lasers, transparent contacts	+
ZnSe	Low power blue LEDs and lasers	Progress in current applications	+
PbS, PbSe	IR detectors	Optical and electro-optical modulators/switches	+

BOX 2.1 ADVENT OF GaN PHOTONICS

Gallium nitride, GaN, is now the second most important semiconducting material after silicon, and probably is the number one photonic semiconductor material. In 1995 Shuji Nakamura and co-workers reported on the first blue LEDs based on a GaInN single quantum well structure. These devices have been significantly improved and extended both to shorter and longer wavelengths from the green to the ultraviolet. These diodes are used in large area screen displays, traffic lights, colored automobile lights, and backlighting for mobile phones and iPads.

Using yellow phosphors with blue GaInN LEDs has allowed developing high-intensity white LEDs that are expected to revolutionize residential lighting in a decade. Blue GaN-based laser diodes are commercially available and have enabled the advent of Blu-ray disks and HD-DVD (high-density DVD) in daily life. In recognition of this decisive contribution to science and technology, Shuji Nakamura together with Isamu Akasaki and Hirosi Amano received the Nobel Prize in 2014 for the invention of blue LEDs, which has enabled bright and energy-saving white light sources.

Isamu Akasaki

Hirosi Amano

Shuji Nakamura

2.6 QUASIPARTICLES: ELECTRONS, HOLES, AND EXCITONS

Electrons in the conduction band of a crystal can be described as "free" particles with charge $-e$, spin $\pm^1/_2$, effective mass m_e^* (basically variable rather than constant, often anisotropic), and quasi-momentum $\hbar k$ with the specific conservation law. Thus, for an electron in a crystal, only charge and spin remain the same as in free space. By and large, the notation "electron" implies a particle whose properties result from the interactions in a many-bodies system consisting of a large number of positive nuclei and negative electrons. Notably, it is the standard and a very efficient approach in the theory of many-bodies systems to replace the large number of interacting particles with a small number of non-interacting *quasiparticles*. These quasiparticles are viewed as elementary excitations of the system consisting of a large number of real particles. Thus, an electron in c-band is a quasiparticle corresponding to the primary elementary excitation in the electron subsystem of a crystal.

The "hole" is another quasiparticle in crystals. It is introduced to describe an ensemble of electrons in the valence band from which one electron has been removed (e.g., by means of a transition to the conduction band). In a sense, the hole can be viewed as a void in air or water to which we can apply notions like size, speed, acceleration, etc. While many pieces of matter drop down, voids apparently go up as if they acquire "negative" mass. Similarly, the empty place in the v-band is viewed as a quasiparticle with charge $+e$ (opposite to electron), effective mass m_h, depending on the v-band shape, spin $\pm\frac{1}{2}$, and kinetic energy rising from the top of the v-band downward (opposite to electron). Note, the top of the valence band typically features at least two branches, giving rise to the notion of *light* and *heavy* holes.

Electrons and holes in semiconductors possess properties of ideal gases at low concentrations. When their concentration rises up, electron–hole plasma develops, evolving sometimes into electron–hole liquid (found in a few crystals, including Si and Ge, under extreme concentrations and low temperatures). Electron and hole gases acquire temperatures depending on the energy distribution functions in the c- and v-bands, respectively; their temperatures are often different from each other and from that of a parent crystal (determined by atomic vibrations).

Using the hole notion, a transition of an electron from the v-band to the c-band can be viewed as the creation of an *electron–hole pair*. Such transitions may occur due to photon absorption (Figure 2.14), with the energy and momentum conservation

$$
\begin{aligned}
\hbar\omega &= E_g + E_{e\,kin} + E_{h\,kin} \\
\hbar\mathbf{k}_{phot} &= \hbar\mathbf{k}_e + \hbar\mathbf{k}_h
\end{aligned}
\quad ,
\tag{2.76}
$$

where $E_{e\,kin}$ ($E_{h\,kin}$) and $\mathbf{k}_e$ ($\mathbf{k}_h$) are the electron (hole) kinetic energy and wave vector, respectively.

Only photons whose energy exceeds E_g can be absorbed. If this condition is met, photons can be readily absorbed provided that energy and momentum conservation laws are met simultaneously. Thus, E_g value defines the spectral position of the optical absorption edge after which absorption spectrum is continuous. If the photon energy equals E_g then the electrons and holes that take part in the photon absorption have zero kinetic energies. The band gap energy equals the minimal energy for creation of a single pair of free charge carriers (an electron and a hole). This statement can serve as the definition of E_g.

If the photon energy exceeds E_g, the excess energy breaks into the kinetic energy of an electron and that of a hole. So a high-energy photon, when absorbed, generates a faster electron–hole pair. Notably, the photon momentum is rather small (Figure 2.14(a)), e.g., for the visible range ($\hbar\omega = 1.7$–3.1 eV), it is two orders of magnitude lower than the electron momentum (Problem 2.12). As the photon momentum is negligibly small, we simply have a nearly vertical transition in the diagram shown in Figure 2.14(b). The energy versus momentum and wave number law for photons plotted in Figure 2.14(a) directly follow from Eqs. (2.2) and (2.3), taking into account $E = \hbar\omega$ and $p = \hbar k$.

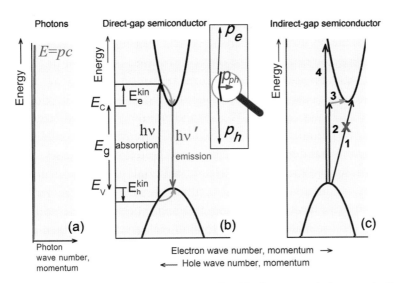

Figure 2.14 Energy versus momentum for (a) photons and electrons in (b) direct-gap and (c) indirect-gap semiconductors.

The insert in Figure 2.14(b) explains the momentum conservation law in photon absorption. Very small photon momentum enables generation of an electron–hole pair when the electron and the hole move in nearly opposite directions to make the vectorial sum of the momenta fit the small value of the photon momentum. This consideration accords with the nearly vertical transition shown in Figure 2.14(b) by the upward blue arrow, if we agree that the momentum axis for holes is in the opposite direction to that for electrons. Interband optical transitions result in high absorption coefficients measuring of the order of 10^4 cm^{-1}.

The reverse process, i.e., a downward radiative transition, is called an electron–hole *recombination* process. An electron–hole pair converts into a photon, again, with simultaneous energy and momentum conservation. Electron and hole energies and momenta may experience relaxation toward extreme points in the c- and v-band, respectively, before the recombination happens. Such relaxation occurs by means of e–e, h–h, and e–h collisions, as well as by interaction with the crystal lattice. The e–e processes modify electron energy distribution and result in electron gas temperature T_e, the h–h processes define hole gas temperature T_h, and the e–h processes make these temperatures equal to each other, whereas electron and hole interaction with the lattice makes the crystal warmer resulting in $T_e = T_h = T_c$, where T_c is the crystal lattice temperature.

Figure 2.14(c) shows possible optical transitions in the case of an indirect-gap semiconductor. The E_g value now corresponds to different points on the k axis and a one-step transition from the top of the v-band to the bottom of the c-band is not allowed (transition 1) since the momentum conservation cannot be met. In this case, the lack of momentum can be taken from the crystal lattice by means of simultaneous absorption of a photon and a single or a number of lattice oscillations energy portions, *phonons*. This many-bodies process has low probability, and therefore indirect-gap semiconductors exhibit low absorption coefficients unless the photon energy enables direct transitions (transition 4 in Figure 2.14). The photon energy threshold for the direct transitions is referred to as the *optical band gap* energy. The recombination rates for indirect transitions are also low. Low absorption and recombination rates are among the issues for indirect-gap semiconductors used in photonic detectors and solar cells (e.g., Si, Ge) and emitters (e.g., GaP-based LEDs).

Table 2.3 summarizes parameters of "photonically" essential semiconductor crystals including lattice structure, band gaps, and electron and hole effective masses.

At zero temperature, semiconductor and dielectric materials have fully occupied v-bands and completely empty conduction bands. This becomes possible in many materials because *s*- and *p*-states in atoms can have no more than two and six electrons, respectively. Many semiconductor and dielectric materials have a total of eight electrons in their constituting atoms. For elemental materials this condition is met for group IV elements only (diamond C, or graphite, silicon, germanium, as well as SiC compound). For binary materials a pair of elements of I–VII (NaCl, LiF, CuCl, AgBr, etc.), II–VI (ZnO, ZnS, ZnSe, ZnTe, similar Cd and Hg compounds), and III–V (BN, GaN, GaP, GaAs, GaSb, similar Al and In compounds) couples to complete eight valence electrons for every pair of atoms in chemical bonds. These types of solids are presented as separate groups in Table 2.3.

There is an important correlation between properties of constituting elements and band gap of crystals formed. First, the band gap decreases within every group with increasing element numbers, i.e., number of electronic shells. Consider, for example, band gaps in the rows C → SiC → Si → Ge, AlN → GaN → InN, GaN → GaP → GaAs. Continue I–VII compounds (see also Problem 2.14). This happens because a higher number of electron shells results in screening of Coulomb potential in the lattice and makes potential wells shallower. This relationship seems to be fulfilled unless crystal lattice structure remains the same. It fails, e.g., for hexagonal ZnO versus cubic ZnS, or cubic HgS versus hexagonal CdS, either of which with shorter crystal lattice will have a bigger gap. Another regularity occurs with respect to the horizontal position of elements in the Periodic Table. For the same total number of shells, the band gap typically rises in the row IV → III–V → II–VI → I–VII. Examples from Table 2.3 are Ge → GaAs → ZnSe → CuBr; others are LiF → BN → C, NaCl → AlP → Si, and there are many more. The reason is stronger polarity of lattice structure with growing charge difference. For a smaller valence difference, bonding is close to covalent, whereas for a larger valence difference it is strongly ionic, making potential alteration in space higher. Thus, to summarize the two above regularities, one can see that *stronger bonding makes shorter lattices and bigger gaps in the electron spectrum.*

2.6.1 Excitons

Electrons and holes form gases and plasma in semiconductors. Not only do they experience multiple scattering resulting in Boltzmann-like energy distributions within the c-band for electrons and v-band for holes, but every e–h pair may form a bound hydrogen-like state

Table 2.3 **Parameters of semiconductors used in photonic components and devices**

Material, lattice type, band gap type	Band gap (eV) at 300 K	Band gap wavelength (nm)	m_e/m_0	m_h/m_0	a_B* (nm)	Ry* (meV)	Lattice constant (nm)
Group IV elements and IV−IV compounds							
C diamond, dia, i	5.47	226	0.36(t) 1.4 (l)	1.1(hh) 0.36 (lh)	–	80	0.357
SiC (6H), i, w	3.02	410	0.48(t) 2(l)	0.66 (t) 1.85 (l)	–	–	0.308 (a) 1.512 (c)
Si, i, dia	i 1.12 d 3.4	1090	0.08 (t) 1.6 (l)	0.3 (hh) 0.43 (lh)	4.3	15	0.543
Ge, i, dia	i 0.67 d 0.80	1850 1550	0.19 (t) 0.92 (l)	0.54 (hh) 0.15 (lh)	24.3	4.1	0.566
III−V compounds							
α-GaN, w, d	3.44	360	0.22	0.3 (lh) 1.4 (hh)	2.1	28	0.319 (a) 0.518 (c)
β-GaN, z, d	3.17	390	0.19	0.2 (lh) 0.7 (hh)	–	26	0.453
GaP, z, i	i 2.27 d 2.79	546 444	0.21	0.17 (lh) 0.67 (hh)	7.3	22	0.545
GaAs, z, d	1.42	872	0.064	0.08 (lh)(111) 0.09 (lh)(100) 0.34 (hh)(100) 0.75 (hh)(111)	12.5	4.6	0.565
InP, z, d	1.34	924	0.07	0.12 (lh) 0.45 (hh)	16.8	4	0.586
InN, w, d	0.7	1770	0.08	–	8	15.2	0.354 (a) 0.570 (c)
InAs, z, d	0.35	3540	0.02	0.35 (100) 0.85 (111)	36	1.5	0.605

Table 2.3 **(Cont.)**

Material, lattice type, band gap type	Band gap (eV) at 300 K	Band gap wavelength (nm)	m_e/m_0	m_h/m_0	a_B^* (nm)	Ry^* (meV)	Lattice constant (nm)
InSb, z, d	0.18	6880	0.01	0.01 (lh) 0.4 (hh)	–	0.5	0.647
AlN, w, d	6.13	202	0.4	3.53 (hh l) 10.4 (hh t) 3.53 (lh l) 0.24 (lh t)	1.2	70	0.311 (a) 0.498 (c)
AlP, z, i	i 2.53 (6 K) d 3.63 (4 K)	490 341	–	–	1.2	25	0.546
AlAs, z, i	i 2.15 d 3.03	576 409	0.19 (t) 1.1 (l)	0.4 (hh)(100) 1.0 (hh)(111) 0.15 (lh)(100) 0.11 (lh)(111)	2.0	20	0.566
II−VI compounds							
ZnO, w, d	3.37	367	0.27	0.59	1.8	63	0.325 (a) 0.520 (c)
ZnS, z, d	3.72	333	0.22	0.23 (lh) 1.76 (hh)	2.5	38	0.541
ZnSe, z, d	2.68	462	0.15	0.75 (hh) 0.14 (lh)	3.8	21	0.567
ZnTe, z, d	2.35	527	0.12	0.6	6.7	13	0.609
CdS, w, d	2.48	499	0.14	0.7 (t) 5 (l)	2.8	29	0.413 (a) 0.675 (c)
CdSe, w, d	1.73	716	0.11	0.45 (t) 1.1 (l)	4.9	16	0.430 (a) 0.701 (c)
CdTe, z, d	1.47	840	0.1	0.4	7.5	10	0.647
HgTe, z	0 (semimetal)	8200	0.03	0.3			0.645

Table 2.3 **(Cont.)**

Material, lattice type, band gap type	Band gap (eV) at 300 K	Band gap wavelength (nm)	m_e/m_0	m_h/m_0	a_B^* (nm)	Ry^* (meV)	Lattice constant (nm)
I−VII compounds							
CuCl, z, d	3.2	390	0.5	2	0.7	190	0.542
CuBr, z, d	2.9	430	0.2	1.1 (lh) 1.5 (hh)	1.2	108	0.453
IV−VI compounds							
PbS, r	0.41	3020	0.040 (t)	0.034 (t)	18	2.3	0.593
PbSe, r, d	0.28	4420	0.070 (l)	0.068 (l)	46	2.0	0.613

Notations: direct-gap structure (d), indirect-gap structure (i); predominant crystal structure: wurtzite (w), zinc blende (z), diamond (dia), rocksalt (r); light holes (lh), heavy holes (hh), transverse (t), longitudinal (l), (a) and (c) lengths as shown in Figure 2.12. Sources: Klingshirn (2004), Madelung (2004), and Gaponenko (2010).

owing to Coulomb interaction. The relevant quasiparticle is termed "exciton." Figure 2.9 gives an idea of *s*-, *p*-, and *d*-states of excitons. Similar to the hydrogen atom, the lowest states feature spherical symmetry and can be characterized by *exciton Bohr radius*, a_B^* (compare to Eq. (2.49)), which in SI units reads

$$a_B^* = 4\pi\varepsilon_0 \frac{\varepsilon\hbar^2}{\mu_{eh}e^2} = \varepsilon\frac{\mu_H}{\mu_{eh}} \cdot 0.053 \text{ nm} \approx \varepsilon\frac{m_0}{\mu_{eh}} \cdot 0.053 \text{ nm,} \qquad (2.77)$$

where μ_H is the electron–proton reduced mass defined as

$$\mu_H^{-1} = m_0^{-1} + m_{\text{proton}}^{-1} \approx m_0^{-1}, \qquad (2.78)$$

the result of which with good accuracy (10^{-3}) equals the electron mass m_0 because $m_0 \ll m_{\text{proton}}$, and μ_{eh} is the electron–hole reduced mass,

$$\mu_{eh}^{-1} = m_e^{*-1} + m_h^{*-1}. \qquad (2.79)$$

Similar to a hydrogen atom, *exciton Rydberg energy Ry** can be written (in SI units) as

$$Ry^* = \frac{1}{4\pi\varepsilon_0}\frac{e^2}{2\varepsilon a_B^*} = \left(\frac{1}{4\pi\varepsilon_0}\right)^2\frac{\mu_{eh}e^4}{2\varepsilon^2\hbar^2} = \frac{\mu_{eh}}{\mu_H}\frac{1}{\varepsilon^2} \cdot 13.60 \text{ eV} \approx \frac{\mu_{eh}}{m_0}\frac{1}{\varepsilon^2} \cdot 13.60 \text{ eV.} \qquad (2.80)$$

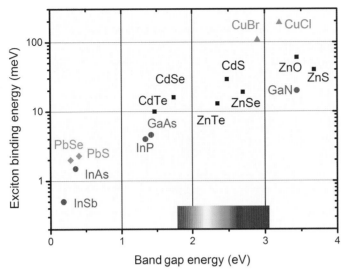

Figure 2.15 Exciton binding energy (Ry^*, meV) versus band gap energy (E_g, eV) for a number of direct-gap semiconductors. A general trend of higher Ry^* for bigger E_g is apparent and can be roughly approximated as a straight line in a semi-logarithmic plot.

The reduced electron–hole mass is smaller than the electron mass m_0, and the *dielectric constant* ε is of the order of 10. This is why the exciton Bohr radius is significantly bigger, and the exciton Rydberg energy is significantly smaller than the relevant values of the hydrogen atom. Absolute values of a_B^* for the common semiconductors range in the interval 10–100 Å, and the exciton Rydberg energy takes the values from approximately 1 to 100 meV (Table 2.3). An exciton can be described as an *elementary excitation in the electron subsystem of a crystal that does not contribute to charge transfer.*

There is a definite correlation between the band gap values, E_g, and exciton binding energy, Ry^*: bigger band gaps correlate with bigger Ry^*. This correlation can be traced in Figure 2.15. It results both from the tendencies of higher electron effective mass and lower dielectric constants for wider-band gap materials. In other words, bigger gaps inherent in materials with strong atomic bonding provides stronger Coulomb interactions of electrons and holes therein. Stronger Coulomb interaction, in turn, gives rise to a small Bohr radius. For the hydrogen-like consideration of electron–hole coupling it is of principal importance that the calculated Bohr radius should necessarily exceed many times the crystal lattice constant, i.e., $a_B \gg a_L$. Otherwise, presentation of electron and hole interacting through a medium with dielectric constant ε is incorrect. Notably, for most of the crystals listed in Table 2.3, the condition $a_B \gg a_L$ is met. However, it is not perfectly met for copper halides, especially CuCl, where a_B is only slightly bigger than the lattice constant. For crystals with wider gaps (more ionic) – like NaCl, KCl, and KBr – the exciton is localized at a given crystal site and does not possess hydrogenic features. In fact, it is this type of exciton that was predicted in 1931 by Ya. Frenkel (USSR) and is known as the *Frenkel* *exciton*. On the contrary, the hydrogenic excitons inherent in semiconductors are referred to as *Wannier–Mott excitons*. The latter type of excitons were observed for the first time in 1951 by E. F. Gross and co-workers (USSR).

Excitons, holes, and electrons in a crystal are viewed as gas consisting of atoms (excitons) that can ionize to give an ion (hole) and an electron. Coupled and free electron–hole pairs exist in dynamic equilibrium, depending on temperature. The relevant equation is known as the equation of ionization equilibrium, or *Saha equation*. It reads:

$$n_{eh}^2 = n_{exc} \left(\frac{\mu_{eh} k_B T}{2\pi\hbar^2} \right)^{3/2} \exp\left(\frac{-Ry^*}{k_B T} \right), \tag{2.81}$$

where n_{exc} is the exciton concentration and n_{eh} is the concentration of free electrons (holes). Note that in an intrinsically neutral crystal, the number of electrons, n_e, will always be equal to the number of holes, n_h. This is emphasized by the n_{eh} notation. Eq. (2.81) has a very general field of application for the relation between neutral and ionized atoms at a certain temperature. It was derived in 1920 by M. Saha, an Indian physicist, when considering ionization of atoms in astrophysics. The Saha equation states that at a low temperature ($k_B T \ll Ry^*$) electron–hole pairs exist in the form of excitons, i.e., $n_{eh} \ll n_{exc}$ holds, whereas at higher temperatures ($k_B T > Ry^*$) most excitons ionize to give free electrons and holes.

Excitons at rest ($p = 0$, $k = 0$) feature the hydrogenic energy spectrum defined by Eq. (2.48):

$$E_n = E_g - \frac{Ry^*}{n^2}, \qquad n = 1, 2, 3, ..., \tag{2.82}$$

with the lowest state at $E_g - Ry^*$ followed by converging series of levels to give final continuum at $E > E_g$, like the generic set for Coulomb potential presented in Figure 2.10. In every state, an exciton can perform translational motion as a particle with mass $M = m_e + m_h$. Therefore the $E(k)$ dependence has an infinite number of parabolic branches (Figure 2.16) in accordance with the relation

$$E(K) = E_g - \frac{Ry^*}{n^2} + \frac{\hbar^2 K^2}{2M}, \qquad M = m_e + m_h, \qquad n = 1, 2, 3, ... \tag{2.83}$$

Recalling $E(k)$ dependence for photons (red curve in Figure 2.16), one can see that every time this curve crosses those of the exciton, energy and momentum conservation laws allow for photon absorption to occur. Since the photon wave number is very small, we can consider these absorption lines to occur exactly at $\hbar\omega_n = E_g - Ry^* / n^2$. The reasonable theoretical consideration of exciton absorption spectra was proposed by R. Elliot in 1957. He predicted that (1) the principal absorption line ($n = 1$) should have intensity compared to the atomic line intensity in proportion with (atomic size/exciton size)³; (2) intensity of exciton lines falls with n as $1/n^3$; (3) infinite number of states with a higher n gives rise to constant absorption for higher $\hbar\omega$; and (4) when $\hbar\omega$ exceeds E_g

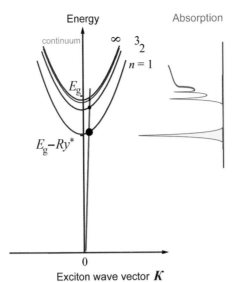

Energy   Absorption

continuum   ∞ 3
             2
             $n = 1$

E_g

$E_g - Ry^*$

0

Exciton wave vector K

Figure 2.16 Energy versus wave number for excitons (blue) and for photons (red). Every point at which the photon curve crosses the exciton curve gives rise to the absorption band shown in the right panel.

noticeably, absorption coefficient rises up as $(\hbar\omega - E_g)^{1/2}$. The absolute value of absorption coefficient in the continuum can be as high as $10^4 \, \text{cm}^{-1}$, i.e., transmission of a 1 μm film will be $1/e = 0.36$.

The semi-classical exciton theory by Elliot implies a Lorentzian shape for every exciton line, but does not provide a way to calculate line widths. The latter is defined by exciton–phonon interaction, i.e., exciton dephasing from scattering on vibrating ions in the crystal lattice. The dephasing rate simply then equals the observed linewidth at the half-maximum on the frequency scale. It is roughly close to the $k_B T$ value. Emission of photons is possible from every exciton ns-state ($1s$-, $2s$-, ...) as a result of the electron recombination with the hole within an exciton, an event which is often referred to as exciton *annihilation*.

To evaluate manifestation of excitonic sharp lines in absorption spectra, exciton binding energies for semiconductors from Table 2.3 should be compared to $k_B T$ for different temperatures, $(k_B T)_{300\,\text{K}} = 26$ meV, $(k_B T)_{77\,\text{K}} = 6.7$ meV, $(k_B T)_{4\,\text{K}} = 0.3$ meV.

Pronounced exciton peaks cannot be observed at room temperature for most typical semiconductors because of strong temperature-induced broadening that smears resonance peaks and also because most excitons ionize into electrons and holes. The ground exciton state typically is readily observed at liquid nitrogen temperatures (77–100 K) in high-quality monocrystals of many II–VI (CdS, CdSe, ZnSe, ZnS) and III–V (GaN) compounds with a film thickness of the order of 1 μm fabricated by means of epitaxial growth on monocrystalline substrates. Two representative examples are given in Figure 2.17.

Owing to their large binding energy, GaN monocrystals feature a well-resolved exciton line with absorption coefficient α about $10^5 \, \text{cm}^{-1}$ at maximum. Evaluation of α values in these conditions can be performed with extremely thin films only (0.4 μm) with carefully accounting for light interference at the low-energy tail of the absorption spectrum. At

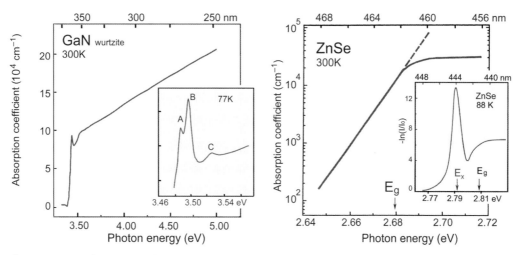

Figure 2.17 Optical absorption spectra at room and liquid nitrogen temperatures for two representative semiconductors, GaN (wurtzite type) and ZnSe (cubic). GaN was epitaxially grown by the MOCVD technique on a sapphire (Al_2O_3) substrate, which is transparent within the spectral range of measurements. ZnSe was epitaxially grown on a GaAs substrate with perfect matching of crystal lattices and transmission was studied through an etched window in the substrate since GaAs features strong intrinsic interband absorption in the visible. Data for GaN are reprinted from Muth et al. (1997) with permission from APS.

liquid nitrogen temperature, three exciton peaks dominate in the band edge spectrum, resulting from the ground states of excitons formed by three hole types from various v-band branches. Excited states of excitons are not resolved at these temperatures but are pronounced very well at liquid helium temperatures.

ZnSe monocrystals show a well-defined exciton band whose low-energy half obeys the Lorentzian shape with a full-width at half-maximum 7 meV, i.e., equal to k_BT for 88 K. At room temperature, exciton band(s) are completely smeared but Coulomb interaction of electrons and holes enhances absorption below E_g. To examine this in detail, samples of different thicknesses have been used. An exponential absorption tail is clearly seen that can be described by the law

$$\alpha(h\nu, \alpha) = \alpha_0 \exp\left(\frac{\sigma(h\nu - E_0)}{kT}\right) \tag{2.84}$$

where σ and E_0 are parameters defining the steepness of the tail. This exponential law was reported for the first time for AgBr crystals by Franz Urbach at Eastman Kodak in Rochester, USA in 1953, and then studied in more detail by Werner Martienssen in 1957. Since then it has been found to occur in pure monocrystals (including GaN); heavily doped, disordered crystalline materials, and glasses. In the case of pure monocrystals it was shown to result from exciton–phonon interaction (mainly elastic scattering), whereas in doped and disordered materials it comes from the lattice disorder. The exponential low-energy absorption tail in semiconductors is referred to as the *Urbach–Martienssen rule*.

2.6.2 Spin effects for quasiparticles in crystals

Electrons and holes possess spin ½ and therefore these are *fermions*. Only one fermion may occupy every available state. They obey the *Fermi–Dirac statistic* with the probability *f* for a state with energy *E* being occupied that reads:

$$f_{FD}(E, k_B T) = \frac{1}{1 + \exp\left(\dfrac{E - E_F)}{k_B T}\right)}.$$

(2.85)

Fermi–Dirac distribution varies from 1 for $E \ll E_F$ to 0 for $E \gg E_F$. Its steepest portion corresponds to the interval of approximately $k_B T$ around E_F. The latter is called the *Fermi level* or *Fermi energy*. It is often said that Fermi energy is the energy level whose occupation probability is ½. It is mathematically correct, but one should keep in mind that there might be no real energy state at $E = E_F$. Only heavily doped or heavily pumped samples feature the Fermi level in the conduction band. Intrinsic semiconductors have the Fermi level in the middle of the band gap. More sound physical interpretation is that Fermi energy stands for *chemical potential*, which is the free energy of an ensemble calculated per particle. It defines how much the total free energy will change if one electron is added or removed from an ensemble.

Excitons consisting of two particles with half-integer spins are *bosons*. Therefore, any number of excitons can occupy the same state. Their energy distribution function is described by the *Bose–Einstein statistic*, which reads

$$f_{BE}(E, k_B T) = \frac{1}{\exp\left(\dfrac{E - \mu)}{k_B T}\right) - 1},$$

(2.86)

with μ being the chemical potential of a system under consideration (see the definition above).

2.6.3 Impurities in semiconductors

Semiconductors can be purposefully doped to modify their electrical and also optical properties. An impurity atom, when substituting an intrinsic one, creates a set of hydrogenic-like energy states within the band gap, and if this state is close to the bottom of the c-band with respect to the $k_B T$ value, then thermal release of an impurity electron provides an extra free electron to the c-band. These impurities are termed *donors*. Donor atom valency should exceed the parent atom(s) valency. For example, group V (e.g., P) elements are donors to Ge and Si, group III elements are donors to II–VI compounds when substituting a group II element in the parent lattice (e.g., Ga in ZnSe when substituting zinc). A semiconductor possessing a high number of free electrons at room temperature is referred to as an *n-type* semiconductor. Its conductivity owing to equilibrium electrons (i.e., without external pumping or injection) exceeds by many orders of magnitude the conductivity of intrinsic – i.e., undoped – crystal.

In a similar way, an impurity atom creating a level in the band gap close to the top of the v-band gives a hole to the v-band. Such an impurity is termed an *acceptor*. These are atoms whose valency is lower than that of the parent crystal. Examples are group III elements in Si, and group II elements in III–V (like Mn in GaN). These types of doped crystals are referred to as *p-type*. Such crystals show high room-temperature conductivity without optical or electrical pumping owing to intrinsic free holes in their v-band.

Heavily doped semiconductors show blue shift of absorption onset since a portion of the c-band (in n-type crystals) or v-band (in p-type crystals) is occupied with charge carriers and is not involved in the absorption process. This phenomenon is known as the *Burstein–Moss shift*.

In binary compounds, the same type of impurity atom can be either donor or acceptor, depending which site it enters in the binary compound. One can see that the group IV element is a donor when substituting Ga, but it can also serve as an acceptor when substituting N in GaN or As in GaAs.

Crystals typically have intrinsic defects like interstitial atoms or vacancies. Their number may rise when donor or acceptor impurity is added. There is a non-trivial phenomenon found in II–VI compounds when an impurity atom itself creates a donor state whereas its complex with a vacancy or an interstitial parent atom develops an acceptor state. Vice versa, impurity thought to be the acceptor often generates undesirable donor states when coupling with an intrinsic defect. Then the situation occurs that a heavily doped semiconductor simultaneously has a high concentration of donors and acceptors but its conductivity remains low since concentration of free carriers is low. In this case, electrons from donor atoms are captured by acceptors instead of becoming free charge carriers in the c-band. This phenomenon is called *self-compensation*. It becomes a serious obstacle in making *p−n junctions* in II–VI compounds, such as when ZnSe is to be used for light-emitting or laser diodes. In the 1980s, most efforts by researchers toward development of visible lasers and LEDs were concentrated on groups II–VI, but it was the self-compensation phenomenon that did not allow these compounds to compete with the recently emerged GaN-, GaInN-, and AlGaN-based light-emitting devices.

In heavily doped compensated semiconductors, discrete levels inherent in impurity atoms spread into wide tails below the c-band and over the v-band because of dense chaotic distribution of potential wells in space. These tails form an Urbach-like exponential absorption wing on the low-energy side of the $\hbar\omega = E_g$ point.

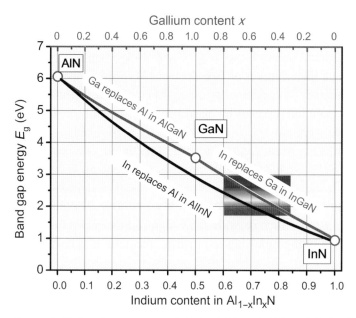

Figure 2.18 Band gap energy of AlN, GaN, and InN and their alloys. Lines are plotted according to calculations by Pelá et al. (2014), which have been shown to describe reasonably experimental data reported by many groups. The spectrum box highlights the $Ga_xIn_{1-x}N$ range, where band gap energy enters the visible, thus enabling LED design. This range is used in GaN-based visible LEDs.

Ternary compounds, like GaInN and AlGaN, are called *solid solutions*. These are developed from binary compounds by substitution of one of the parent atom type with an *isovalent* impurity. That means that a parent group III atom should be substituted by a group III impurity, or a parent group V element should be replaced by another group V atom. For GaN as a starting material, a compound with a wider gap, AlN, can be developed using $Ga_xAl_{1-x}N$ as well as compounds with a smaller band gap, InN, namely $Ga_xIn_{1-x}N$ can be fabricated. Ternary compounds feature the band gap energy and lattice constant intermediate between the starting and ending binary ones (Figure 2.18). Groups II–VI and IV compounds can also make solid solutions, e.g., $CdSe_xS_{1-x}$, $Zn_xCd_{1-x}Se$, Si_xGe_{1-x}. In a similar way, *quaternary solid solutions* can be developed – e.g., $Ga_xAl_{1-x}N_yAs_{1-y}$, $Zn_xCd_{1-x}Se_yTe_{1-y}$. Solid solutions enable tuning of band gaps as well as crystal lattices. Tuning band gaps is crucially important in development of photonic devices as it is the band gap that defines the absorption onset and emission wavelengths. Tuning the lattice period enables better matching of lattices in epitaxial growth when the crystal lattice of a growing semiconductor develops on top of the crystalline substrate lattice.

Conclusion

- Electron properties of solids define both their electronic and optical characteristics since it is electronic excitation that determines light interaction with matter.

- Electrons and other quantum particles in potential wells feature discrete spectra, the spacing between levels being constant only in the case of a harmonic oscillator (parabolic well).

- In potential wells steeper than parabolic ones (e.g., rectangular), energy-level spacing of a quantum particle rises for the growing level numbers, whereas if the wells are smoother than the parabolic one (e.g., Coulomb), energy levels get closer for the growing level numbers.

- Tunneling occurs in the case of a potential barrier with finite height and width.

- Resonant tunneling with high transmission probability is possible in the case of two barriers if a particle's energy coincides with any of the possible steady states in the well between barriers.

- Periodic potential gives rise to energy bands separated by energy gaps.

- Elementary excitations in the electron subsystem of crystals are electrons, holes, and excitons, the latter being the hydrogenic state of electrons and holes.

- Semiconductor crystals obey either a direct-gap or indirect-gap structure.

- The absorption spectrum of semiconductors features high transparency for photon energies below exciton energies and high absorption for photon energies higher than exciton energies, with sharp multiple peaks in the range $E_g - Ry^* < \hbar\omega < E_g$ and continuous growth of absorption for $\hbar\omega > E_g$. For higher temperatures ($k_B T \gg Ry^*$), sharp exciton peaks evolve to the exponential absorption tail.

Problems

2.1 Recall the electron–volt energy scale and compare it to joules.

2.2 Calculate the de Broglie wavelength of an electron whose kinetic energy is 10 eV. Compare this with the same values for a proton. Discuss the difference.

2.3 Calculate the de Broglie wavelength of an electron whose kinetic energy equals $k_B T$ at room temperature.

2.4 In obsolete cathode-ray tubes used in TV sets and computer monitors by your parents and grandfathers, electrons emitted from the cathode experience acceleration between plates at 10 kV voltage. Find the electron de Broglie wavelength and make a judgment about the role of wave properties of electrons in this case.

2.5 Explain why the quantum mechanical problem for a well with infinite potential walls is often referred to as the "hard-wall problem." Hint: consider the acoustic analogy.

2.6 Using Eq. (2.9), prove that inter-level spacing grows with level numbers.

2.7 Prove that the Taylor series allows for parabolic approximation for U-like and V-like potentials.

2.8 Calculate a few of the lowest energies of an electron in a spherical box with infinite walls with radii $a = 1, 2, 3, 4, 5$ nm.

2.9 Explain why the Bohr radius and Rydberg energy only very slightly differ from the atomic length and energy units.

2.10 Compare tunneling probabilities for an atom versus an electron.

2.11 Look at the resonance tunneling presentation and consider modification of wave function on the way from the well to the right-hand side of the right barrier. Compare with Figure 2.7. Observe and explain the analogy.

2.12 Calculate photon momentum for the visible range and compare it with momentum of an electron possessing kinetic energy equal to photon energy.

2.13 Based on the data in Table 2.3, examine the correlation between semiconductor band gap energy and effective masses of electrons and holes.

2.14 Consider modification of the band gap energy for a number of compounds (Table 2.3) within the series CdS → CdSe → CdTe → HgTe; AlN → AlP → AlAs → InSb; CuCl → CuBr. Try to predict band gaps for HgS, AlSb, AgBr, and CuI. Compare your predictions with the reference data.

2.15 Compare data on E_g versus composition (Figure 2.18) and E_g versus lattice constant (Table 2.3) for AlN–GaN–InN. Make a conclusion on correlation between lattice constant and composition.

2.16 Suggest semiconductor compounds to develop light emitters at 400, 500, and 600 nm.

2.17 Consider and summarize similarities and differences of an exciton and a hydrogen atom.

Further reading

Chichibu, S., Mizutani, T., Shioda, T., et al. (1997). Urbach–Martienssen tails in a wurtzite GaN epilayer. *Appl Phys Lett*, **70**(25), 3440–3442.

Harrison, W. A. (2000). *Applied Quantum Mechanics*. World Scientific.

Klingshirn, C. F. (2012). *Semiconductor Optics*. Springer Science & Business Media.

Klingshirn, C. F., Waag, A., Hoffmann, A., and Geurts, J. (2010). *Zinc Oxide: From Fundamental Properties Towards Novel Applications*. Springer Science & Business Media.

Levinshtein, M. E., Rumyantsev, S. L., and Shur, M. S. (2001). *Properties of Advanced Semiconductor Materials: GaN, AlN, InN, BN, SiC, SiGe*. John Wiley & Sons.

Miller, D. A. B. (2008). *Quantum Mechanics for Scientists and Engineers*. Cambridge University Press.

Peyghambarian, N., Koch, S. W., and Mysyrowicz, A. (1994). *Introduction to Semiconductor Optics*. Prentice-Hall, Inc.

Schmitt-Rink, S., Haug, H., and Mohler, E. (1981). Derivation of Urbach's rule in terms of exciton interband scattering by optical phonons. *Phys Rev B*, **24**(10), 6043.

Yariv, A. (2013). *An Introduction to Theory and Applications of Quantum Mechanics*. Courier Corporation.

Yu, P. Y., and Cardona, M. (1996). *Fundamentals of Semiconductor Optics*. Springer.

References

Gaponenko, S. V. (2010). *Introduction to Nanophotonics*. Cambridge University Press.

Klingshirn, C. F. (2004). *Semiconductor Quantum Structures. Part 2: Optical Properties*. Springer Science & Business Media.

Madelung, O. (2004). *Semiconductors: Data Handbook*. 3rd edition. Springer.

Muth, J. F., Lee, J. H., Shmagin, I. K., et al. (1997). Absorption coefficient, energy gap, exciton binding energy, and recombination lifetime of GaN obtained from transmission measurements. *Appl Phys Lett*, **71**(18), 2572–2574.

Pelá, R. R., Caetano, C., Marques, M., et al. (2011). Accurate band gaps of AlGaN, InGaN, and AlInN alloys calculations based on LDA-1/2 approach. *Appl Phys Lett*, **98**(15), 151907.

3 Quantum confinement effects in semiconductors

Absorption and emission of light by atoms, molecules, and solids arise from electron transitions. Electron confinement phenomena in solids with restricted geometry like nanoparticles, nanorods, or nanoplatelets gives rise to the modification of optical absorption and emission spectra and transition probabilities in semiconductor nanostructures. These phenomena are direct consequences of the wave properties of electrons. In this chapter we describe size-dependent optical properties of semiconductor nanostructures related to quantum confinement.

3.1 DENSITY OF STATES FOR VARIOUS DIMENSIONALITIES

Density of states is among the basic attributes of quantum particles that directly follow from their wave properties. This notion is used for the function D defining the number of available states in a unit volume calculated per unit interval of any of the following parameters: energy E, momentum p, wave number k, or wavelength λ. Since all of these arguments can be expressed through any one of the others, knowing the density of states versus any of the arguments allows the calculation of the density of states for every other argument as well.

For classical waves, states are called *modes*. This notion implies a definite type of oscillation featuring a certain wavelength, wave vector, frequency, and polarization in the case of transverse waves. To understand density of states in quantum physics, let us first consider density of modes for classical waves.

Recall Figure 2.2. Consider standing waves in a cube with size a having hard walls – say, acoustic waves. One can see that the wavelengths obey a series $a/2, 2a/2, \ldots, na/2$ in which $n = 1,2,3, \ldots$ Accordingly, wave numbers will form three series for the three-dimensional space,

$$k_x = n_x \frac{\pi}{a}, \ k_y = n_y \frac{\pi}{a}, \ k_z = n_z \frac{\pi}{a}. \tag{3.1}$$

Modes form a discrete set of points in k-space, and every pair of neighboring modes has the spacing

$$\Delta k_x = \Delta k_y = \Delta k_z = \frac{\pi}{a}. \tag{3.2}$$

One can say that every mode occupies in k-space the volume

$$V_k = \left(\frac{\pi}{a}\right)^3.$$

(3.3)

Let us count the number of modes for all directions within the interval $[k, k + dk]$, i.e., the number of modes contained in a spherical shell between a sphere with radius k, and a sphere with radius $k + dk$. Taking the volume of such a layer,

$$dV_k = 4\pi k^2 dk$$

(3.4)

and keeping only positive k (i.e., 1/8 of the whole-layer volume) by dividing the volume into the volume of a single mode (Eq. (3.3)), we arrive at the number of modes in the interval $[k, k + dk]$:

$$\frac{dV_k}{V_k} = \frac{a^3}{2\pi^2}k^2 dk.$$

(3.5)

To get the number of modes per unit volume, one has to divide (Eq. (3.5)) by a^3. Then we can introduce density of modes $D(k)$ through the number of modes dN within the interval dk as

$$dN = D(k)dk \equiv \frac{dV_k}{V_k L^3} = \frac{k^2}{2\pi^2}dk.$$

(3.6)

Applying similar considerations to two- and one-dimensional problems (Problem 3.1), we can finally write

$$D_3(k) = \frac{k^2}{2\pi^2}, \quad D_2(k) = \frac{k}{2\pi}, \quad D_1(k) = \frac{1}{\pi}.$$

(3.7)

In Eq. (3.7), subscripts 1,2,3 denote space dimensionality. It has been rigorously mathematically shown that the final result in finding mode density does not depend upon the shape of the box. Moreover, functions D_1, D_2, and D_3 are the common laws for all waves, including acoustic and electromagnetic, and also including electrons and other quantum particles (protons, holes, excitons, atoms, molecules, etc.), since these are viewed as waves. For transverse (e.g., electromagnetic) waves, a factor of 2 should be added to account for the two possible orthogonal polarizations. For electrons, holes, and other *fermions*, it is known that owing to their half-integer spin value, two particles with opposite spins may exist in the same state.[1] Therefore, a factor of 2 has to be added in each of the parts of Eq. (3.7). For bosons (excitons, atoms, and molecules), Eq. (3.7) should be used as presented.

When making a transfer from classical waves to quantum particles, the notation "modes" should be replaced by "states." Therefore, starting from mode counting we arrive at the density of states (DOS) in quantum physics.

---

[1] Strictly speaking, only one fermion is allowed to be in every state, but the state acquires spin orientation as the necessary attribute in addition to energy and momentum values.

Knowing the $D(k)$ function, one can readily derive quantum particle density of states versus energy, E, and momentum, p. The relevant relationships read as

$$D(E) = D(k)\frac{dk}{dE}, \quad D(p) = D(k)\frac{dk}{dp}, \quad D(p) = D(E)\frac{dE}{dp}. \qquad (3.8)$$

Recalling $E(k)$ and $p(k)$ (Eqs. (2.4) and (2.5)), we arrive at the following important formulas for different dimensionalities, d

$$d = 3: \qquad D_3(E) = \frac{8\pi m^{3/2}E^{1/2}}{2^{1/2}h^3}, \quad D_3(p) = \frac{4\pi p^2}{h^3}, \qquad (3.9)$$

$$d = 2: \qquad D_2(E) = \frac{2\pi m}{h^2}, \quad D_2(p) = \frac{2\pi p}{h^2}, \qquad (3.10)$$

$$d = 1: \qquad D_1(E) = \frac{\sqrt{2}m^{1/2}E^{-1/2}}{h}, \quad D_1(p) = \frac{1}{h}. \qquad (3.11)$$

All three formulas contain dimensionality-dependent factors that can be written as

$$D_d(E) \propto \frac{m^{\frac{d}{2}}E^{\frac{d}{2}-1}}{h^d}, \qquad D_d(p) \propto \frac{p^{d-1}}{h^d}.$$

However, different pre-factors in Eqs. (3.9)–(3.11) do not allow one to suggest the common formula for all three dimensionalities ($d = 1,2,3$). Such a formula can be readily written for $d = 1,2$, namely:

$$D_d(E) = \frac{2^{\frac{d}{2}}\pi^{d-1}m^{\frac{d}{2}}E^{\frac{d}{2}-1}}{h^d}, \qquad D_d(p) \propto \frac{(2\pi)^{d-1}p^{d-1}}{h^d} \qquad \text{for } d = 1, d = 2. \qquad (3.12)$$

Eqs. (3.9)–(3.12) are valid for every type of real quantum particles and quasiparticles (electrons, holes, excitons, atoms, molecules), but for fermions (electrons and holes) a factor of 2 should be added to account for spin. Note that in the 2D space, density of states appears to be energy-independent and to depend only on particle mass, m. For electrons in a vacuum it reads $D_2(E) = 4.17 \cdot 10^6/(\text{eV} \cdot \mu\text{m}^2)$. All results are summarized in Eq. (3.13) and in Figure 3.1:

$$D_3(E) = \frac{8\sqrt{2}\pi m^{3/2}E^{1/2}}{h^3}\left[\frac{1}{J \cdot m^3}\right], \qquad D_2(E) = \frac{4\pi m}{h^2}\left[\frac{1}{J \cdot m^2}\right], \qquad (3.13)$$

$$D_1(E) = \frac{2\sqrt{2}m^{1/2}E^{-1/2}}{h}\left[\frac{1}{J \cdot m}\right].$$

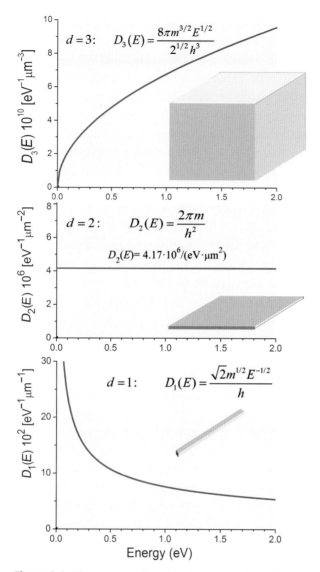

Figure 3.1 Electron density of states for electrons in vacuum for different dimensionalities of space $d = 3, 2, 1$. To account for the two possible spin orientation the factor of 2 should be added.

3.1.1 Density of states in terms of elementary h^d cells in phase space

In quantum mechanics, every state is believed to occupy the same and the finite portion of phase space, namely h^d. Let us remember that phase space is space of momentum and spatial coordinates. It has $2 \cdot d$ dimensions, i.e., six for three-dimensional space, four for two-dimensional space, and two for one-dimensional space. The idea on quantization of phase space for quantum particles was advanced by M. Planck in 1916, well prior to the Schrödinger equation (1926), and conforms to the uncertainty relation introduced in quantum theory by W. Heisenberg in 1927. So, unlike classical mechanics, in which states with definite momentum and coordinate(s) are described by a point in phase space, in the quantum world every state necessarily occupies the finite phase space volume h^d. With this assumption, the density of states versus momentum can be immediately derived. To calculate $dN = D(p)dp$ for $d = 3$ space, one has to divide a spherical layer volume in p-space, whose radius is p and thickness is dp, by h^3, i.e., $4\pi p^2 dp/h^3$, to arrive at the right-hand formula in Eq. (3.9). Results for $d = 2$ and $d = 1$ can be derived accordingly.

Which approach to density of states is more correct: counting modes or counting phase space cells? The first approach contains only the assumption of wave properties of the object in question, then using the $p = \hbar k$ relation to move from classical modes to quantum states. The second approach seems to be shorter, but it uses an assumption on the elementary phase space cell, which in turn should be thoroughly proven by derivation of the uncertainty relation. Noteworthy is that in the first approach, cells in k-space appear naturally without any additional assumptions.

3.1.2 Where density of states matters

Density of states plays an important role in every quantum process in which the final state of a quantum particle belongs to a continuous spectrum, which is correct for the case for electrons in the c-band and holes in the v-band in solids. The probability of a quantum process occurring in this case is proportional to the density of the final states. This statement is the consequence of the quantum mechanical perturbation theory and was called the "golden rule" by E. Fermi to emphasize its significance. Therefore, it is often referred to as the *Fermi golden rule*, though E. Fermi himself never claimed authorship for it.

Optical *absorption* by a transition from the v-band to the c-band should be (and is) proportional to the density of the final electron states. In the simple model of parabolic bands without Coulomb electron–hole interaction, the absorption coefficient of a semiconductor equals zero below E_g and grows as $(\hbar\omega - E_g)^{1/2}$ for $\hbar\omega > E_g$ in proportion with the $D(E)$ function. In a real semiconductor, interband absorption is strongly enhanced by electron–hole Coulomb interaction.

Electron–electron, hole–hole, and electron–hole *scattering*, as well as their scattering by lattice vibrations (phonons), are proportional to density of final states. In optics, scattering defines relaxation times in ultrafast phenomena. In the charge transfer processes in conductivity, current depends on the electron's ability to increase energy in the c-band and for the hole to do so in the v-band, in the electric field. Again, density of states will enter all expressions for charge carrier dynamics.

The latter example leads to a remarkable property of an ideal quantum wire. In such a wire, the electron is supposed to have no scattering between two edges, and so-called *ballistic transport* occurs. The density of states appears to be inversely proportional to the square root of charge carrier energy, i.e., to its velocity, v. Eq. (3.10) can be rewritten as $D_2(E) = 1/(hv)$. To have higher current in such a wire, the electron should move faster. However, since the product of its velocity and density of states is constant (which means faster electrons are less likely to exist), the current appears to be independent of wire length. Moreover, the current–voltage relation in this case leads to the notion of *conductance quantum*:

$$G_0 = \frac{2e^2}{h} \approx 7.75 \cdot 10^{-5} \text{siemens, or resistance } R \approx 12.9 \text{ k ohm/s.} \tag{3.14}$$

It is remarkably defined by the two fundamental constants and does not depend on parameters of the problem. This property describes an ideal wire, whereas in the real case a factor from 0 to 1 should be added to account for non-ballistic contributions to the charge carrier motion. Quantum conductance was advanced by R. Landauer from IBM in 1957 and has recently become the subject of active research in nanoelectronics. Over the past decades its manifestations have been reported in many experiments; however, these are beyond the scope of this book.

The equation for *ionization equilibrium* (the Saha equation, Eq. (2.81)) includes the contribution from the electron–hole density of states, since a quantum event of exciton dissociation into an electron and a hole is proportional to the density of final states for these particles. Without going into detail, dimensionality affects interplay of free-to-bound e–h pairs, which in turn enhances the role of exciton states in low-dimensional structures. Eq. (3.15) summarizes ionization balance for various dimensionalities.

$$n_{\text{eh}}^2 = n_{\text{exc}} \frac{1}{\hbar^d} \left(\frac{2\mu_{eh}k_{\text{B}}T}{\pi} \right)^{d/2} \exp\left(\frac{-E_b(d)}{k_{\text{B}}T} \right) \quad for \quad d = 1, d = 2. \tag{3.15}$$

Here, $E_b(d)$ is exciton binding energy for low-dimensional space that is dimensionality-dependent and exceeds the Ry^* value. It is the subject of Sections 3.3 and 3.4, where we consider properties of quantum wells and quantum wires. Similar to density-of-states formulas (Eq. (3.12)), universal d-dependent expressions can be written for $d = 2$ and $d = 1$, whereas for $d = 3$ numerical factor(s) fall out from the general formula, though power numbers for mass, energy, and Planck constant fall in line with $d = 1$ and $d = 2$.

3.2 ELECTRON CONFINEMENT IN SEMICONDUCTORS AND IN METALS

In intrinsic semiconductors, i.e., semiconductors without impurities, in the limit of low temperatures there are no free charge carriers in c- and v-bands. Charge carriers appear upon heating and optical or electrical pumping. For large and reasonable temperature range and optical and electrical excitation conditions, the number of free carriers remains much lower than the number of available states. In this case the free electrons and holes

Table 3.1 **Lattice constants and electron de Broglie wavelength at room temperature for different semiconductor crystals**

Material	Electron effective mass, m_e	Electron de Broglie wavelength λ_e (nm)	Lattice constant a_L (nm)	First energy level in a 5 nm box (meV)
GaN	$0.22\,m_0$	16	0.518	68
GaAs	$0.067\,m_0$	29	0.564	224
InN	$0.08\,m_0$	26	0.570	187
ZnSe	$0.15\,m_0$	19	0.567	100
ZnO	$0.27\,m_0$	15	0.520	55
CdSe	$0.11\,m_0$	23	0.701	136
PbS	$0.04\,m_0$	38	0.593	375
Si	$0.08\,m_0$	26	0.543	187
Ge	$0.19\,m_0$	16	0.564	79
In vacuum	m_0	7.6	–	15

tend to occupy the lowest available state and their energy in c- and v-band can be roughly taken to be of the order of $k_B T$.

Let us now compare the de Broglie wavelength of free electrons whose kinetic energy equals $k_B T$ with the crystal lattice constant, a_L. We use Eq. (2.6) and data from Table 2.3. In every typical semiconducting material, at room temperature

$$a_L \ll \lambda \tag{3.16}$$

holds (Table 3.1). This means that crystalline materials can be restricted in dimensions to get spatial confinement of electrons whereas crystal lattice properties will only slightly be disturbed.

This makes possible development of a variety of size-dependent phenomena, which are referred to as *quantum confinement phenomena* since these come from wave properties of quantum particles. Table 3.1 shows that for common semiconductor materials used in electronics and optoelectronics, electron de Broglie wavelength at room temperature exceeds lattice constant by more than one order of magnitude. Though being somewhat shorter, de Broglie wavelength for holes still measures many times the lattice period. This circumstance offers the possibility of implementing quantum confinement phenomena in semiconductor nanostructures with electron confinement effects dominating over hole confinement because of lower mass and therefore longer wavelength. The right-hand column shows that the lowest energy state (Eq. (2.9)),

$$E_1 = \pi^2 \hbar^2 / 2ma^2,$$

in a5 nm box (i.e., approximately ten-fold lattice period) with infinite barrier will exceed by many times the room temperature $k_B T$ value, and becomes non-negligible when compared to the band gap energy E_g. Quantum confinement then changes the energy spectrum of electrons and holes, moves onset of interband transitions in the absorption spectrum to higher energies, modifies electron and hole motion, and thus affects relaxation processes and scattering rates. Modification of density of states in low-dimensional systems further modifies the spectral shape of the absorption spectrum, relaxation and recombination rates, and balance between free and bound (excitons) e–h pairs.

These relations form the decisive prerequisite for *low-dimensional structures* to occur. These imply that in some directions electronic excitations experience quantum confinement, whereas in other direction(s) they have free motion. These effects give rise to the notion of semiconductor structures with lower *dimensionality*, which are referred to as low-dimensional structures. If in one dimension a crystal has size of the order of the electron de Broglie wavelength, a two-dimensional structure develops. It is referred to as a *quantum well*. If in two dimensions the crystal size approaches the de Broglie wavelength, then the notation of *quantum wire* is applied. Finally, in the case of size restrictions in three dimensions, the term *quantum dot* is used.

Another physical dimension in the electron subsystem of a crystal that can be drastically invaded in nanostructures is the exciton Bohr radius, a_B^*. In most common semiconductors, $a_B^* \gg a_L$ holds, i.e., it is possible to restrict exciton motion without serious effect on the crystal lattice. This changes the exciton spectrum drastically and gives rise to higher binding energy and sharper exciton features both in absorption and emission spectra.

3.2.1 Technological approaches

There are two basically different approaches in synthesis of semiconductor low-dimensional structures, namely *epitaxial* and *colloidal* ones.

Epitaxy offers development of a thin monocrystalline film on top of the monocrystalline substrate whose crystal lattice period matches the lattice period of the growing film. It is the powerful technique that enables mass production of semiconductor lasers and light-emitting diodes (LEDs). In epitaxial growth, quantum well structures can be developed as very thin films between thicker films of wider-band gap materials; quantum wires can be developed in narrow grooves of wider-band gap materials; and quantum dots can be developed in the course of spontaneous self-organization in the submonolayer epitaxy.

Alternatively, using *colloidal techniques*, quantum well structures can be fabricated as nanoplatelets; quantum wires can be developed as nanorods; and quantum dots can be grown as nanocrystals with nearly spherical shape or crystalline faceting. Additionally, quantum dots can be readily developed in glasses using colloidal-like aggregation of ions promoted by their diffusion in viscous glassy matrix at elevated temperatures. Colloidal techniques are still at the beginning of wide commercial applications in photonic devices, mainly these are used in research laboratories rather than in factories. Nevertheless, development of colloidal techniques for nanophotonic research has enabled many challenging properties of quantum dots to be discovered that otherwise need enormous efforts in the epitaxial approach. There are already a few examples of colloidal nanostructures entering display component production in tablets and cell phones, which is considered as the beginning of a newly emerging versatile technological platform for nanophotonic engineering.

3.2.2 Why quantum confinement is impossible in metals

In metals, an enormous number of free electrons (of the order of 10^{22} cm^{-3}) exists in the conduction band (Table 3.2). Unlike semiconductors, in metals the Fermi level is located in the c-band (Figure 3.2). The Fermi energy for given electron density N can be found by means of statistical physics based on the statement that the given total density of electrons within the conduction band should be equal to the integral

$$N = \int_0^{E^*} \bar{N}(E,T)D(E)\,dE \tag{3.17}$$

where $D(E)$ is the electron density of states defined by Eq. (3.8). The integration range here starts at the bottom of the conduction band where kinetic energy equals zero and extends high enough to account for all occupied states. At $T = 0$, the Fermi energy is the highest energy an electron possesses and then the E_F value can be readily elucidated, being the upper point of the integration range. It reads

$$E_F = \frac{\hbar^2}{2m_e^*}\left(3\pi^2 N\right)^{2/3} \tag{3.18}$$

and measures several electronvolts in real metals. Because of the *Pauli exclusion principle* for fermions (only one fermion can occupy any given state), only those electrons that have energy close to the Fermi level can participate in any dynamical process (scattering, charge transfer, etc.). One can see that a few electronvolts means less than 1 nm de Broglie wavelength, which in turn means that in realistic nanostructures electrons in metal do not

Table 3.2 **Parameters of a few common metals**

Metal	m_e	E_F eV	n_e cm^{-3}	λ_e (nm)
Cu	1.01 m_0	7.0	$8.45 \cdot 10^{22}$	0.46
Ag	0.99 m_0	5.5	$5.85 \cdot 10^{22}$	0.52
Au	1.10 m_0	5.5	$5.9 \cdot 10^{22}$	0.52

Source: Kasap 2002.

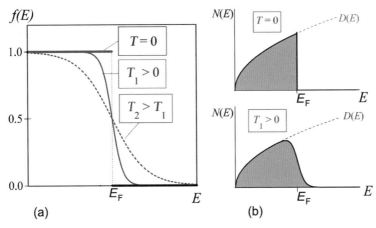

Figure 3.2 Free electrons in a metal. (a) Fermi–Dirac distribution function $f(E)$ at different temperatures. (b) Electron density of states $D(E)$ (dashes) and the product $f(E)D(E)$ (solid line). The shaded area under the curve $f(E)D(E)$ equals the total number of electrons in the c-band in accordance with Eq. (3.17).

experience strong confinement until the size of a nanostructure noticeably exceeds the lattice period. If the latter condition is not met, we arrive at cluster physics, where quantum chemistry, similar to molecular science, is to be applied instead of solid-state physics.

3.3   QUANTUM WELLS, NANOPLATELETS, AND SUPERLATTICES

In modern electronics and optoelectronics, high-quality monocrystalline layers are obtained by means of epitaxial growth in which a new thin monocrystal develops on top of a monocrystalline substrate whose lattice period should match the lattice period of the newborn crystal. Not every desirable monocrystal is available in bulk form to serve as a substrate. Many that are available appear to be too expensive for commercial device development. Therefore, researchers and engineers are often faced with the problem of finding the proper substrate. Figure 3.3 shows the lattice constants for important III–V compounds along with possible II–VI substrate candidates.

3.3.1   The double heterostructure concept

In fact, every epitaxial layer on top of a monocrystalline matched substrate represents a *heterostructure*, i.e., a structure in which two different materials form a single crystal lattice. In the early 1960s, Zh. Alferov and R. Kazarinov in the USSR and H. Kremer in the USA independently suggested and implemented a *double heterostructure* (Figure 3.4). In this structure, the middle, thinner layer of a semiconductor with a smaller band gap is placed between the two thicker layers of a wider-gap semiconductor. In a p–n junction, a potential well between thick *p*- and *n*-layers collects electrons and holes so efficiently that not only was it possible to enhance efficiency of semiconductor laser diodes and light-emitting

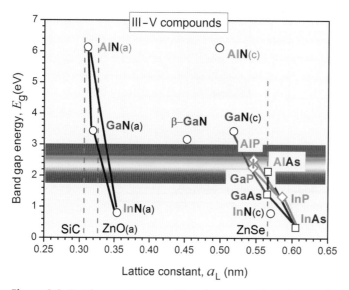

Figure 3.3 Lattice constants and band gap energies of a number of III–V compounds. SiC and ZnO lattice constants are shown as possible substrates for GaN-based monocrystals and ZnSe – for GaAs-based monocrystals.

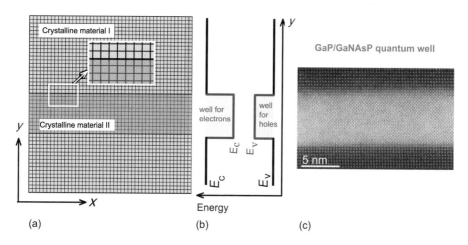

Figure 3.4 A double heterostructure and a single quantum well structure. (a) A sketch of a double heterostructure highlighting the lattice match condition. (b) Energy diagram with a potential well for both electrons and holes in the middle. (c) Scanning electron microscopy image of a real GaP/GaNAsP/GaP quantum well. GaP is the material on the top and bottom, and appears darker, GaNAsP is the material sandwiched in the center and appears lighter. This structure is used in quantum well lasers. Image reprinted from Straubinger et al. (2016).

diodes, but it appears to be possible to obtain efficient light emission by direct injection of charge carriers without making p–n junctions with heavily doped semiconductors.

If a well in a double heterostructure has thickness of the order of a few nm, it represents a *quantum well* for electrons and/or holes. Such a quantum well structure is shown in the right-hand part of Figure 3.4. For more than three decades quantum well structures have been fabricated and examined in detail. Quantum well structures are used today in semiconductor lasers (in optical communication, CD-ROMs, laser printers, laser pointers, etc.), and in LEDs, including the outstanding invention of GaInN-based efficient violet and blue LEDs. N. Holonyak at General Electric was the first to develop a quantum well laser in 1977 by liquid-phase epitaxy; later he initiated gas-phase epitaxy research, nowadays referred to as MOVPE (metallo-organic vapor-phase epitaxy) or MOCVD (metallo-organic chemical vapor deposition). The latter dominates today in mass production of semiconductor LEDs.

3.3.2 Electrons and excitons in two dimensions and in quantum wells

For an electron in a quantum well with thickness L_z and infinite potential barriers, the wave function can be written as

$$\psi(\mathbf{r}) = \sqrt{\frac{2}{L_x L_y L_z}} \sin(k_n z) \exp(i\mathbf{k}_{xy} \cdot \mathbf{r}) \tag{3.19}$$

BOX 3.1 DOUBLE HETEROSTRUCTURES AND QUANTUM WELLS IN PHOTONICS

Double heterostructures and quantum wells are used today in commercial mass production of laser diodes and LEDs, including the recently developed blue GaN diodes. In 1963, Zhores Alferov, Rudolf Kazarinov, and Herbert Kroemer suggested a double heterostructure would be advantageous in a sense of non-equilibrium carrier collection in potential wells, enabling more efficient electron–hole recombination. In 1977, Nick Holonyak and co-workers at General Electric advanced fine epitaxial techniques for quantum well fabrication and developed the first quantum well laser. In 1962 he developed the first semiconductor LED. Alferov and Kroemer in 2000 became Nobel Prize winners for "developing semiconductor heterostructures used in high-speed- and optoelectronics."

| Zhores Alferov | Herbert Kroemer | Nick Holonyak |

where wave number

$$k_n = \frac{\pi}{L_z} n , \qquad n = 1, 2, 3, \ldots \qquad (3.20)$$

takes discrete values determined by the well width, whereas $\mathbf{k}_{xy}$ takes continuous values and corresponds to infinite in-plane motion of an electron. Accordingly, the electron energy is the sum of the kinetic energy relevant to infinite motion $E = \hbar^2 k_{xy}^2 / 2m_e^*$ and the discrete set of energy values $E_n = \hbar^2 k_n^2 / 2m_e^*$ relevant to quantized states of an electron, i.e.,

$$E = \frac{\hbar^2 k_{xy}^2}{2m_e^*} + \frac{\pi^2 \hbar^2}{2m_e^* L_z^2} n^2 , \qquad n = 1, 2, 3, \ldots \qquad (3.21)$$

Recalling that density of states for electrons (and holes) in 2D space is energy-independent (Figure 3.1), one can expect as a first approximation that the absorption spectrum of a

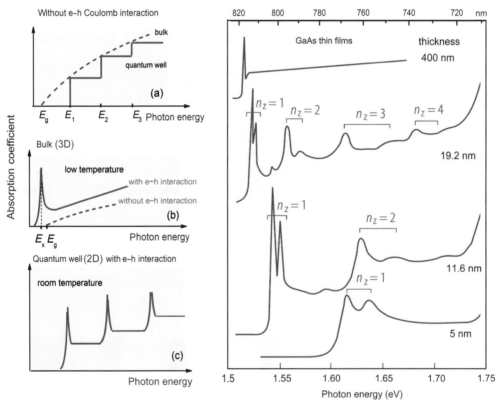

Figure 3.5 A sketch of the absorption spectrum in bulk (3D) and quantum well (2D) semiconductors (left panel) and the low temperature optical absorption spectra of GaAs films of different thicknesses between $Al_{0.25}Ga_{0.75}As$ barriers (right panel). n denotes energy state numbers in the quantum well. Splitting of the absorption band comes from electron coupling to heavy and light holes. Experimental data are adapted from Göbel and Ploog (1990) with permission from Elsevier.

semiconductor quantum well will evolve to a step-like dependence provided that the bulk parabolic band model (without e–h Coloumb interaction) features $(\hbar\omega - E_g)^{1/2}$ dependence. This simplified consideration for 3D and 2D semiconductors is shown in Figure 3.5(a).

However, as we saw in Section 2.6 (Figure 2.16), electron–hole Coulomb interaction gives rise to excitonic absorption band(s) at $\hbar\omega < E_g$ and strong absorption enhancement for $\hbar\omega > E_g$. This effect is sketched in Figure 3.5(b). Note that for most semiconductors the exciton line is pronounced at low temperatures only.

Introducing Coulomb interaction in the 2D semiconductor model gives rise to a strong increase in exciton binding energy and in many cases exciton lines become well pronounced at room temperatures (Figure 3.5(c)). This is the important result of spatial confinement of excitons.

To understand what will happen with excitons in 2D spaces, one needs to reconsider the Schrödinger equation for relative motion of a Coulombically coupled electron and a hole:

$$-\frac{\hbar^2}{2\mu_{eh}}\nabla^2\Psi(r) - \frac{e^2}{\varepsilon r} = E\Psi(r) \tag{3.22}$$

for the cases of 2D and 1D space. Here, μ_{eh} is the electron–hole reduced mass and ε is the crystal dielectric permittivity.

A 2D Schrödinger equation for an electron and a hole with Coulomb interaction has solutions with the energy spectrum for excitonic s-states in the form (Ralph 1965)

$$\boxed{E_n = -\frac{Ry^*}{(n-\frac{1}{2})^2}, \quad n = 1, 2, 3, \dots} \tag{3.23}$$

which means that

$$E_1 = -4Ry^*, \quad E_2 = -\frac{4Ry^*}{9}, \quad E_3 = -\frac{4Ry^*}{25}, \quad \dots \tag{3.24}$$

Remarkably, the lowest energy state is now four times lower than in the 3D case, and the spacing between the first and the second levels appears to be even more than four times greater than for the 3D case. Enhanced interaction in 2D space results not only in greater binding energy, but also gives rise to a smaller exciton Bohr radius in 2D space, namely

$$a_B^{2d} = \tfrac{1}{2}a_B^*, \tag{3.25}$$

where a_B is the 3D value expressed by Eq. (2.77). By and large, this property enormously enhances excitonic phenomena in two-dimensional space, bringing the liquid nitrogen temperature scale into the room-temperature range.

Experimental studies revealed reasonable agreement with the theoretical predictions (Figure 3.5, right panel; Figure 3.6). Nanometer-thick GaAs quantum wells epitaxially grown between wider-gap $Al_xGa_{1-x}As$ monocrystalline layers were extensively examined experimentally to test electron–hole confinement phenomena in quantum wells. It was

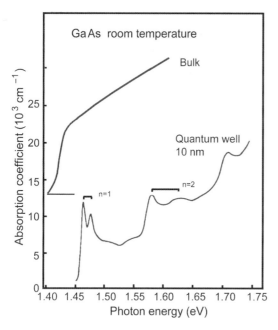

Figure 3.6 Comparison of optical absorption spectra for room temperature for GaAs 1.2 μm (bulk) monocrystalline film and a 10 nm-thick quantum well. In both cases GaAs is grown between Al$_x$Ga$_{1-x}$As layers. Adapted from Sell and Casey (1974), with the permission of AIP Publishing, and from Chemla and Miller (1985) with the permission of OSA Publishing.

found that (1) quantum wells exhibit sharp multiple exciton peaks even at room tempera-ture, whereas parent bulk crystal shows no exciton peak because of $k_B T > Ry^*$; (2) exciton peaks split because heavy and light holes possess different energy shift (in accordance with Eq. (2.9)); (3) absorption between exciton peaks varies only slightly, which is supposed to confirm constant-like 2D density of states.

However, though the 2D consideration gives reasonable trends and therefore is instructive for experimenters, the accurate theoretical description and full understanding of optical properties of real quantum well *structures* needs the finite barriers and the finite thicknesses to be taken into consideration. For the 2D consideration to be true it needs, first, infinite barriers and, second, thickness which is negligibly small as compared to $a_B^{2d} = \frac{1}{2} a_B^*$, rather than a_B^*. Both conditions are not perfectly met in real *structures*. Actually, when starting from the thick intermediate layer and going to a thinner one the band gap energy and exci-ton binding energy rise, but note that in a three-layer sandwich structure the extreme case of the well width going to zero does not arrive to the pure 2D space but instead leads to the 3D exciton of the surrounding material. This effect is qualitatively explained in Figure 3.7.

To account for possible intermediate behavior between ideal 3D and ideal 2D cases in finite-width quantum well structures with finite potential barriers, He (1991) provided a general solution of the Schrödinger equation for a hydrogenic system in fractional space with dimensionality ranging from 1 to 3. He arrived at instructive formulas for the energy spectrum and mean radius of a hydrogenic system as a function of space dimensionality d:

$$E_n(d) = E_g - \frac{Ry^*}{\left(n + \dfrac{d-3}{2}\right)^2} \quad , \quad a_n(d) = \left(n + \frac{d-3}{2}\right)^2 a_B^* \, , \quad n = 1, 2, 3, \ldots \qquad (3.26)$$

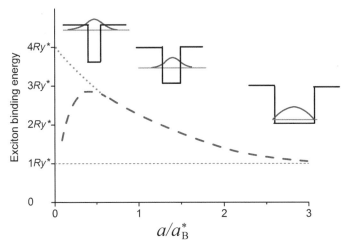

Figure 3.7 Explanation for non-monotonic dependence of exciton binding energy versus well width, a, in real quantum well structures with finite barriers. In the case of infinite barrier exciton binding energy tends to the $4Ry^*$ value inherent in the extreme 2D case, whereas in the case of a finite barrier, penetration of the wave function into the barrier material turns binding energy toward that of the 3D barrier material.

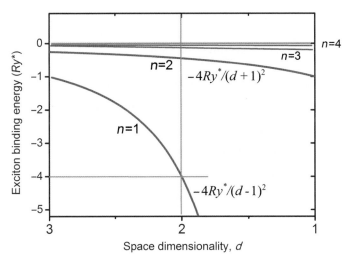

Figure 3.8 Energy levels of a hydrogenic system for the space dimensionality continuously changing from $d = 3$ to $d = 2$ and then to $d = 1$, based on He's formula (Eq. (3.26)).

One can see these expressions contain formulas for 3D and 2D space as special cases. Energy levels according to He's formula are presented in Figure 3.8.

The enhanced role of excitons in quantum wells is also pronounced in luminescence as a strong emission band from annihilation of an exciton in its ground state. This remarkable property of quantum wells is used in commercial InGaN-based LEDs. Figure 3.9 shows the design and emission spectrum of a single quantum well LED invented by S. Nakamura

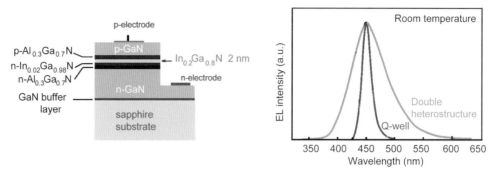

Figure 3.9 Design (left) and operation (right) of the first blue GaN-based quantum well structure advanced by S. Nakamura and co-workers in 1995. As compared to a standard double heterostructure, the quantum well diode showed a narrower emission spectrum and nearly twice the efficiency at the same current values. Adapted from Nakamura et al. (1995) with the permission of AIP Publishing.

in the 1990s. The quantum well LED, in spite of a tiny (2 nm) active layer, exhibited nearly twice the efficiency compared to a thick layer in a double heterostructure, with the emission spectrum being considerably sharper for the well.

Recently, advanced colloidal chemistry offered an approach to development of very thin monocrystalline sheets measuring a few nanometers in one direction and micrometers otherwise. These have been termed *nanoplatelets* and can be treated as colloidal counterparts of epitaxial quantum wells. Nanoplatelets of II–VI compounds have been successfully grown and thoroughly examined. Similar to epitaxial quantum wells, nanoplatelets exhibit thickness-dependent multiple exciton peaks in absorption and show strong excitonic emission from the ground state of the confined exciton (Figure 3.10(a)). A representative example shows well-pronounced exciton bands at room temperature, huge high-energy shift of band gap energy (from $E_g = 1.73$ eV inherent in bulk parent crystal to about 2.8 eV), and luminescence quantum yield of about 50%. In remarkable agreement with the constant density of states in 2D space, there is a well-defined plateau in absorption at which interband transitions occur (>2.9 eV).

There is one important issue, the dielectric confinement effect, that becomes crucially important for colloidal quantum wells but is less pronounced in epitaxial ones. When two charges, an electron and a hole, are within a thin layer in real space (Figure 3.9(b)), Coulomb interaction is not restricted by the well but extends beyond, to space outside the well. Unlike quantum well heterostructures, in the case of nanoplatelets the surrounding medium (air, liquid, or polymer) has much lower dielectric constant value ($\varepsilon_{out} = 1...5$) as compared to $\varepsilon_{in} > 10$ in the well region. Accordingly, interaction energy outside the well is $e^2 / \varepsilon_{out} r$, whereas inside it is $e^2 / \varepsilon_{in} r$. Therefore, in thin nanoplatelets binding energy of confined excitons will be strongly enhanced. Then exciton binding energy, E_B (Ry^* notation is no longer appropriate as we have departed from the hydrogen model) should rise

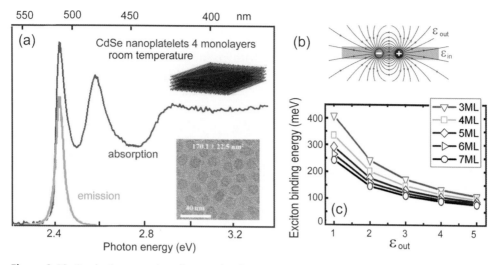

Figure 3.10 Optical properties of nanoplatelets. (a) Sample absorption and emission spectra from an ensemble of four-monolayer thick CdSe nanoplatelets. (b) Coulomb interaction scheme. (c) Modeling of exciton binding energy for various environments with 3D eight-band $k \cdot p$-theory by A. Achtstein et al. (2012). Note, for bulk CdSe, band gap energy E_g = 1.73 eV, and exciton binding energy Ry^* = 16 meV.

and exciton radius a_{ex} should fall accordingly. For the extreme thickness limit $d \ll a_{ex} < a_B^*$, L. Keldysh (1979) derived the following formulas (SI units):

Exciton binding energy

$$E_B = 2Ry^* \frac{a_B^*}{d} \left\{ \ln\left[\left(\frac{\varepsilon_{in}}{\varepsilon_{out}} \right)^2 \frac{d}{a_B^*} \right] - 0.8 \right\}, \quad (3.27)$$

Exciton radius

$$a_{ex} = \frac{1}{2}\sqrt{a_B^* d} \ , \quad a_B^* = 4\pi\varepsilon_0 \frac{\varepsilon_{in}\hbar^2}{\mu_{eh}e^2}. \quad (3.28)$$

One can see that these formulas predict manifold *increase in binding energy* versus both bulk value Ry^* and 2D value $4Ry^*$, as well as manifold *decrease in radius* again, as compared to the bulk value a_B^* and 2D value $a_B^*/2$. It is worth noting that the formulas are valid only when *d is small as compared to resulting* a_{ex} value rather than a_B^*. If this criterion is met, Eq. (3.27) predicts stronger exciton features in nanoplatelets as compared to quantum wells. This prediction has further been confirmed by numerical modeling (Figure 3.10(c)).

The structures with multiple quantum wells can readily be grown and are widely used in electronic and optoelectronic devices. In the case of a couple or a few wells, energy levels are split (see Figures 2.5 and 2.6) and an electron (and a hole) occupies all wells simultaneously. Multi-well structures with tens, or even hundreds, of identical wells possess minibands (see Figure 2.5), and electron and hole move therein mainly by means of resonance tunneling (see Figure 2.8). Since an electron and a hole experience another periodic potential superimposed over the crystal potential, these structures acquired notation *superlattices*. Superlattices play a role in ultrafast nanoelectronics and in quantum cascade deep-infrared lasers.

3.4   QUANTUM WIRES AND NANORODS

For an electron in a quantum wire with rectangular cross-section that allows infinite motion in the z direction, with infinite potential barriers in the x,y directions, wave function can be written as

$$\psi(\mathbf{r}) = 2\sqrt{\frac{1}{L_x L_y L_z}} \sin(k_n^{(x)}x)\sin(k_m^{(y)}y)\exp(ik_z z) \qquad (3.29)$$

where wave number k_z takes continuous values and $k_n^{(x)}$, $k_m^{(y)}$ are discrete:

$$k_n^{(x)} = \frac{\pi}{L_x}n, \quad n = 1,2,3,\dots, \quad k_m^{(y)} = \frac{\pi}{L_y}m, \quad m = 1,2,3,\dots \qquad (3.30)$$

The energy spectrum consists of a continuous term for motion along the x axis and the two discrete sets resulting from confinement in the y and z directions:

$$E_{nm} = \frac{\hbar^2 k_z^2}{2m_e^*} + \frac{\pi^2\hbar^2}{2m_e^* L_x^2}n^2 + \frac{\pi^2\hbar^2}{2m_e^* L_y^2}m^2, \quad n = 1,2,3,\dots, \quad m = 1,2,3,\dots \qquad (3.31)$$

These properties of electrons in the c-band of a one-dimensional semiconductor are supposed to give rise to an absorption spectrum with sharp absorption steps corresponding to discrete levels and $1/(\hbar\omega - E_{nm})$, resulting from the density of states (Figure 3.1) in accordance with the Fermi golden rule and in a similar way to the discrete energy levels and density of states spectrum discussed in Section 3.3 in the context of the quantum well absorption spectrum. However, this simple consideration cannot provide a reasonable explanation for experimentally observed absorption spectra for the 1D-like semiconductor structure.

The correct consideration should involve electron–hole Coulomb interaction. One-dimensional space in this context was found to differ drastically from 3D and 2D analog. The one-dimensional Schrödinger equation for a couple of particles interacting via potential $U(z) = -e^2/z$ gives a diverging solution for the ground state $E_1 \rightarrow -\infty$. This means that the true 1D hydrogenic system has no stable ground state. A signature of this property is seen in Figure 3.8 and in Eq. (3.26).

To model the properties of real 1D-like structures, two approaches are used. In the first approach, the 2D problem is solved numerically with finite cross-sectional dimensions and aspect ratio of real elongated wire-like nanocrystals. In the second, the 1D Schrödinger equation is solved with a modified interaction potential $U(z) = -e^2/(z + z_0)$ to prevent $U(z)$ from going to $-\infty$ when z tends to zero. Then z_0 is used as a flexible parameter that should be chosen to get the best agreement with experimental observations.

Quasi-1D structures can be fabricated by means of epitaxial growth combined with lithographically defined etching and by means of colloidal chemistry. Epitaxial quantum

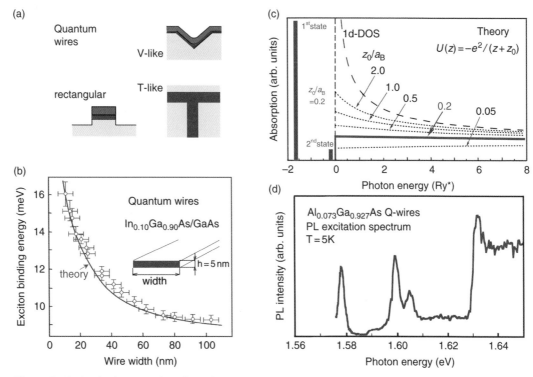

Figure 3.11 Optical properties of semiconductor quantum wires. (a) Different geometries of quantum wire structures developed by epitaxial growth and etching. (b) Experimentally observed (dots) and calculated (solid line) exciton binding energy for $In_{0.1}Ga_{0.9}As$ rectangular quantum wire with 5 nm height in a $GaAs/In_{0.1}Ga_{0.9}As/GaAs$ heterostructure. Adapted from Bayer et al. (1998) with permission, copyright APS. (c) Calculation of continuum absorption spectrum for quantum wire with adjustable parameter z_0 in the 1D Coulomb interaction potential resolving the discontinuity problem; two first exciton states are also shown. Adapted from Ogawa and Takagahara (1991) with permission, copyright APS. (d) Experimentally measured low-temperature exciton photoluminescence excitation spectrum for AlGaAs wires which shows constant absorption features between sharp excitonic peaks that correlates with the theory presented in panel (c). Adapted from Akiyama et al. (2003), with the permission of AIP Publishing.

wires are developed on a semiconductor substrate as narrow heterostructures in which barriers are formed by means of wider-gap layers. Figure 3.11(a) shows the possible geometries (rectangular, V-like, and T-like); Figure 3.11(b,c) gives examples of size-dependent optical absorption.

Figure 3.11(b) shows width-dependent exciton binding energy modeled (solid line) and measured (dots) for a rectangular planar quantum wire-like structure of 5 nm in height. In the limit of big width (100 nm) the structure resembles properties of a quantum well because of a very large width-to-height ratio. Exciton binding energy in this case tends to 8 meV, which is twice that of 3D exciton binding energy for the ternary compound examined. In the opposite limit, for width close to 10 nm, binding energy shows a rapid increase to 16 meV (i.e., four times greater than the 3D case) and bears witness to even higher growth for smaller widths. These thorough theoretical and experimental analyses clearly show that when going from the bulk to quasi-2D films (quantum wells), exciton binding energy rises but typically no more than 2–3 times, whereas reduction of a thin film to a wire-like geometry gives rise to further increases in exciton binding energy, up to $4Ry^*$ or maybe even more.

Figure 3.11(c) shows a set of calculated optical absorption spectra corresponding to continuum states with electron–hole interaction expressed as $U(z) = -e^2/(z + z_0)$. One can see that $1/(\hbar\omega - E_{nm})$ behavior expected from 1D density of states is not the case. Depending on z_0 value, continuum absorption can even obey energy-independent behavior (bright red curve in Figure 3.11(c)). In this case, the 1D structure exhibits an absorption spectrum like the 2D one, for which energy-independent absorption comes from constant 2D density of states. Energy-independent absorption features have been experimentally found in photoluminescence (PL) excitation spectra of quantum wires (Figure 3.11(d)). Note that for quantum wires measurements of the absorption spectrum by means of transmission detection is not always possible. In this case, analysis of PL excitation spectra becomes very instructive since it reproduces to a large extent the main absorption features.

Elongated nanocrystals with different aspect ratio (*nanorods*) can be developed by means of colloidal chemistry in liquids or in polymers. Nanorods show high-energy shifts with respect to the parent 3D crystal and feature absorption peaks depending on size and shape (Figure 3.12). Nanorods with higher aspect ratio represent a colloidal counterpart of quantum wire-like structures.

To date, 1D-like semiconductor nanostructures, including both epitaxial quantum wires and colloidal nanorods, have not found commercial applications in photonics, but their size- and shape-dependent absorption and emission properties can be used in display

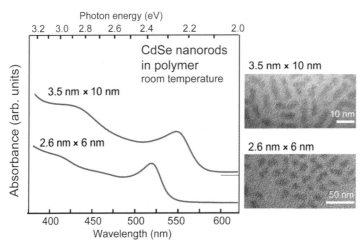

Figure 3.12 Optical absorption spectra and images of CdSe nanorods in polymer. Photo courtesy of M. Artemyev.

components, e.g., backlight sources for LCD-based devices where strong anisotropy of rods may become advantageous for polarized light emission. The latter is crucially important for all LCD-based displays since liquid crystals need strictly linearly polarized light for backlighting.

3.5 NANOCRYSTALS, QUANTUM DOTS, AND QUANTUM DOT SOLIDS

3.5.1 From cluster to crystal

In the context of quantum confinement phenomena in semiconductors, nanocrystals – i.e., nanometer-size crystalline particles – represent a kind of low-dimensional structures complementary to quantum wells (two-dimensional structures) and quantum wires (one-dimensional structures). However, nanocrystals have a number of specific features that are not inherent in the two- and one-dimensional structures. Quantum wells and quantum wires feature translational symmetry in two or one dimensions, and a statistically large number of electronic excitations can be created therein. In nanocrystals, the translational symmetry is totally broken and only a finite number of electrons and holes can be created within the same nanocrystal. Therefore, the concepts of the electron–hole gas and quasi-momentum fail in nanocrystals. It is instructive to consider evolution from single atoms to bulk solids (Figure 3.13).

A few atoms can form stable configurations called *clusters.* Clusters have no periodicity in atom arrangement; every type of atom possesses a specific set of stable configurations that can be predicted in terms of quantum chemistry. Probably the most famous cluster is *fullerene* C_{60}, in which 60 carbon atoms form a spherical surface with equal distance between every pair of atoms.

Sets of numbers relevant to stable clusters are referred to as *magic numbers* and these numbers differ for different atoms. The smallest clusters start from 3–4 atoms and extend to

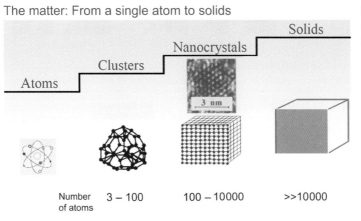

Figure 3.13 Evolution of matter from a single atom to bulk solid.

approximately 10^2 atoms. Upon further increasing the number of atoms, when the nanoparticle measures a few nanometers, the crystalline lattice develops. The electron microscope image in Figure 3.13 shows the early stage of crystal lattice emergence in a CdTe nanoparticle. At the nanocrystalline stage, from approximately 2–3 to 20–30 nm, electrons and holes can be viewed as quantum particles in a potential well where well size is defined by nanocrystal size and well depth is defined by a potential barrier at the nanocrystal/host matter interface. Liquids, polymers, glasses, and crystalline or polycrystalline matrices can be the host material for nanocrystals. In the range of approximately 2–10 nm optical properties of nanocrystals feature pronounced size dependence because of the quantum confinement effect on electron and hole energy spectra. This property was discovered and highlighted by Alexei Ekimov and Alexander Efros and co-workers in 1982 and Louis Brus in 1983 (see Box 3.2). Strong size-dependent properties of nanocrystals in correlation with planar quantum wells and rod-like quantum wires has given rise to notation of *quantum dots*.

BOX 3.2 EMERGENCE OF COLLOIDAL OPTOELECTRONICS

Size-dependent optical properties of semiconductor nanocrystals were discovered and explained in 1982–1983 by independent researchers in terms of the quantum confinement effect on electron and hole energy spectra based on glass technology – Alexei Ekimov and Alexander Efros and co-workers (Leningrad, USSR) – and based on colloidal chemistry – Louis Brus (Murray Hill, USA). Their pioneering publications established the beginning of the novel interdisciplinary field bridging solid-state physics, colloidal chemistry, and optics to give rise to emerging colloidal nanophotonics and optoelectronics.

| Alexei Ekimov | Alexander Efros | Louis Brus |

3.5.2 Synthesis of semiconductor nanocrystals

Semiconductor nanocrystals can be made by means of colloidal chemistry, glass technology, and epitaxial growth. Colloidal chemistry offers versatile techniques for a wide variety of compounds, typically II–VI compounds (CdTe, CdSe, CdS, ZnSe, ZnS, ZnO, HgSe, etc.), a few III–V compounds (e.g., InP, GaAs, GaP), IV–VI compounds (PbS, PbSe), and also I–VII compounds (e.g., AgBr) (Figure 3.14). The main advantages of colloidal

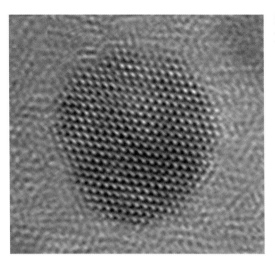

Figure 3.14 CdSe single nanocrystals in a polymer matrix. Courtesy of O. Chen.

techniques is cheap synthesizing processes (no high vacuum, no high temperatures) along with flexible surface and interface features providing, e.g., control of potential barrier height and profile, surface traps/defects, and guest–host effects. Sizes from cluster regime (1–2 nm) to 10–20 nm are affordable. The possibility exists to control size dispersion to a certain extent. The colloidal technique offers the widest possible range of nanocrystal concentration, from strongly diluted to close-packed arrangements. The latter represents an example of *colloidal crystals* and has gained the notation of *quantum dot solids*.

Optical glass technology did actually involve development of semiconductor nanocrystals in glass matrices for many decades, well before the notion of quantum confinement arose. Remarkably, in many cases quantum confinement contributes to and in a few cases has a decisive effect on the optical absorption spectra of commercial cutoff (based on binary and ternary II–VI compounds) and photochromic (based on I–VII compounds) filters. Growth of nanocrystals occurs in viscous glassy matrices at ~600 °C over several hours by means of diffusion-controlled aggregation of ions dissolved in the viscous glassy matrix. It can be considered as a kind of colloidal technique contrary to epitaxial growth based on lattice-matched deposition of atoms in high vacuum. Glass technology offers nanocrystals of II–VI compounds (mainly CdSe, CdS and their solid solutions), I–VII compounds (CuCl, CuBr, and others), and IV–VI compounds (PbS, PbSe) in production scale, and some other semiconductors including Si and Ge for research purposes. Glasses with semiconductor nanocrystals possess stable optical absorption spectra but show poor luminescence yield because of the lack of control of interface properties and guest–host phenomena (e.g., compressive strain from the glass matrix, surface traps, etc.). Sizes from the nucleation stage (about 2 nm) to opaque glass (about 50 nm) are available. Glasses feature noticeable nanocrystal size dispersion resulting in broadening of absorption spectra. The maximal possible concentration of semiconductor phase in glass is typically less than 1%.

Epitaxial growth can lead to nanocrystalline islands under conditions of lattice mismatch in *submonolayer heteroepitaxy*. This growth mode is referred to as the Stranski–Krastanov regime. It occurs by means of compressive strain in growing the epitaxial layer, which results in transformation of a 2D layer in a number of 3D islands. It was proposed for semiconductor nanocrystals growth in the 1990s and is extensively used for III–V compounds (InAs, InGaAs, InGaN, InP), including the first commercial quantum dot diode lasers. Si/Ge and CdSe/ZnSe quantum dot structures can also be fabricated. Heteroepitaxial quantum dots feature a specific shape (dome- or hut-like) and limited size range because of the complex interplay of several factors, primarily compensation of strain from dot–substrate lattice mismatch by means of internal mechanical tension at a dot. Certain possibilities of mean size variation and concentration control exist, but the size is typically more than 5 nm and concentration is well below the close-packed arrangement. Multilayer arrangement of quantum dots separated by thin planar interfacial layers is possible with certain quantum dot 3D self-ordering through the interplay of mechanical strain within and between layers. Self-ordering occurs because strain-induced distortions of a substrate material do not allow development of another dome- or hut-like crystallite very close to an existing one.

3.5.3　Optical absorption spectra

For semiconductor nanocrystals dispersed in a solid matrix or in a liquid with average distance much larger than crystallite size, the model of a single spherical potential well with rectangular barrier at the boundaries is very useful. If a nanocrystal size, i.e., a box radius a, is considerably larger than the exciton Bohr radius a_B^*, then the size-dependent optical absorption spectrum can be described in terms of exciton confinement. The energy shift is smaller or close to exciton binding energy and therefore does not change essentially the optical properties of nanocrystals. The formula for a quantum particle in a spherical box (Eq. (2.47)) is applicable with respect to the exciton (Efros and Efros, 1982) as a reasonable approximation for size-dependent absorption peaks,

$$E_{nml} = E_{g} - \frac{Ry^*}{n^2} + \frac{\hbar^2 \chi_{ml}^2}{2Ma^2}, \quad n, m, l = 1, 2, 3, \dots \tag{3.32}$$

with the roots of the Bessel function χ_{ml} (Table 2.1 and Figure 2.9). Here, $M = m_e^* + m_h^*$ is the exciton mass, which equals the sum of electron m_e^* and hole m_h^* effective masses. Thus an exciton in a spherical quantum dot is characterized by the quantum number n describing its *internal* states arising from the Coulomb electron–hole interaction (1S; 2S, 2P; 3S, 3P, 3D; ...), and by the two additional numbers, m and l, describing the states related to the center-of-mass motion in the presence of the *external* potential barrier featuring spherical symmetry (1s, 1p, 1d ..., 2s, 2p, 2d ...). Here, S(s), P(p), D(d), F(f), etc. label states with $l = 0,1,2,3$, ..., respectively as it is used in atomic spectroscopy. To distinguish the "internal" and the "external" states, we shall use capital letters for the former and lower case for the latter.

For the lowest 1S1s state ($n = 1$, $m = 1$, $l = 0$) the energy is expressed as

$$E_{1S1s} = E_g - Ry^* + \frac{\pi^2 \hbar^2}{2Ma^2},\tag{3.33}$$

or in another way as

$$E_{1S1s} = E_g - Ry^* \left[1 - \frac{\mu}{M} \left(\frac{\pi a_B^*}{a} \right)^2 \right],\tag{3.34}$$

where μ is the electron–hole reduced mass $\mu = m_e^* m_h^* / (m_e^* + m_h^*)$. In Eqs. (3.33) and (3.34) the value $\chi_{10} = \pi$ was used. Therefore, the first exciton resonance in a spherical quantum dot experiences a high energy shift by the value

$$\Delta E_{1S1s} = \frac{\mu}{M} \left(\frac{\pi a_B^*}{a} \right)^2 Ry^* < Ry^*.\tag{3.35}$$

Note that this value remains small compared to Ry^* since we consider the case $a \gg a_B^*$. This is the quantitative justification of the term "weak confinement."

Taking into account that photon absorption can create an exciton with zero angular momentum only, the absorption spectrum will consist of a number of lines corresponding to states with $l = 0$. Then the absorption spectrum can be derived from Eq. (3.32), with $a \gg a_B^*$, (see Table 2.1), i.e.:

$$E_{nm} = E_g - \frac{Ry^*}{n^2} + \frac{\hbar^2 \pi^2}{2Ma^2} m^2, \quad n, m = 1, 2, 3, \ldots\tag{3.36}$$

A weak confinement regime is feasible in wide-band semiconductors of I–VII compounds (copper halides) featuring small exciton Bohr radius (about 1 nm) and large exciton Rydberg energy (about 100 meV, see Table 2.3). Such nanocrystals are incorporated into certain commercial photochromic glasses.

3.5.4 Strong confinement regime

For many semiconductors – e.g., Si, Ge, II–VI (e.g., CdSe, CdTe, HgSe, HgS,), III–V (e.g., GaAs, InP), IV–VI (e.g., PbS, PbSe) – a condition

$$a_L < a < a_B^*\tag{3.37}$$

can be met for nanocrystals whose radius falls in the range of approximately 1.5–5 nm, depending on the specific semiconductor type. In this case, both absorption and emission spectra exhibit such crucial changes that color observed by the naked eye becomes strongly size-dependent.

In this regime, the hydrogenic exciton state can no longer exist and absorption occurs through optical transitions between fully discrete energy levels in valence and conduction bands and then continuous edge-like absorption of a bulk crystal reduces to a set of discrete sharp lines (Figure 3.15). When applying Eq. (2.47) to an electron and a hole in a spherical dot, one has

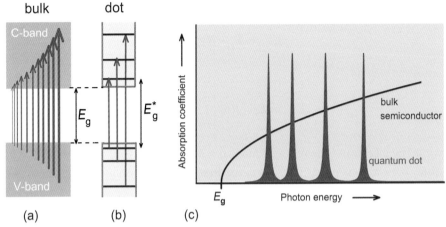

Figure 3.15 Optical absorption by semiconductor nanocrystals in the strong confinement regime. A sketch of the energy spectrum, optical transitions, and absorption spectra in a simple quantum dot model versus simple bulk semiconductor model. Unlike continuous optical transitions in a bulk crystal (a), electron and hole energy levels obey a series of states inherent for a particle in a spherical box (b). Selection rules allow optical transitions coupling the hole and the electron states with the same quantum numbers. Therefore, the optical absorption spectrum of the parent bulk crystal reduces to a number of discrete bands (c).

$$E_{nl}^{\text{electron}} = \frac{\hbar^2 \chi_{nl}^2}{2m_e^* a^2}, \quad E_{nl}^{\text{hole}} = \frac{\hbar^2 \chi_{nl}^2}{2m_h^* a^2}, \quad n,l = 1,2,3,\dots \tag{3.38}$$

Taking into account that selection rules allow only transitions between states with the same nl numbers, we arrive at the absorption spectrum (Efros and Efros, 1982), which has peaks at

$$E_{nl} = E_g + \frac{\hbar^2 \chi_{nl}^2}{2\mu a^2}, \quad \mu = \frac{m_e^* m_h^*}{m_e^* + m_h^*}, \quad n,l = 1,2,3,\dots \tag{3.39}$$

For χ_{nl} numbers, see Table 2.1. Notably, though hydrogenic electron–hole binding is not possible in the case under consideration, electron–hole Coulomb attractive interaction is still present and gives rise to lower energy of optical transitions with the Coulomb correction factor scaling as $e^2/(\varepsilon a)$ (Brus 1984). Since $a < a_B$ holds, the Coulomb correction term for an e–h pair in a dot can readily exceed the exciton binding energy of the parent bulk crystal. In more detail, the correction factors depend on n value; namely, the first peak in optical absorption takes the form

$$E_{1s1s} = E_g + \frac{\pi^2 \hbar^2}{2\mu a^2} - 1.786 \frac{e^2}{\varepsilon a}, \quad \mu = \frac{m_e^* m_h^*}{m_e^* + m_h^*}, \tag{3.40}$$

the second peak is

$$E_{1p1p} = E_g + \frac{4.49^2 \hbar^2}{2\mu a^2} - 1.884 \frac{e^2}{\varepsilon a}, \quad \mu = \frac{m_e^* m_h^*}{m_e^* + m_h^*}, \tag{3.41}$$

and further peaks can be calculated using χ_{nl} numbers from Table 2.1 and the numerical factor in the Coulomb term ranging from –1.6 to –1.8 (Shmidt and Weller 1986).

The simple "particle-in-a-box" model is very instructive. However, real semiconductors possess complex valence band structures, typically with three branches, and complicated confinement effects on holes restrict application of the simplified model. It is typically used as a quick and reasonable estimate for size-dependent absorption blue shift that needs calculation of the first absorption peak, and also as an estimate for emission spectrum, which features small low-energy shifts (so-called Stokes shift) with respect to the first absorption maximum.

In what follows we briefly discuss size-dependent absorption properties of colloidal II–VI nanocrystals which were examined and synthesized in recent decades and are considered as promising technological platforms for colloidal nano-optoelectronics, including lasers, LEDs, and photovoltaics.

Figure 3.16 gives two representative examples of size-dependent optical absorption spectra for nanocrystals in glass matrix and in solutions. For decades, the glass industry has used semiconductor-doped glasses to produce cutoff filters for the visible range. In these filters, nanocrystals of CdS_xSe_{1-x} are responsible for the absorption in the visible spectrum and in many cases their size-dependent properties make decisive contributions to the filter spectrum. For a typical application with 2–3 mm thickness, filters show a sharp decrease in transmission when wavelength becomes shorter. But if thinner samples, like 0.1 mm, are examined, the pronounced absorption maxima are apparent, indicating a discrete optical

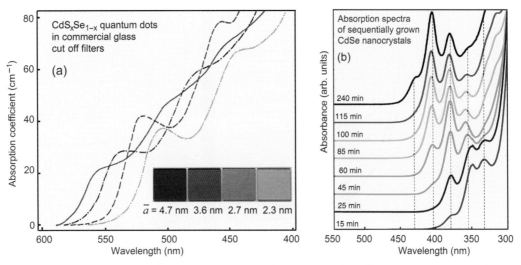

Figure 3.16 Representative size-dependent absorption spectra of semiconductor nanocrystals in (a) glass and (b) a solution. Nanocrystal size decreases from left to right (a) and from top to bottom (b). (b) Adapted from Kudera et al. (2007) with permission of John Wiley and Sons. © 2007 WILEY-VCH Verlag GmbH & Co. KGaA, Weinheim.

transition contrary to continuous absorption inherent in bulk crystals (Figure 3.16(a)). In the ultimate limit, nanocrystals reduce to clusters with a magic-size arrangement of atoms. However, in the intermediate size range, when the crystal lattice is present but the number of atomic shells and/or lattice period takes discrete values, an ensemble of nanocrystals with different sizes show a discrete set of absorption bands (Figure 3.16(b)).

Figure 3.16 gives an idea of how far downsizing of semiconductor nanocrystals can shift their absorption edge in terms of the first absorption peak. Apparently, the quantum confinement effect on absorption is more pronounced for narrow-band semiconductors. This is the consequence of smaller electron effective mass and higher dielectric permittivity inherent in narrow-band crystals (not only II–VI but also other types). See Table 2.3 for numerical data. Notably, for crystals like CdTe and CdSe, the whole visible range can be run through. The same takes place for other narrow-band semiconductors like GaAs, InSb, PbS, and PbSe.

Since the quantum confinement effect expressed by Eq. (3.40) and Figure 3.17 is dependent on electron–hole reduced mass and material dielectric permittivity, it correlates with the properties of excitons in the parent bulk crystals. Recalling Table 2.3 and Figure 2.15, one can see that exciton binding energy typically rises for wider-gap materials, whereas exciton Bohr radius falls accordingly. This allows for rigorous and elegant mathematical expression of Eq. (3.40) in terms of dimensionless energy E/Ry^* and dimensionless radius a/a_B^*,

$$(E_{1s1s} - E_g)/Ry^* = \pi^2 \left(\frac{a_B^*}{a}\right)^2 - 1.786\frac{a_B^*}{a}. \tag{3.42}$$

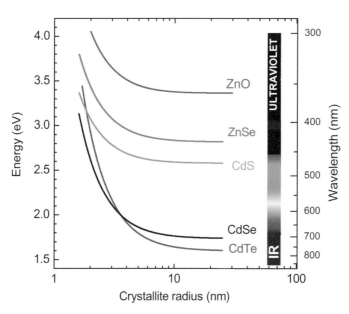

Figure 3.17 Dependence of the photon energy for the first optical transition versus quantum dot radius for the simple model of a spherical dot with infinite potential barriers for a number of common semiconductors.

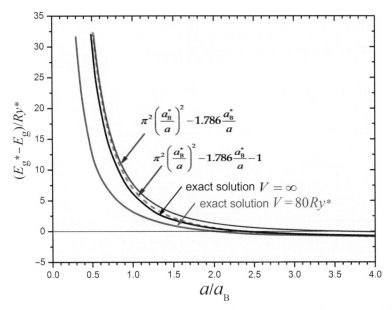

Figure 3.18 The results of the simple particle-in-a-box model presented in material-independent dimensionless energy and length scales. Position of the first absorption maximum E_g^* versus box radius a: Eq. (3.42) (black), its modification (dashes) and exact solutions of the two-particle Schrödinger equation for infinite ($V = \infty$) and finite ($V = 80\ Ry^*$) potential barrier by Thoai et al. (1990). E_g^* stands for E_{1s1s} energy.

Figure 3.18 summarizes results of size-dependent absorption blue shift for semiconductor nanocrystals in terms of dimensionless material-independent energy and length scales. The simple analytical solution is seen to reasonably coincide with the exact solution of the Schrödinger equation for an electron and a hole in a spherical box with infinite barrier (blue line). This solution gives continuous high-energy shift of the first absorption maximum denoted as E_g^* with descending box radius. There is no need to introduce weak and strong confinement limits. Note that Eq. (3.42) reasonably agrees with the accurate solution and this agreement can be made even better if a (–1) term is added to start with $E_g - Ry^*$ instead of E_g. Data for the finite barrier (red line) show overall lowering of energy in accordance with the general properties of a quantum particle in a box (see Figure 2.3).

Experimental studies of size-dependent optical absorption spectra for colloidal semiconductor quantum dots reveal that the simple particle-in-a-infinite-box model does predict the right trend in the size dependence of the energy for the first absorption band. Figure 3.19 shows a representative example. However, for smaller size the energy shift predicted by the simple model gives bigger values than the experimentally observed ones. The better agreement of particle-in-a-box theory with the experiment can be obtained when the finite potential barrier, nonparabolicity of the c-band, and complex structure of the v-band are involved. All these factors result in lower energy shift for smaller sizes, providing reasonable agreement down to 2 nm radii. The finite potential barrier effect is clear from Figure 2.3 and from calculations presented in Figure 3.18. Nonparabolicity of

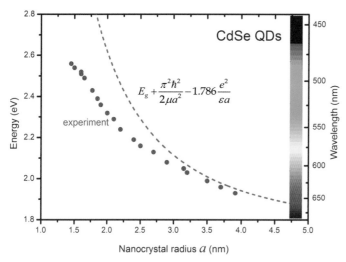

Figure 3.19 Size dependence of the first absorption band maximum for colloidal CdSe quantum dots (red dots, Norris and Bawendi 1996) and the simple particle-in-a-box theory (blue dashes).

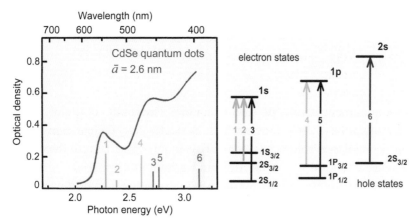

Figure 3.20 Experimental absorption spectrum for CdSe nanocrystals in glass and identification of optical transitions. Courtesy of Al L. Efros.

the c-band occurs since the electron has kinetic energy corresponding to a high k value, far from $k = 0$. Then nonparabolicity can be described as higher effective mass and, accordingly, energy goes down. This effect occurs mainly for electrons since their effective mass is originally much smaller than hole mass and electrons depart from parabolicity in bigger boxes. Complex v-band structure gives rise to a multitude of optical transitions which are presented in Figure 3.20, where the experimentally measured absorption spectrum and calculated optical transitions for CdSe nanocrystals are shown.

When quantum dot size exceeds exciton Bohr radius, a_B^*, there are free and bound electron–hole pairs and optical properties of quantum dots only slightly deviate from those of a parent bulk crystal. The main effect is the blue shift of exciton peak(s), which is of the order of Ry^* and therefore is much smaller than E_g. When size becomes noticeably smaller than a_B^*, the hydrogenic coupling of electrons and holes is no longer possible since a box offers no space for electron and hole hydrogenic arrangement. However, electron–hole Coulomb interaction is present and, moreover, size-dependent attraction energy $1.786e^2 / \varepsilon a$ always exceeds the corresponding bulk value $Ry^* = e^2 / (\varepsilon a_B^*)$. Furthermore, if only optical excitation is considered, in nanocrystals an electron and a hole are always created simultaneously. Therefore, the notation of *exciton in a quantum dot* has become common and instructive for an elementary electronic excitation in nanocrystals. It emphasizes the role of Coulomb interaction and pair-like creation and recombination of electrons and holes in quantum dots in the course of absorption and emission of light (Woggon and Gaponenko 1995).

Important properties of quantum dots are the discrete number of electron–hole pairs (excitons) and discrete number of trap states. Unlike bulk crystals and two- and one-dimensional nanostructures, the notion of concentration with respect to electrons, holes, excitons, and defects (traps) cannot be applied to nanocrystals. A discrete number of e–h pairs gives rise to intensity-dependent optical properties. Roughly speaking, one e–h pair per dot gives rise to complete bleaching (absorption saturation) for a quantum dot ensemble, and two e–h pairs per dot result in optical gain. A discrete number of traps results in statistical lack of trap states in every quantum dot ensemble. This property enables enhancing the luminescence efficiency of nanocrystalline luminophores.

Glasses provide durable absorption properties which enable optical filter and shutter (modulators) developments. However, glass technology does not allow for the surface and interface properties to be altered or controlled. Basically, only the mean size and to some extent the nanocrystal quality can be controlled by avoiding the so-called competitive growth stage for the sake of nucleation and normal growth processes (Gaponenko et al. 1993). However, these growth regimes require a combination of low temperature and a long time in thermal treatment; crystallites with a mean size of the order of 5 nm need a few days of thermal processing at ~600 °C to grow properly. The temperature should carefully be kept constant, with accuracy better than 1% to get reproducible results. Furthermore, in most cases cooling down to room temperature after growth gives rise to strong compressive strain since the thermal expansion (i.e., cooling compression) coefficient of glassy matrix exceeds that of the crystalline material. I–VII crystallites (like CuCl) are possibly the only case in which strain is not essential because of low melting temperature of those crystals.

3.5.5 Luminescence of nanocrystals

Luminescent properties of nanocrystals define important fields of their potential applications from light converters for blue or UV LEDs to fully colloidal display optoelectronic devices. Figure 3.21 shows typical examples of photoluminescence spectra for

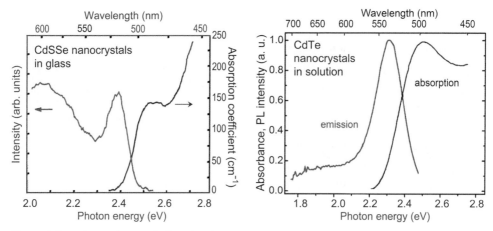

Figure 3.21 Absorption and photoluminescence emission spectra for typical II–VI nanocrystals in the strong confinement regime in glass and in a solution.

small nanocrystals of II–VI compounds in glass and in solution. Along with the intrinsic emission peaking around 530 nm, an additional band presents that results from defect states, mainly surface trap states. In glasses, luminescence properties cannot be controlled; moreover, luminescence in glasses features pronounced degradation upon continuous illumination because of the photoionization phenomenon. Therefore, semiconductor-doped glasses cannot be considered as light-emitting materials. Fortunately, colloidal chemistry offers fine surface and interface engineering techniques that are capable of minimizing the number of traps and taming photoionization. In this context, the concept of the *core–shell colloidal quantum dot* structure coined by M. Bawendi and co-workers in 1997 was extremely helpful. In this structure, the core of the narrow-band semiconductor is covered by the shell of a wider-band semiconductor, the absolute levels of c- and v-bands enabling an additional potential barrier both for electrons and for holes (Figure 3.22). These core–shell nanocrystals are called "type I" quantum dots and represent the main trend in colloidal quantum dot luminophores resulting in high quantum yield and superior photostability. Alternatively, it is possible to develop core–shell structures in which an electron will have the potential well whereas a hole will not (Figure 3.22, right-hand panel). These so-called "type II" quantum dot structures represent the principal approach to efficient colloidal *quantum dot lasers* and will be considered in detail in Part II.

Modern colloidal nanotechnologies demonstrate a versatile approach to development of stable luminescent core–shell nanocrystals (Figure 3.23) with high quantum yield and the spectrum tunable from near-IR (PbS, PbSe) through the whole visible spectrum (CdTe, CdSe) to UV (ZnSe). Many of these are commercially available in the form of solutions. When embedded in polymers, these nanocrystals represent emerging technological platforms for display devices.

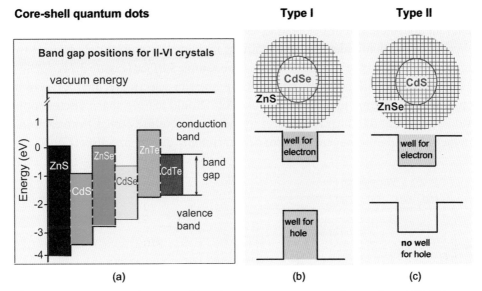

Figure 3.22 (a) Energy diagram of band gaps versus vacuum of a number of II–VI compounds and (b) the two types of core–shell quantum dots.

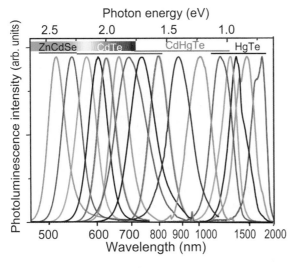

Figure 3.23 Variation of size enables wide-range tuning of photoluminescence emission spectra. Courtesy of N. Gaponik, TU Dresden.

Every nanocrystal ensemble has *inhomogeneously broadened absorption and emission spectra* due to distribution of sizes, defect concentration, shape fluctuations, environmental inhomogeneities, and other features. Therefore, the most efficient way to examine the properties of a single nanocrystal smeared by *inhomogeneous broadening* is to use selective techniques. At the early stages of quantum dot studies, selective absorption saturation and spectrally selective photoluminescence excitation spectroscopy were applied. The relevant optical phenomena are referred to as *spectral hole burning* and *fluorescence line narrowing* (see Gaponenko

1998 for details). Later, a single-molecule photoluminescence technique was introduced into quantum dot studies. In this technique, spectrally selective excitation is applied to a spatially selective area so that under certain conditions in diluted ensembles only one dot will exhibit luminescence within the probed portion of the sample surface (Figure 3.24).

Thus, fine spectrally selective, size-selective, and site-selective techniques unambiguously confirmed the original idea that quantum dots should feature sharp discrete optical transitions. Smearing of sharp resonances at room temperature comes from exciton–phonon interactions and inhomogeneous broadening, mainly due to size dispersion.

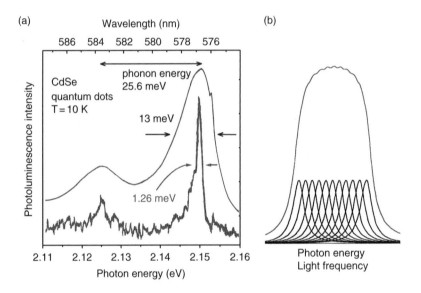

Figure 3.24 Single-dot photoluminescence emission spectrum (red) along with the ensemble-averaged spectrum (blue) for colloidal CdSe nanocrystals at low temperature, and explanation for inhomogeneous broadening. Adapted from Empedocles et al. (1996); Copyright 1996 by the American Physical Society.

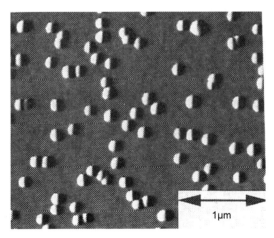

Figure 3.25 InP nanocrystals on a GaInP substrate (atomic force microscope image). Reprinted from Carlsson et al. (1995), with permission from Elsevier.

Epitaxial quantum dots (Figure 3.25) became the subject of extensive research in the mid-1990s when the principal size-dependent optical properties of quantum dots featured by nanocrystals in colloidal and glass environments was established. Therefore, studies of epitaxial dots had been from the very beginning targeted at practical applications, first in semiconductor injection lasers. Nowadays InAs quantum dots on GaAs substrate are used in commercial lasers for optical communications at a wavelength range around 1.3 μm. Recent progress in InGaN epitaxial dots promises development of blue lasers. Quantum dot lasers feature lower threshold current density and less pronounced undesirable temperature effects as compared to quantum well lasers (Ledentsov 2011). Quantum dot lasers are considered in detail in Part II.

3.5.6 Quantum dot solids

Semiconductor nanocrystals can be organized in close-packed periodic or in dense random solid-like structures. The first case is possible if very narrow size distribution is reached. Then, a face-centered cubic lattice of nanocrystals can be fabricated. These structures represent a type of *colloidal crystal* which was first identified in science in 1957. In periodic structures, formation of minibands and resonant tunneling of electrons and holes over large distances is possible (see Figures 2.6 and 2.8), as well is the enhanced energy transfer phenomenon. In the context of semiconductor low-dimensional structures, colloidal nanocrystalline crystals represent three-dimensional superlattices. These structures can be use in optoelectronic devices, including electroluminescent structures and photodetectors.

Several groups have reported on successful realization of periodic two- and even three-dimensional arrays of nanocrystals using the self-organization effect in strained heterostructures under the condition of submonolayer epitaxy. Two-dimensional self-organization within a layer is possible because strain beneath a given quantum dot and around it does not allow for development of another dot very close to the first one. Three-dimensional arrangement is performed by means of multiple-layer growth. In this case, however, the typical size of nanocrystals is equal to or larger than 10 nm, the inter-dot spacing being of the order of 10 nm.

Random but dense arrangement of nanocrystals should give rise to delocalized electron and hole states, provided certain critical density is achieved (by analogy with *Anderson transition* in disordered solids). This behavior was observed for very small nanocrystals belonging to the so-called magic-size regime, and results in systematic evolution of an optical absorption spectrum from sharp peaks to band edge absorption which is interpreted as development of delocalized electron and/or hole states in a densely packed subsystem of spherical quantum wells. Figure 3.26 presents experimental and computational data on this phenomenon.

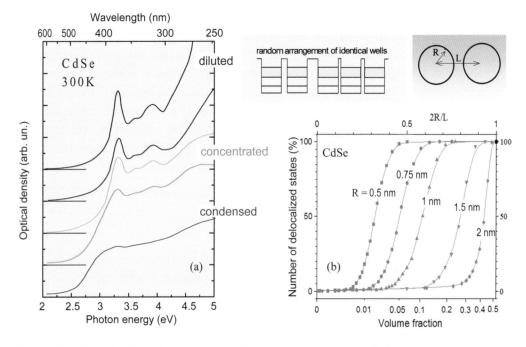

Figure 3.26 Modification of optical absorption spectra in a dense colloidal quantum dot ensemble. (a) Experimental optical absorption spectra for CdSe nanocrystals with 1.6 nm mean radius for diluted and condensed systems. Solvent concentration from top to bottom is 37%, 8%, 3%, 1%, and 0%. (b) Calculated number of delocalized states versus volume fraction and dimensionless inter-dot spacing $2R/L$ for various dot radii in the model of randomly arranged 3375 ($=15^3$) spherical potential wells. Figures adapted with permission from Artemyev et al. (1999), copyrighted by the American Physical Society, and Artemyev et al. (2001), copyrighted by John Wiley and Sons. © 2001 WILEY-VCH Verlag Berlin GmbH, Fed. Rep. of Germany.

Conclusion

- Spatial confinement results in a number of principal effects on electrons, holes, and excitons in semiconductors.

- When confinement extension becomes comparable with de Broglie wavelength (of the order of 10 nm) for electrons and holes but remains considerably bigger than the lattice period (of the order of 1 nm), which is feasible in many crystalline materials, the lattice properties remain only slightly changed whereas electron, hole, and exciton properties change drastically since these quasiparticles find themselves in quantum boxes and in spaces with lower dimensionality.

- Absorption, emission spectra, recombination rates, relaxation times, and luminescence quantum yields in low-dimensional structures become size- and shape-dependent.

- Very thin films (quantum wells) obey to a large extent properties of a two-dimensional space for electrons, holes, and excitons. Needle-like crystals (quantum wires) feature to a large extent properties of a one-dimensional space for the abovementioned quasiparticles.
- Properties of quantum wires and quantum wells in many cases can be satisfactorily understood when bulk-like equations are rewritten for lower dimensionality.
- Semiconductor nanocrystals (quantum dots) represent what are often termed "zero-dimensional objects." However, one must understand that neither property of a quantum dot can be obtained by simple scaling of 3D, 2D, or 1D equations to $d = 0$. Nanocrystals therefore represent a different case of low-dimensional structures in which ultimate space restriction can be understood only in terms of the finite number of e–h pairs, electron and hole traps, etc.
- Quantum wells, quantum wires, and quantum dots can be fabricated by epitaxial and colloidal techniques.
- Epitaxial quantum wells are already successfully used in commercial LEDs and lasers.
- Quantum wires are the subject of research and are fabricated mainly in labs rather than in factories.
- Quantum dots when epitaxially grown are used in commercial lasers. Colloidal quantum dots have recently entered commercial display components in TV sets, tablets, and cell phones. Colloidal quantum dots, as well as the recently emerged colloidal quantum wires (nanorods), and colloidal quantum wells (nanoplatelets) promise breakthroughs in nano-optoelectronics based on cheap and versatile colloidal fabrication techniques.

Problems

3.1 Derive density of modes versus wave number for one- and two-dimensional space (Eq. (3.7)).
3.2 Derive Eq. (3.18) to calculate the Fermi energy for metals and arrive at numerical values based on free electron concentration data (Table 3.1).
3.3 Derive Eq. (3.42) based on Eq. (3.40) and formulas for exciton parameters.
3.4 Calculate and plot the energy of the first optical transition versus radius for PbS and PbSe quantum dots, assuming infinite potential barriers.
3.5 Suggest optical cutoff filters for 300, 400, 500, 600, and 700 nm based on nanocrystals in a dielectric matrix. A cutoff filter has transmission close to 1 for $\lambda > \lambda_{crit}$ and close to 0 for $\lambda < \lambda_{crit}$.

Further reading

Alferov, Z. I. (1998). The history and future of semiconductor heterostructures. *Semiconductors*, **32** (1), 1–14.

Bányai, L., and Koch, S. W. (1993). *Semiconductor Quantum Dots*. World Scientific.

Bastard, G. (1988). *Wave Mechanics Applied to Semiconductor Heterostructures*. Les Editions de Physique.

Bimberg, D., Grundmann, M., and Ledentsov, N. N. (1999). *Quantum Dot Heterostructures*. John Wiley & Sons.

Carlsson, N., Georgsson, K., Montelius, L., et al. (1995). Improved size homogeneity of InP-on GaInP Stranski–Krastanow islands by growth on a thin GaP interface layer. *J Cryst Growth*, **156**, 23–29.

Dabbousi, B. O., Rodriguez-Viejo, J., Mikulec, F. V., et al. (1997). (CdSe) ZnS core–shell quantum dots: synthesis and characterization of a size series of highly luminescent nanocrystallites. *J Phys Chem B*, **101**(46), 9463–9475.

Elliott, R. J. (1957). Intensity of optical absorption by excitons. *Phys Rev*, **108**, 1384–1389.

Gaponenko, S. V. (1998). *Optical Properties of Semiconductor Nanocrystals*. Cambridge University Press.

Gaponenko, S. V. (2010). *Introduction to Nanophotonics*. Cambridge University Press.

Gaponik, N., Hickey, S. G., Dorfs, D., Rogach, A. L., and Eychmüller, A. (2010). Progress in the light emission of colloidal semiconductor nanocrystals. *Small*, **6**, 1364–1378.

Guzelturk, B., Martinez, P. L. H., Zhang, Q., et al. (2014). Excitonics of semiconductor quantum dots and wires for lighting and displays. *Laser Photonics Rev*, **8**, 73–93.

Harrison, P. (2009). *Quantum Wells, Wires and Dots: Theoretical and Computational Physics of Semiconductor Nanostructures*. John Wiley & Sons.

Kalt, H., and Hetterich, M. (eds.) (2013). *Optics of Semiconductors and Their Nanostructures*. Springer Science & Business Media.

Klimov, V. (ed.) (2010). *Nanocrystal Quantum Dots*. CRC Press.

Klingshirn, C. (ed.) (2001). *Semiconductor Quantum Structures: Optical Properties. Part 1*. Springer.

Klingshirn, C. (ed.) (2004). *Semiconductor Quantum Structures: Optical Properties. Part 2*. Springer.

Markov, I. V. (1995). *Crystal Growth for Beginners: Fundamentals of Nucleation, Crystal Growth, and Epitaxy*. World Scientific.

Pelá, R. R., Caetano, C., Marques, M., et al. (2011). Accurate band gaps of AlGaN, InGaN, and AlInN alloys calculations based on LDA-1/2 approach. *Appl Phys Lett*, **98**, 151907.

Rogach, A. (ed.) (2008). *Semiconductor Nanocrystal Quantum Dots*. Springer.

Ustinov, V. M. (2003). *Quantum Dot Lasers*. Oxford University Press.

Woggon, U. (1997). *Optical Properties of Semiconductor Quantum Dots*. Springer.

References

Achtstein, A. W., Schliwa, A., Prudnikau, A., et al. (2012). Electronic structure and exciton–phonon interaction in two-dimensional colloidal CdSe nanosheets. *Nano Lett*, **12**, 3151–3157.

Akiyama, H., Yoshita, M., Pfeiffer, L. N., West, K. W., and Pinczuk, A. (2003). One-dimensional continuum and exciton states in quantum wires. *Appl Phys Lett*, **82**, 379–381.

Artemyev, M. V., Bibik, A. I., Gurinovich, L. I., Gaponenko, S. V., and Woggon, U. (1999). Evolution from individual to collective electron states in a dense quantum dot ensemble. *Phys Rev B*, **60**, 1504–1507.

Artemyev, M. V., Bibik, A. I., Gurinovich, L. I., et al. (2001). Optical properties of dense and diluted ensembles of semiconductor quantum dots. *Physica Status Solidi (b)*, **224**, 393–396.

Bayer, M., Walck, S. N., Reinecke, T. L., and Forchel, A. (1998). Exciton binding energies and diamagnetic shifts in semiconductor quantum wires and quantum dots. *Phys Rev B*, **57**, 6584–6591.

Brus, L. E. (1984). Electron–electron and electron–hole interactions in small semiconductor crystallites: the size dependence of the lowest excited electronic state. *J Chem Phys*, **80**, 4403–4409.

Chemla, D. S., and Miller, D. A. (1985). Room-temperature excitonic nonlinear-optical effects in semiconductor quantum-well structures. *J Opt Soc, Amer B*, **2**, 1155–1173.

Efros A.L., and Efros A.L. (1982). Interband absorption of light in a semiconductor sphere. *Soviet Physics Semiconductors-USSR*, **16**, 772–775.

Empedocles, S. A., Norris, D. J., and Bawendi, M. G. (1996). Photoluminescence spectroscopy of single CdSe quantum dots. *Phys Rev Lett*, **77**, 3873–3876.

Gaponenko, S. V. (1998). *Optical Properties of Semiconductor Nanocrystals*. Cambridge University Press.

Gaponenko, S., Woggon, U., Saleh, M., et al. (1993). Nonlinear-optical properties of semiconductor quantum dots and their correlation with the precipitation stage. *J Opt Soc Amer B*, **10**, 1947–1955.

Göbel, E. O., and Ploog, K. (1990). Fabrication and optical properties of semiconductor quantum wells and superlattices. *Progress in Quantum Electronics*, **14**, 289–356.

He, X. F. (1991). Excitons in anisotropic solids: the model of fractional-dimensional space. *Phys Rev B*, **43**, 2063–2069.

Kasap, S. O. (2002). *Principles of Electronic Materials and Devices*, 2nd edn. McGraw-Hill.

Keldysh, L. V. (1979). Coulomb interaction in thin semiconductor and semimetal films. *J Exp Theor Phys*, **29**, 658–662.

Kudera, S., Zanella, M., Giannini, C., et al. (2007). Sequential growth of magic size CdSe nanocrystals. *Adv Mater*, **19**, 548–552.

Ledentsov, N. N. (2011). Quantum dot laser. *Semicond Sci Technol*, **26**, 014001.

Nakamura, S., Senoh, M., Iwasa, N., and Nagahama, S. I. (1995). High-power InGaN single-quantum-well-structure blue and violet light-emitting diodes. *Appl Phys Lett*, **67**, 1868–1870.

Norris, D. J., and Bawendi, M. G. (1996). Measurement and assignment of the size-dependent optical spectrum in CdSe quantum dots. *Phys Rev B*, **53**, 16336–16342.

Ogawa, T., and Takagahara, T. (1991). Optical absorption and Sommerfeld factors of one-dimensional semiconductors: an exact treatment of excitonic effects. *Phys Rev B*, **44**, 8138–8144.

Ralph, H. I. (1965). The electronic absorption edge in layer type crystals. *Solid State Commun*, **3**, 303–306.

Schmidt, H. M., and Weller, H. (1986). Quantum size effects in semiconductor crystallites: calculation of the energy spectrum for the confined exciton. *Chem Phys Lett*, **129**(6), 615–618.

Sell, D. D., and Casey, Jr., H. C. (1974). Optical absorption and photoluminescence studies of thin GaAs layers in GaAs–Al$_x$Ga$_{1-x}$As double heterostructures. *J Appl Phys*, **45**(2), 800–807.

Straubinger, R., Beyer, A., and Volz, K. (2016). Preparation and loading process of single crystalline samples into a gas environmental cell holder for in situ atomic resolution scanning transmission electron microscopic observation. *Microsc Microanalysis*, **22**, 515–519.

Thoai, D. T., Hu, Y. Z., and Koch, S. W. (1990). Influence of the confinement potential on the electron–hole-pair states in semiconductor microcrystallites. *Phys Rev B*, **42**, 11261–11270.

Woggon, U., and Gaponenko, S. V. (1995). Excitons in quantum dots. *Phys Stat Sol (b)*, **189**, 286–343.

Lightwaves in restricted geometries

The subject of this chapter is propagation, reflection, and evanescence of electromagnetic waves in continuous and complex media and structures to introduce the basics of wave optics and its implementation in novel nanophotonic conceptions. Electromagnetic waves propagate in a medium with positive permittivity, reflect at every permittivity step, and evanesce in a metal featuring negative permittivity. Combination of different dielectrics allows for wave confinement, tunneling, and energy storage, and gives rise to the photonic crystals notion. Combination of a metal with a dielectric material in nanostructures allows for optical material design like stained glass and very high local concentration of incident field to arrive at the notion of nanoplasmonics and optical antennas.

4.1 LIGHT AT THE INTERFACE OF TWO DIELECTRICS

4.1.1 Dispersion law and refractive index

The simple relation occurs between wavelength, λ, velocity, ε, and period of oscillations, T:

$$\lambda = vT, \tag{4.1}$$

which simply states that distance covered by a wave during a single period of oscillations equals the product of velocity and time. Recalling the notion of wave vector $\mathbf{k}$ (see Eq. (2.3)),

$$k = |\mathbf{k}| = \frac{2\pi}{\lambda}, \tag{4.2}$$

where k is wave number, v is frequency, and ω is circular frequency:

$$\omega = 2\pi v = \frac{2\pi}{T} \tag{4.3}$$

one can arrive at the important relation

$$\omega = vk \tag{4.4}$$

or

$$\boxed{v = \frac{\omega}{k}.} \tag{4.5}$$

It is important to distinguish between *phase velocity* and *group velocity* of waves. Phase velocity describes the speed of motion of a plane with constant phase. It is phase velocity that enters Eqs. (4.1), (4.4), and (4.5). Group velocity, ε_g describes speed and direction of energy transfer in the course of a wave process. It reads as a vector in a 2D and 3D case:

$$\mathbf{v}_g = \frac{d\omega}{d\mathbf{k}}, \tag{4.6}$$

and reduces to a scalar value in a 1D problem, i.e.,

$$v_g = \frac{d\omega}{dk}. \tag{4.7}$$

In a particular case of linear $\omega(k)$ relation, phase and group velocities coincide. In isotropic media, group velocity has direction coinciding with the wave vector. For electromagnetic waves in a vacuum, the group and phase velocities are equal to $c = 299,792,458$ m/s. In a general case, a medium features frequency-dependent phase velocity and is characterized by the *refractive index* $n(\omega)$, which shows how many times phase velocity of light in a given medium is smaller than c. Then the dispersion law reads

$$\omega = \frac{c}{n(\omega)} k. \tag{4.8}$$

Relation between frequency and wave number or wave vector is very important in the physics of waves and oscillations, and is referred to as the *dispersion law*. Figure 4.1 shows a few simple cases relevant to optics. Note that in optics the term dispersion means dependence of light speed (and refractive index) on frequency.

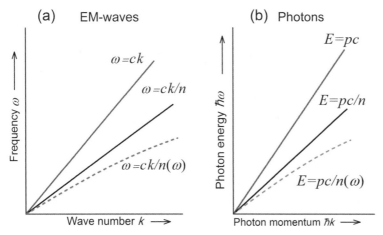

Figure 4.1 Dispersion law (frequency versus wave number) for (a) electromagnetic waves and (b) photons. Red lines – in vacuum or in air ($n = 1$); blue lines – in dielectric medium with frequency-independent refractive index $n > 1$; dashed green lines – in a dielectric medium whose refractive index n grows with frequency.

It is possible to change scales in Figure 4.1(a) by multiplying frequency and wave number by the factor of $\hbar$. Then we arrive at the *photon dispersion curve* (Figure 4.1(b)) in accordance with the relations for photon energy $E = \hbar\omega$ and photon momentum $p = \hbar k$. It is common in theoretical physics to consider the photon dispersion law $E(p)$ instead of electromagnetic wave function $\omega(k)$. Then, photon and electron properties look alike except for parabolic $E(p)$ dependence for electrons in a vacuum versus linear $E(p)$ dependence for photons in a vacuum. However, every mass particle including an electron can be disturbed by means of impact resulting from interactions, e.g., collisions, gravity, Coulomb interaction for charged particles, etc. Therefore its momentum can experience continuous change along with its energy in accordance with the $E(p)$ dependence. For photons, the situation is different. Neither impact can continuously move a photon along its $E(p)$ dependence. The photon can only be absorbed or emitted in the course of light–matter interactions. Then the $E(p)$ dependence for the photon simply gives a way to calculate photon energy when its momentum is known and, *vice versa,* to calculate photon momentum based on its energy.

In the transparency spectral range for dielectrics and semiconductors, i.e., for photon energies $\hbar\omega < E_g$, the refractive index is always more than 1 and typically does not exceed 4. There is a general tendency for the refractive index to rise with frequency for every given dielectric or semiconductor crystal, and then the dispersion law typically has the form shown by green dashes in Figure 4.1. For a given material, relative n variation typically measures about 10% within the optical range, provided the material is transparent.

There is another important trend: wider-gap materials feature lower n values (Figure 4.2). For example, Ge ($E_g = 0.7$ eV) below the band gap has $n = 4$, ZnSe below band gap ($E_g = 2.8$ eV) has $n = 2.5$, and SiO_2 ($E_g = 8$ eV) has $n = 1.45$. This trend means that stronger atomic bonding responsible for a wider-band gap does not allow ions in crystal lattice sites to react strongly to electromagnetic oscillations. Table 4.1 gives refractive indices for a number of common semiconductors and dielectrics.

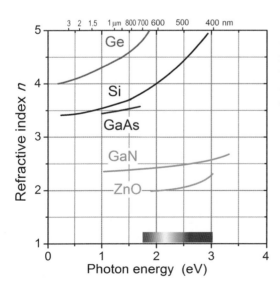

Figure 4.2 Refractive index for a few semiconductors. For every material refractive index rises with photon energy. There is also the general trend that lower band gap crystals (Ge, Si, GaAs) possess higher refractive index than wider-band gap compounds (GaN, ZnO).

Table 4.1 **Refractive index for selected solids**

Material	n	Material	n
Na_3AlF_6	1.34	CdS	2.47
MgF_2	1.37	ZnSe	2.50
LiF	1.39	CdSe 1 μm	2.55
CaF_2	1.43	InN 1 μm	2.56
Fused silica (SiO_2)	1.46	SiC	2.64
Mica	1.58	CdTe 10.6 μm	2.69
Polystyrene	1.59	TiO_2 rutile	2.80
MgO	1.74	ZnTe	2.98
Al_2O_3	1.77	GaP	3.31
ZnO	2.0	GaAs 1.15 μm	3.37
TiO_2 polycrystalline	2.22	InAs 10.6 μm	3.42
Nb_2O_5	2.26	Si 10.6 μm	3.42
GaN	2.38	GaSb 10.6 μm	3.84
ZnS	2.35	InSb 10.6 μm	3.95
Diamond	2.41	Ge 10.6 μm	4.00

Wavelength is 632 nm if not specified. Materials with specified wavelengths absorb light in the visible spectrum. Materials are ordered according to growing n.

Source: Yariv and Yeh (1984) and Madelung (2012).

4.1.2 Helmholtz equation

Electromagnetic field is exhaustively described by *Maxwell's equations*, which mathematically express the existence of electrical charges and absence of magnetic charges, as well as mutual interplay of alternative magnetic and electric fields. The electromagnetic field is described by a set of four vectors: electric field vector $\mathbf{E}$, magnetic field vector $\mathbf{H}$, electric displacement vector $\mathbf{D}$, and magnetic induction vector $\mathbf{B}$. Vectors $\mathbf{E}$ and $\mathbf{H}$ are independent, whereas vectors $\mathbf{D}$ and $\mathbf{B}$ are related to the first two vectors via material equations:

$$\mathbf{D} = \varepsilon\varepsilon_0\mathbf{E} = \varepsilon_0\mathbf{E} + \mathbf{P}, \ \ \mathbf{B} = \mu\mu_0\mathbf{H} = \mu_0\mathbf{H} + \mathbf{M}. \tag{4.9}$$

Here, ε is dimensionless relative dielectric permittivity of the medium under considera-tion, μ is dimensionless relative magnetic permeability of the medium, ε_0 and μ_0 are the basic constants that are often referred to as permittivity and permeability of a vacuum, respectively, and **P** and **M** are electric and magnetic polarizations, respectively. In a general case of an anisotropic medium, ε and μ are tensors. For isotropic media, these tensors reduce to scalars. Vectors **E**, **H**, **D**, **B** satisfy Maxwell's equations:

$$\nabla \times \mathbf{E} = -\frac{\partial \mathbf{B}}{\partial t}, \quad \nabla \times \mathbf{H} = \frac{\partial \mathbf{D}}{\partial t} + \mathbf{J},$$

$$\nabla \cdot \mathbf{D} = \rho, \quad \nabla \cdot \mathbf{B} = 0 \tag{4.10}$$

where **J** is the electric current density (A/m^2) and ρ is the electric charge density (C/m^3).

For a medium without charges and currents (**J** = 0, ρ = 0), Eqs. (4.8)–(4.10) reduce to a couple of *wave equations* for **E** and **H**:

$$\nabla^2 \mathbf{E} - \mu\varepsilon\mu_0\varepsilon_0 \frac{\partial^2 \mathbf{E}}{\partial t^2} = 0, \quad \nabla^2 \mathbf{H} - \mu\varepsilon\mu_0\varepsilon_0 \frac{\partial^2 \mathbf{H}}{\partial t^2} = 0 \tag{4.11}$$

with the known solutions in the form of plane waves,

$$\psi = e^{i(\omega t - \mathbf{k}\cdot\mathbf{r})}. \tag{4.12}$$

Here, wave number k reads

$$k = |\mathbf{k}| = \omega\sqrt{\varepsilon\varepsilon_0\mu\mu_0}. \tag{4.13}$$

Then phase velocity of the wave (Eq. (4.5)) takes the form

$$\upsilon = \frac{\omega}{k} = \frac{1}{\sqrt{\varepsilon\varepsilon_0\mu\mu_0}}. \tag{4.14}$$

For a vacuum $\varepsilon = 1$ and $\mu = 1$ hold, and phase velocity equals

$$\upsilon_{\text{vacuum}} = c = \frac{1}{\sqrt{\varepsilon_0\mu_0}}. \tag{4.15}$$

For media other than a vacuum, we can write

$$\upsilon = c/n, \quad n = \sqrt{\varepsilon\mu} \tag{4.16}$$

Finally, in the optical frequency range, $\mu = 1$ holds and then one simply has

$$\boxed{n = \sqrt{\varepsilon}.} \tag{4.17}$$

In the case of a plane monochromatic wave with frequency ω, the time-dependent part (tracing the electric field only for convenience) has the form

$$E(\mathbf{r}, t) = E(\mathbf{r})e^{i\omega t}, \tag{4.18}$$

and the wave equation can be reduced to the time-independent equation for $E(\mathbf{r})$ function,

$$\nabla^2 E(\mathbf{r}) + \frac{n^2(\mathbf{r})}{c^2} \omega^2 E(\mathbf{r}) = 0 \tag{4.19}$$

or

$$\boxed{\nabla^2 E(\mathbf{r}) + k^2 E(\mathbf{r}) = 0, \quad k = \frac{n(\mathbf{r})}{c}\omega.} \tag{4.20}$$

This equation is called the *Helmholtz equation*. It describes the space distribution of the electric field when an electromagnetic wave with frequency ω propagates in a medium with index of refraction $n(\mathbf{r})$. It is clear that similar consideration holds for the $\mathbf{H}(\mathbf{r},t)$ counterpart.

Many practical problems can be reduced to 1D considerations. In this case, the 1D Helmholtz equation is to be examined. It reads

$$\boxed{\frac{d^2 E(x)}{dx^2} + k^2 E(x) = 0, \quad k = \frac{n}{c}\omega.} \tag{4.21}$$

This equation enables analysis of electromagnetic wave propagation in complex media, with refractive index depending on x.

4.1.3 Reflection at the interface of two dielectric materials at normal incidence

This is a simple but instructive and practically important problem (Figure 4.3). We can describe the incident and reflected waves as

$$E_I(x) = A \exp(ik_1 x) + B \exp(-ik_1 x) \text{ for } x < 0 \tag{4.22}$$

$$E_{II}(x) = C \exp(ik_2 x) \text{ for } x > 0 \tag{4.23}$$

where

$$k_1 = \frac{n_1}{c}\omega, \quad k_2 = \frac{n_2}{c}\omega \tag{4.24}$$

are wave numbers for every medium. One can see immediately that Eq. (4.24) means that wavelength has been changed at the interface from λ_1 to λ_2 as follows:

$$\lambda_1 = \frac{2\pi}{k_1} = \frac{c}{n_1 v}, \quad \lambda_2 = \frac{2\pi}{k_2} = \frac{c}{n_2 v}. \tag{4.25}$$

The latter equation simply means that wavelength in a medium with refractive index n is n times shorter than in a vacuum.

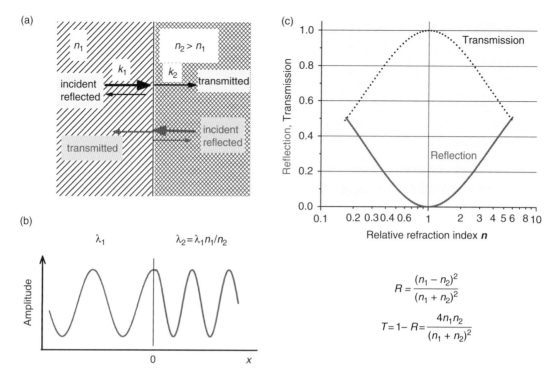

Figure 4.3 Transmission and reflection of electromagnetic waves at the border of two dielectric materials with different refractive indices, n_1 and n_2. (a) incident, transmitted, and reflected waves; (b) amplitudes of waves in two media; (c) transmission and reflection coefficients versus relative refractive index $n = n_1/n_2$ or $n = n_2/n_1$ along with the formulas. Propagation properties are the same when waves move from right to left or from left to right, which is also seen in symmetry of T and R plots with respect to lg n in (c) and T and R formulas with respect to changes $n_2 \leftrightarrow n_1$.

To calculate reflection coefficient, $R = I_{\text{reflected}} / I_{\text{incident}}$ and transmission coefficient, $T = I_{\text{transmitted}} / I_{\text{incident}}$ for intensity $I \propto E^2$, we shall first calculate reflection coefficient for field amplitude $r = B / A$, then take $R = r^2$, and $T = 1 - R$. When finding A, B values we shall use conditions of continuous $E(x)$ and $dE(x)/dx$ at the interface, i.e., at $x = 0$. Then we arrive at

$$r = \frac{k_1 - k_2}{k_1 + k_2} = \frac{n_1 - n_2}{n_1 + n_2}, \tag{4.26}$$

$$R = \frac{\left(n_1 - n_2\right)^2}{\left(n_1 + n_2\right)^2}, \quad T = 1 - R = \frac{4n_1 n_2}{\left(n_1 + n_2\right)^2}. \tag{4.27}$$

Notably, in Eq. (4.27) reflection and transmission coefficients depend on the relative refraction index $n = n_1 / n_2$, but not on the absolute values of n_1, n_2. Using n notation, one can rewrite Eqs. (4.27) as

$$R = \frac{(n-1)^2}{(n+1)^2}, \quad T = 1 - R = \frac{4n}{(n+1)^2}.$$ (4.28)

Reflection coefficients for interfaces of dielectric materials with air in the near-IR, visible, and near-UV range (from a few micrometers to approximately 300 nm) range from a few percent (e.g., silica) to 40–50% (e.g., Ge, Si, GaAs). Table 4.2 gives reflection coefficient values for silicon and gallium nitride interfaces with air in the near-IR and the visible spectra.

Table 4.2 **Reflection coefficient for different wavelengths for Si–air and GaN–air interfaces at normal incidence**

Wavelength, λ (nm)	Photon energy, $\hbar\omega$ (eV)	Refractive index, n	Reflection coefficient $R = (n-1)^2/(n+1)^2$
Si–air			
400	3.10	4.95	0.43
500	2.48	4.35	0.39
600	2.06	3.9	0.35
700	1.77	3.8	0.34
800	1.55	3.75	0.33
900	1.38	3.65	0.32
1000	1.24	3.60	0.32
GaN–air			
380	3.26	2.7	0.21
400	3.10	2.6	0.20
500	2.48	2.5	0.18
600	2.06	2.4	0.17
700	1.77	2.37	0.165
800	1.55	2.35	0.16
1000	1.24	2.32	0.16

4.1.4 Oblique incidence: Snell's law, total reflection phenomenon, Fresnel formulas, and the Brewster angle

In the case of oblique incidence, the problem of electromagnetic wave transformation at the interface of two dielectrics becomes three-dimensional. Its thorough analysis is based on Maxwell equations that account for boundary conditions for electric and magnetic fields at the interface. The conditions are: (1) tangential components of $\mathbf{E}$ and $\mathbf{H}$ are equal in both dielectrics and (2) normal components of $\mathbf{D}$ and $\mathbf{B}$ are equal in both dielectrics. The relations between angle of incidence, α, angle of reflection, and angle of refraction, β, obey the following laws:

1 Incident, reflected, refracted (transmitted) light beams and the normal to the interface all lie within the plane of incidence formed by the incident beam and the normal to the interface.

2 Angle of reflection equals angle of incidence.

3 Angle of refraction, β, is related to the α angle via phase velocities v_1, v_2 of light in media and hence via wavelengths λ_1, λ_2 and their refractive indices, n_1, n_2 (*Snell's law*), namely

$$\frac{\sin \alpha}{\sin \beta} = \frac{v_1}{v_2} = \frac{\lambda_1}{\lambda_2} = \frac{n_2}{n_1}. \tag{4.29}$$

When an electromagnetic wave travels from a medium with high n to a medium with lower n (shown in Figure 4.4), the critical angle α_{crit} exists when $\beta = 90°$. It reads

$$\alpha_{crit} = \arcsin\left(\frac{n_2}{n_1}\right). \tag{4.30}$$

When the angle of incidence exceeds the critical value α_{crit}, *total reflection* occurs since no transmitted light enters the medium with lower n (Figure 4.4(c)). For larger n_1/n_2 ratio, total reflection develops more readily, i.e., it starts at a smaller angle of incidence α. For materials interfacing with air, the critical angle value versus the material refractive index is

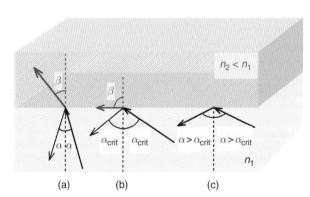

Figure 4.4 Reflection and refraction of light propagating from a medium with higher refractive index n_1 to a medium with lower refractive index $n_2 < n_1$ in cases of (a) $\alpha < \alpha_{crit}$, (b) $\alpha = \alpha_{crit}$, and (c) $\alpha > \alpha_{crit}$.

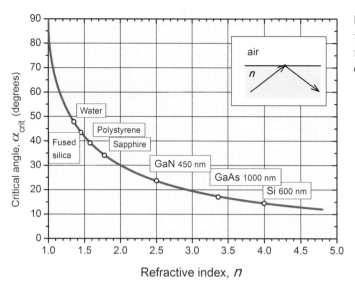

plotted in Figure 4.5. One can see that for GaAs and Si total reflection develops at an angle of incidence below 20°, and for GaN total reflection starts at about 25°. This phenomenon becomes a serious obstacle in light extraction from semiconductor devices. At the same time, total reflection is the basic principle of optical fiber *waveguides* used routinely in optical communication networks.

For oblique incidence at the interface of two media, *polarization* of the electromagnetic wave becomes important. Electromagnetic waves are *transverse* waves, i.e., the direction of the **E** vector oscillation and **H** vector oscillation are both perpendicular to the beam direction. Note that **E** and **H** vectors are orthogonal to each other, and the three vectors **E**, **H**, and that defining propagation direction, form the right-hand set of vectors, i.e., propagation direction obeys the direction of a right-hand screw when rotation is imagined from **E** to **H**. When the electric field oscillates in a certain plane, light is referred to as *linearly polarized*. When the electric field plane of oscillation rotates, it is called *circularly polarized*. *Natural light* is viewed as a combination of linearly polarized electromagnetic waves with randomly oriented **E** planes. The transverse nature of electromagnetic waves gives rise to the two principal cases of light reflection and refraction for linearly polarized beams, which are presented in Figure 4.6.

The first case corresponds to the **E** vector oscillating in the plane of incidence. This case is referred to as *p*-polarized light ("*p*" comes from "parallel"), or $E_\parallel$-polarization to emphasize that the electric field oscillates in the plane of incidence. This case is also sometimes referred to as TM-polarization (from "transverse **m**agnetic") to emphasize that the **H** vector in this case is transverse with respect to the plane of incidence.

The second case corresponds to the **H** vector oscillating in the plane of incidence, whereas the **E** vector oscillates in the plane orthogonal to it. This case is referred to as *s*-polarized light ("*s*" comes from the German *senkrecht*, meaning perpendicular), or $E_\perp$-polarization and TE-polarization (from "transverse **e**lectric") to emphasize that electric field oscillations are transverse with respect to the plane of incidence.

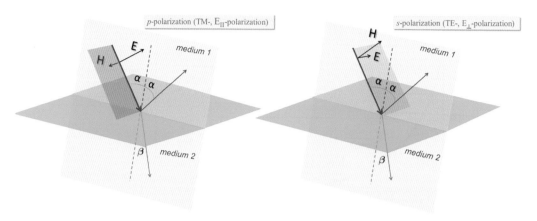

Figure 4.6 The two principal orientations of the **E** vector for oblique incidence at an interface of two dielectrics. The left-hand case (*p*-polarization, or TM- or $E_{\parallel}$-polarization) corresponds to the **E** vector lying in the plane of incidence. The right-hand case represents the **E** vector that is normal to the plane of incidence, whereas the **H** vector is parallel to it (*s*-polarization, or TE- or $E_{\perp}$-polarization).

At oblique incidence, intensity reflection coefficient, R, and intensity transmission coefficient, $T = 1 - R$, both become angle-dependent, with the angular dependencies differing for different polarizations. These values are defined by the *Fresnel formulas* that read as follows:

$$R_p = \frac{\text{tg}^2(\alpha - \beta)}{\text{tg}^2(\alpha + \beta)} \ , \quad T_p = \frac{\sin 2\alpha \sin 2\beta}{\sin^2(\alpha + \beta)\cos^2(\alpha - \beta)} \ , \tag{4.31}$$

$$R_s = \frac{\sin^2(\alpha - \beta)}{\sin^2(\alpha + \beta)} \ , \quad T_s = 1 - \frac{\sin^2(\alpha - \beta)}{\sin^2(\alpha + \beta)} \ , \tag{4.32}$$

where s and p subscripts correspond to the values for s- and p-polarized waves, α is the angle of incidence, and β is the refraction angle to be found from Eq. (4.29); i.e.:

$$\beta = \arcsin\left(\frac{n_1}{n_2}\sin\alpha\right). \tag{4.33}$$

The representative calculated angle-dependent reflection data for sapphire–air and GaN–air interfaces are plotted in Figure 4.7. Sapphire and GaN monocrystals are used in light-emitting diodes (LEDs). Note the difference for s- and p-polarizations. For s-polarization, reflection monotonically grows from the minimal value corresponding to normal incidence to 1 in the limit of α tending either to 90° (when radiation propagates from air to the material in question) or to α_{crit}, defining the total reflection onset when radiation propagates from the material into air. For p-polarization, angular dependence of reflection coefficient is essentially non-monotonic. First, reflection goes down and equals 0 at the angle at which $\alpha + \beta = \pi / 2$ holds. Mathematically zero reflection results from an infinite denominator in the left-hand expression of Eq. (4.31). Physically zero reflection is the direct consequence of the transverse nature of electromagnetic waves. Since the reflection angle equals the incidence angle, the expected reflected beam appears to be orthogonal to the transmitted one, and therefore **E** vector oscillations have to coincide with the reflected beam propagation. The latter is impossible for transverse waves. The phenomenon of zero p-component in reflected light is called the *Brewster law* and the angle α_{Brewster} meeting the condition $\alpha_{\text{Brewster}} + \beta = \pi/2$ is termed the *Brewster angle*. Snell's law (Eq. (4.29)) together with the $\alpha_{\text{Brewster}} + \beta = \pi/2$ condition gives rise to the expression

$$\alpha_{\text{Brewster}} = \text{arctg}\left(\frac{n_2}{n_1}\right). \tag{4.34}$$

The Brewster law is widely used to convert natural light into linearly polarized light.

For angle of incidence $\alpha > \alpha_{\text{Brewster}}$, reflectivity rises up and reaches 1 in the limit of $\alpha = \pi/2 \ (90°)$ in the case $n_2 > n_1$, or $\alpha = \alpha_{\text{crit}}$ in the case $n_2 < n_1$ (right-hand panels in Figure 4.7).

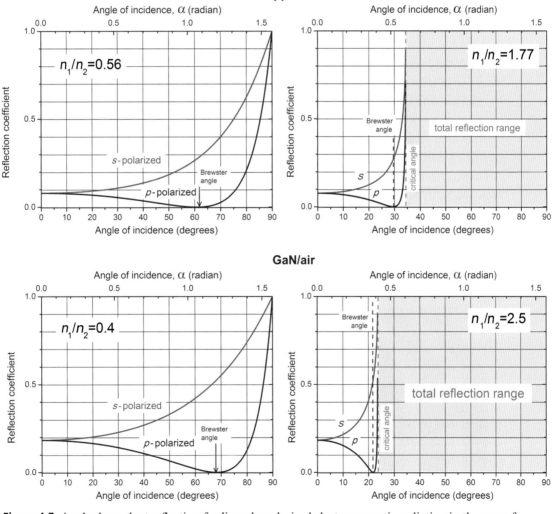

Figure 4.7 Angle-dependent reflection for linearly polarized electromagnetic radiation in the case of sapphire–air and GaN–air interfaces. The left-hand panels correspond to propagation of waves from air to sapphire (GaN) and right-hand panels correspond to propagation of waves from sapphire (GaN) into air.

Natural light is viewed as electromagnetic radiation of random linear polarization, i.e., the **E** vector oscillates in all possible directions within the plane normal to the propagation direction. Then the reflection coefficient for intensity can be calculated as

$$R_{natural} = \frac{1}{2}(R_s + R_p).$$

(4.35)

4.2   LIGHT IN A PERIODIC MEDIUM

The simplest case of a medium with periodic alteration of refractive index $n(x)$ along the propagation direction of the electromagnetic wave, both for theoretical consideration and for experimental implementation, is a two-component multilayer stack with infinite number of layers. A portion of such a structure is shown in Figure 4.8. The problem to be solved is the Helmholtz equation (Eq. (4.21)) with a periodic $n(x)$ term. This problem represents nothing but an optical analog to the known Kronig–Penney model in solid-state physics, which is described in numerous textbooks with respect to an electron in a periodic rectangular potential. Though for an electron in a periodic potential coming from Coulomb interaction, the rectangular potential profile represents the simplest approximation only, in optics this "potential" in terms of refractive index periodicity explicitly corresponds to multilayered structures that can be grown as monocrystals by epitaxial techniques or fabricated as polycrystalline films by cheap vacuum deposition or sol-gel techniques.

Taking into account mathematical isomorphism of Eq. (4.21) and its electron counterpart, the Schrödinger equation (see Section 2.4, Eq. (2.55)), we can summarize the general properties of electromagnetic waves in a periodic medium.

(1) Electric field $E(x)$ is described by *Bloch waves* with the amplitude modulated by period a,

$$E(x) = E_k(x)e^{ikx},$$

(4.36)

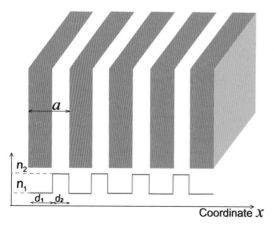

Figure 4.8 A portion of the multilayer periodic structure consisting of two materials with different refractive indices.

where k is the Bloch wave number and $E_k(x) = E_k(x+a)$. Subscript k in $E_k(x)$ means this function depends on k.

(2) The dispersion curve $\omega(k)$ breaks into portions (Figure 4.9) featuring discontinuities at every wave number,

$$k_N = N\frac{\pi}{a} \, , \, N = 1, 2, 3, \ldots \tag{4.37}$$

For every k_N meeting this condition, the half-wavelength equals the integer number of the period, Na. The physical reason for discontinuity is formation of standing waves. Then group velocity expressed by $v_g = d\omega/dk$ equals zero, which means the tangent lines at these points should be parallel to the k axis. There is no solution of the Helmholtz equation in the form of plane waves within the frequency gaps corresponding to the breaks. Only an evanescent electromagnetic field exists in these gaps. For this reason, the wave coming from outside reflects back. These gaps are the correct classical electromagnetic analogs to band gaps for electrons in crystalline solids. In the gaps, only *tunneling* of electromagnetic waves is possible. In optics, these gaps are referred to as *reflection bands, stop bands*, or *photonic band gaps*.

(3) Similar to the case of electrons, the states differing in wave number by the value $k_N = k \pm \frac{2\pi}{a}N$ are equivalent. This is a direct consequence of the translational symmetry of space in the problem under consideration. This property leads to the notion of *Brillouin zones*, i.e., intervals in the k axis with width $2\pi/a$. All branches in the dispersion curve can be shifted by the integer number of $2\pi/a$ to fall into the single Brillouin zone. Usually the interval of k from $-2\pi/a$ to $+2\pi/a$ is used as is shown in Figure 4.9(b).

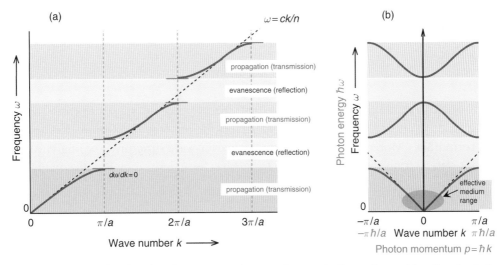

Figure 4.9 Dispersion law for electromagnetic waves in a periodic medium with refractive index profile independent of frequency. The straight line inherent in a homogeneous medium breaks into an infinite number of branches because of the standing waves at every $k = N\pi/a$. These branches allow propagation of waves, whereas in the gaps between branches only an evanescent field occurs.

In Section 2.5 (Figure 2.14) we have seen that reduced presentation of all energy states within a single Brillouin zone and the relevant quasi-momentum conservation law opens a realm of nearly vertical electron transitions between bands which are impossible for a continuous space. In optics, reduced presentation for waves in a periodic medium does not result in such dramatic consequences – it simply helps to overview all possible waves in a periodic medium in a convenient and instructive form.

Notably, the original straight dispersion law does not deviate in a periodic medium for small k (Figure 4.9). This is because smaller wave numbers correspond to larger wavelengths, thus periodicity of the medium is "seen" by long waves as a minor deviation from the continuous medium and can be expressed in terms of the effective medium whose effective refractive index falls between refractive indices of constituting materials.

To get closer to its electronic counterpart $E(p)$, both axes in Figure 4.9(b) can be multiplied by $\hbar$ to arrive at photon energy $E = \hbar\omega$ versus photon momentum $p = \hbar k$ (shown in green). Then one can speak about photon band gaps and photon band structure. However, it is instructive to always keep in mind that the results presented in Figure 4.9 do not imply photons but are essentially derived from the classical electrodynamics based on Maxwell equations reduced to the Helmholtz one.

There is one practically important particular case of multilayer periodic structures appreciated by optical designers. This is alternating layers of two materials with different refractive indices n_1, n_2 and thicknesses d_1, d_2, but with the same optical length nd of every layer, i.e.,

$$n_1 d_1 = n_2 d_2 \equiv nd. \tag{4.38}$$

In this case the central (midgap) frequency ω_0 in the reflection band is determined from the so-called quarter-wave condition,

$$\lambda_0 / 4 = nd \tag{4.39}$$

and reads

$$\omega_0 = 2\pi c / \lambda_0 = \pi c / 2nd. \tag{4.40}$$

Such structures are referred to as quarter-wave periodic stacks. For quarter-wave periodic structures, the transmission spectrum is the periodic function of frequency with the period $2\omega_0$. Moreover, for finite periodic quarter-wave structures, the analytical expression for optical transmission coefficient $T_{QW}(\omega)$ has been derived by Bendikson et al. (1996):

$$T_{QW}(\omega) = \frac{1 - 2R_{12} + \cos\pi\tilde{\omega}}{1 - 2R_{12} + \cos\pi\tilde{\omega} + 2R_{12}\sin^2\left[N\arccos\left(\frac{\cos\pi\tilde{\omega} - R_{12}}{1 - R_{12}}\right)\right]}, \tag{4.41}$$

where $\tilde{\omega} = \omega / \omega_0$ is the midgap-normalized dimensionless frequency, N is the number of periods, and R_{12} is the reflection coefficient at the $n_1 \leftrightarrow n_2$ refractive step (see Eq. (4.28)).

One-dimensional finite-length periodic structures offer a very instructive way to understand band gap formation for wave propagation in a periodic medium. Representative spectra for different n_1 / n_2 values and different numbers of periods N are given in Figures 4.10 and 4.11. The first three periods in the transmission spectrum are shown in the left-hand panel of Figure 4.10, whereas all other graphs show only one period in the transmission spectra. The frequency scale is normalized with respect to ω_0. Within every period on the frequency scale, the transmission spectrum of a quarter-wave structure is symmetrical with respect to points $\omega_0, 3\omega_0, 5\omega_0$, etc. Spectral symmetry can be written in a general form as

$$T(\omega + 2N\omega_0) = T(\omega), \quad N = 1, 2, 3, \ldots \quad (4.42)$$

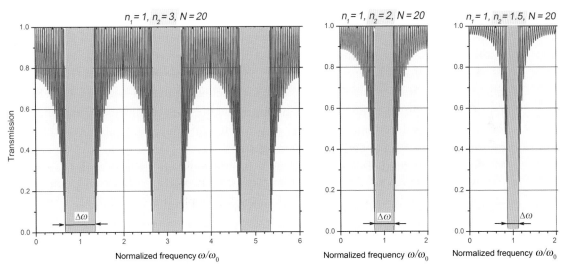

Figure 4.10 Transmission spectra for 20-period quarter-wave structures differing in n_2 values; $n_1 = 1$. Stop bands with zero transmission correspond to reflection coefficient $R = 1$ (highlighted gray). The gaps shrink for smaller n_2/n_1 in accordance with Eq. (4.43).

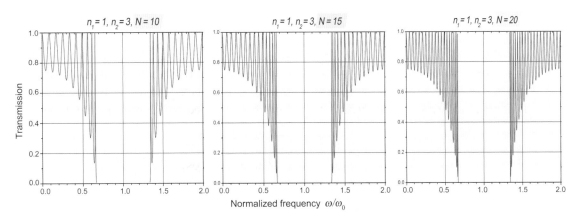

Figure 4.11 Transmission spectrum of a periodic stack consisting of N periods of materials with refractive indices $n_1 = 1$ and $n_2 = 3$. Note that the number of transmission peaks in the transmission bands rises in proportion with N. The gap width remains the same for all structures since the n_1/n_2 value is the same.

The relative width of the reflection band for a quarter-wave periodic structure reads (Yariv and Yeh 1984):

$$\frac{\Delta\omega_{gap}}{\omega} = \frac{4}{\pi}\frac{|n_1 - n_2|}{n_1 + n_2} = \frac{4}{\pi}\left(\frac{1 - n_1/n_2}{1 + n_1/n_2}\right) = \frac{4}{\pi}\sqrt{R_{12}}, \qquad (4.43)$$

i.e., it is unambiguously defined by the n_1/n_2 value.

Eq. (4.41) is helpful for calculation of the spectral position of reflection and transmission bands, the number of transmission bands, and their profiles. However, it fails to describe accurately the buildup of reflection band(s) in the case of a few layers. In this case, numerical calculations should be performed (see, e.g., Gaponenko (2010) for details).

When an electromagnetic wave with frequency corresponding to the band gap enters a periodic medium, intensity falls rapidly over a distance of a few periods. Therefore, transmission rapidly drops to near-zero values as the number of periods grows. However, in a thin slab, finite transmission occurs – e.g., for thickness equal to a few micrometers in Figure 4.12. This property resembles tunneling in quantum mechanics (see Chapter 2) and actually represents the common property of waves to maintain a finite amplitude of oscillations in the case of finite height of potential steps.

Multilayer periodic dielectric mirrors have been extensively used in optical engineering for many decades. Every commercial laser has a resonator formed by two mirrors; typically, multilayer dielectric mirrors are preferable to metallic ones since metallic mirrors have lower reflectivity and non-desirable absorptive losses resulting not only in higher threshold energy but also in poor durability because of heating. At least four different techniques can be outlined for fabrication of periodic structures as mirrors.

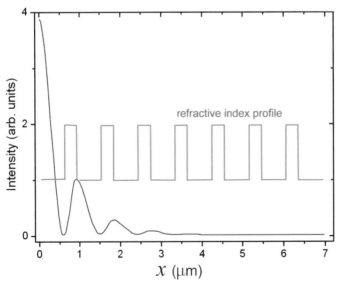

Figure 4.12 The spatial intensity profile in a finite periodic multilayer structure for the frequency value in the center of the band gap ω_0. The gray line shows the stepwise refractive index profile. Courtesy of S. Zhukovsky.

1 **Polycrystalline films by vacuum deposition** (Figure 4.13). In bulky solid-state or gas lasers, from laser pointers to laser scribers or cutting machines, polycrystalline structures are fabricated by means of vacuum deposition using magnetron sputtering or ion beam-assisted deposition. Typical materials for the visible and near-IR are MgF_2, SiO_2, and sometimes Na_3AlF_6 as low-n components, and TiO_2, ZnS, and Nb_2O_5 as high-n counterparts. With these materials, reflection of greater than 95% can be obtained with just a few periods. Figure 4.13 presents the example of a commercial multilayer dielectric mirror, namely quarter-wave SiO_2/TiO_2 layers on a glass substrate. One can see that the reflection spectrum remarkably correlates with the theory. However, there are a few deviations that can be easily accounted for through numerical calculations. First, there is a certain dependence of refractive indices of constituent materials on wavelength and therefore the reflection (and transmission) spectra are not ideally symmetric with respect to the center of the stop band. Second, the contribution from glass substrate reflections gives rise to final reflectance outside the main reflection band and therefore transmission does not reach 1 and reflection does not fall to 0 in the transmission bands.

2 **Epitaxial growth offers a possibility to develop monocrystalline mirrors** integrated with an active medium and contacts. In semiconductor microchip lasers, vertical cavity surface-emitting lasers (VCSELs) are now common. In these lasers, bottom and top multilayer mirrors are made through an epitaxial process together with the intermediate active layer (often semiconductor quantum well). In this case, lattice-matching conditions may restrict the available materials, and because of the low refractive indices ratio the number of periods should exceed 20 to get reflectance close to 100%. Mirrors are often located within the

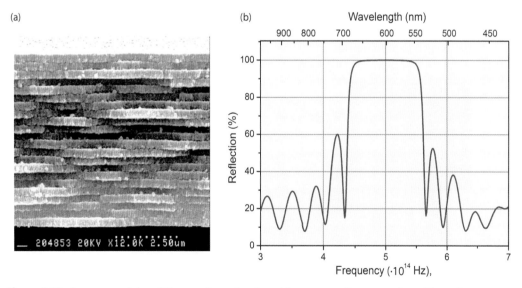

Figure 4.13 A commercial multilayer mirror developed by means of vacuum deposition of polycrystalline films on a glass substrate: (a) cross-section image by means of electron microscopy, (b) reflection spectrum. The photo image is reprinted from Voitovich (2006), with permission of Springer.

current flow area and then the additional condition to be met by constituting materials is the proper conductivity.

3 **Templated etching** enables development of vertical semiconductor layers separated by air spacers (Figure 4.14(a)). This technique gives very high refractive indices ratios (e.g., for Si $n = 3.4$), which is otherwise impossible. Furthermore, air spacers can be impregnated with another material, e.g., liquid crystals whose refraction can be electrically driven. The tricky issue in this approach is deep anisotropic etching, since usually only a thin layer near the surface resembles the template image well (Baldycheva et al. 2011).

4 **Electrochemical etching with alternating porosity** (Figure 4.14(b)). Electrochemical etching of silicon (and many other semiconductors) can result in nanoporous materials whose porosity will be the function of etching conditions such as concentration of etchant, temperature, and current, whereas etching depth will be proportional to processing time. Modulation of current in time changes porosity accordingly. This ingenious approach was proposed by Pellegrini et al. (1995) for making periodic structures and microcavities. Modulation of porosity will give modulation of the refractive index, $n = \sqrt{\varepsilon}$, between silicon and air values. If a medium whose dielectric permittivity is ε contains inclusions of another material whose dielectric permittivity is ε_1 and its volume-filling factor is $0 < f < 1$, then provided the size of inhomogeneities is much smaller than the wavelength, the effective medium approach can be used and dielectric permittivity ε_{mix} can be ascribed to the composite mixture. It can be calculated using the Bruggeman formula (Bruggeman 1935):

$$f \frac{\varepsilon_1(\omega) - \varepsilon_{mix}(\omega)}{\varepsilon_1(\omega) + 2\varepsilon_{mix}(\omega)} + (1 - f) \frac{\varepsilon(\omega) - \varepsilon_{mix}(\omega)}{\varepsilon(\omega) + 2\varepsilon_{mix}(\omega)} = 0. \tag{4.44}$$

Refractive index of the composite medium is always an intermediate value with respect to the constituent materials.

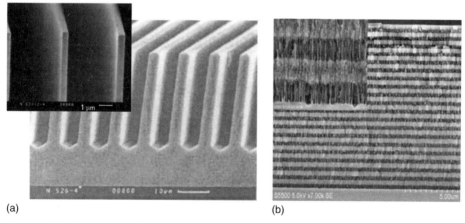

(a)   (b)

Figure 4.14 Silicon periodic structures fabricated by two techniques: (a) vertical anisotropic etching using a lithographic template on the top; (b) electrochemical etching using modulation of current in time to get alternating porosity. Courtesy of A. Baldycheva (a) and L. Pavesi (b).

Recently, the notation of *distributed Bragg reflector* (DBR) or Bragg mirror was coined for multilayer dielectric mirrors. It became popular and commonly used. The notation recalls the pioneering work of W. H. Bragg and W. L. Bragg in 1913 on observation of x-ray diffraction on a three-dimensional crystal lattice. For the sake of historical truth, it is noteworthy that these are multiple scattering and interference rather than diffraction that define the reflectance of multilayer structures in optics. Lord Rayleigh was the pioneer of this design; in 1887 he predicted development of the reflection band for electromagnetic waves in a medium with periodic alteration on the refractive index along wave propagation (Strutt 1887).

4.3 PHOTONIC CRYSTALS

4.3.1 The conception

Photonic crystals are composite structures whose refractive indices feature periodicity in two or three dimensions. This elegant conception was elaborated upon in the 1980s. It can be considered as an extension of the known multilayer periodic stacks to the higher dimensions. From the other side, this conception is the excellent transfer of the electron theory of solids to the electromagnetic phenomena based on the similarities of the Helmholtz equation in electromagnetic theory and the Schrödinger time-independent equation in quantum mechanics. For the optical range, from near-UV to near-IR, components of photonic crystals should be of the order of 100 nm. Figure 4.15 sketches periodic structures in various dimensions. Within the photonic crystals paradigm, multilayer stacks are often referred to as 1D photonic crystals.

The term *photonic crystal* should not be used to mean a crystal for photons, but should rather be considered to mean a periodic structure for electromagnetic waves. Very often the term *photonic crystal* is used as a quick reference or just a label, whereas the correct physical notion is expressed as *electromagnetic crystal structures*.

Figure 4.15 A sketch of 1D, 2D, and 3D periodic structures referred to as photonic crystals. Different colors correspond to the two different materials with differing refractive indices.

The basic properties of photonic crystals are different conditions for propagation of different modes, i.e., type of electromagnetic waves featuring a certain frequency, direction, and polarization. The basic reason for the special properties of photonic crystals is the scattering of waves at the refractive index inhomogeneities and subsequent interference of scattered waves. The theory of ideal infinite photonic crystals resembles to a large extent the electron theory of crystalline solids. The difference is that contrary to Coulomb potential wells inherent for an electron in solids, in photonic crystal theory the refractive index profile should be defined by the photonic crystal topology and constituent materials. Notably, in the case of the optical range, development of omnidirectional photonic band gaps (i.e., gaps in the frequency or wavelength domains at which electromagnetic waves cannot propagate) becomes a challenge. This is because the existing materials severely restrict the range of refractive index variation (see Table 4.1). This restriction is further aggravated by the requirements for technological feasibility and material transparency – e.g., highly refractive materials like Si, GaAs, CdTe are not transparent in the visible spectrum.

4.3.2 Theoretical modeling of photonic band structures

The theory is based on numerical solution of the Helmholtz equation (Eq. (4.19)) for periodic $n(\mathbf{r})$ function. The computational details are beyond the scope of this textbook and can be found in the specialist books by Joannopoulos et al. (2011) and Sakoda (2004). Nowadays, computational software packages are available.

Analysis of electromagnetic wave properties is presented in the form of a frequency versus wave number diagram $\omega(k)$, similar to the electron band structure in solids, where energy versus wave number is calculated. Recall the notion of the Brillouin zone (see Figure 4.9 and comments on it). Owing to periodicity of the space under consideration, all modes can be presented within the first Brillouin zone, but it is important to remember that the Brillouin zone is a portion of k-space whose dimensionality equals dimensionality of the relevant geometrical space. That is, for a 1D periodic structure it is just an interval on the k axis, whereas for a 2D periodic space it extents to a polygon, and in a 3D periodic space it becomes a polyhedron.

Consider a few selected examples of model two-dimensional structures. Figure 4.16 presents the band structure for a two-dimensional triangular lattice of air holes in a dielectric with a relatively large dielectric permittivity, $\varepsilon = 12$, relevant to silicon and also close to GaAs, InAs, and GaP. This is a very favorable combination of media to obtain pronounced modification of electromagnetic wave properties since the refractive indices ratio is quite big. The results are presented in terms of dimensionless frequency $\omega a/2\pi c \equiv a/\lambda_0$, where λ_0 is the light wavelength in a vacuum. Presentation of the band structure in terms of dimensionless frequency offers direct translation of results along the electromagnetic wave scale.

Light propagates in the plane normal to the holes (x–y plane). In this case, all Bloch modes can be separated into two sub-groups: the Bloch modes for which the electric field is parallel to the hole axis (referred to as E-polarized modes); and the Bloch modes for which the magnetic field is parallel to the hole axis (H-polarized modes). In terms of traditional

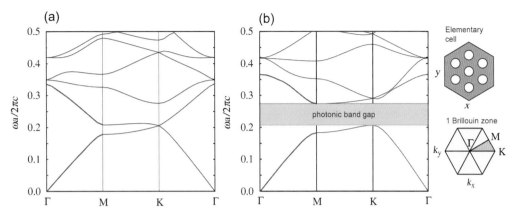

Figure 4.16 Photonic band structure for (a) *E*-polarized and (b) *H*-polarized light in a 2D photonic crystal created by a triangular lattice of air holes ($\varepsilon_1 = 1$) with the radius $r_0 = 0.3a$ (where *a* is the lattice period) in a dielectric material with the permittivity $\varepsilon_2 = 12$. The top-right inset shows a cross-sectional view of the 2D photonic crystal cell. The bottom-right inset shows the first Brillouin zone, with the irreducible zone shaded. Adapted from Busch et al. (2007), with permission of Elsevier.

transverse electric and transverse magnetic notations, *H*-polarized light corresponds to the transverse electric mode and *E*-polarized light corresponds to the transverse magnetic mode.

The band structure diagrams in Figure 4.16 for a 2D lattice represent the dispersion law $\omega(k)$ for the first Brillouin zone, as in the case of a one-dimensional periodic structure. However, two-dimensional space needs three-dimensional multivalued functions $\omega(k_x, k_y)$ to be plotted, which is rather cumbersome. Therefore, the following presentation style is generally adopted for 2D periodic structures. The origin of the coordinate system ω, k corresponds to $\omega = 0$, $k = 0$ and is indicated as the Γ-point in the Brillouin zone. Then the $\omega(k)$ function is presented, with *k* varying along the $\Gamma \rightarrow M$ direction.

Note that the two $\omega(k)$ branches for every polarization with the lowest frequency resemble to a large extent the 1D diagram in Figure 4.9(b). These portions of the dispersion curves are plotted with thicker lines. Note that at the very beginning, for low frequency and wave number, the dispersion curve is nearly linear as it was in the 1D case. This means that propagation of light can be interpreted in terms of an effective medium with constant refractive index. This is the familiar case of the longwave approximation of complex media in optics in which the wavelength is much larger than inhomogeneity sizes.

Afterwards, the **k** vector changes in value and direction between the M and K points at the edge of the first Brillouin zone. According to the theory, there is no need to examine the situation for *k* beyond the ΓMK triangle in *k*-space. Every type of lattice possesses its specific portion of Brillouin zone which gives full data on the band structure. This is an important property of waves in periodic spaces. The omnidirectional forbidden gap on the frequency axis in Figure 4.16 (b) means absence of $\omega(k)$ points within a finite range of frequencies. It is shown in the panel Figure 4.16(b) as a shadowed area. Note, in Figure 4.16

an omnidirectional (2D) band gap develops for *H*-polarized light only. It was shown that the triangular lattice of holes features an omnidirectional band gap for both polarizations only for 0.4 < *r*/*a* < 0.5 (note, the upper limit corresponds to touching holes).

Formation of complete omnidirectional gaps for both types of modes in two dimensions is actually feasible with highly refractive materials, depending on the crystal geometry, shape of an individual scatterer, and volume-filling factor of the scatterers. Table 4.3 presents the widths of complete band gaps in the dimensionless units $\Delta\omega/\omega$ evaluated for GaAs ($\varepsilon = 12.96$) rods in air (Figure 4.17), along with the surface filling factor f (the portion of the square filled with rods). Notably, triangular lattices were found to offer wider gaps as compared with rectangular and honeycomb ones.

Analysis of various topologies for *3D lattices* has revealed the following features of band gap formation. For the same lattice structure (simple cubic, face-centered cubic, and BCC lattices were examined), the structures comprising small volume fraction ($f = 0.2 \dots 0.3$) of a dielectric in air appear to be more advantageous. Among a variety of cubic lattices, the diamond structure, a kind of face-centered cubic lattice, features a wider-band gap. For the diamond lattice consisting of close-packed dielectric spherical particles, the band gap was found to open for all directions when refractive index contrast is $n_{diel}/n_{air} > 2$.

Table 4.3 **Parameters of the selected 2D photonic crystals consisting of the rods with $\varepsilon = 12.96$ in air**

Notation in Figure 4.17	Lattice type	Scatterer type	Band gap $\Delta\omega/\omega$	Surface filling factor f
a	Rectangular	Rectangular	0.15	0.67
b	Rectangular	Circular	0.04	0.71
c	Rectangular	Hexagonal	0.025	0.71
d	Triangular	Rectangular	0.09	0.68
e	Triangular	Circular	0.2	0.85
f	Triangular	Hexagonal	0.23	0.70
g	Honeycomb	Rectangular	0.06	0.43
h	Honeycomb	Circular	0.11	0.2
i	Honeycomb	Hexagonal	0.11	0.2

Source: after Wang et al. (2001).

An example of the calculated band structure for the diamond-like lattice exhibiting an omnidirectional band gap is shown in Figure 4.18 for the silicon–air structures. The dimensionless frequency $\omega a / 2\pi c \equiv a / \lambda$ is used here, with λ being the wavelength value in a vacuum. Note that as in the case of 1D and 2D periodic structures, for low frequencies

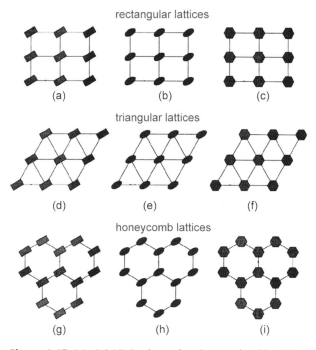

Figure 4.17 Model 2D lattices of rods examined by Wang et al. (2001): (a–c) rectangular; (d–f) triangular; (g–i) honeycomb.

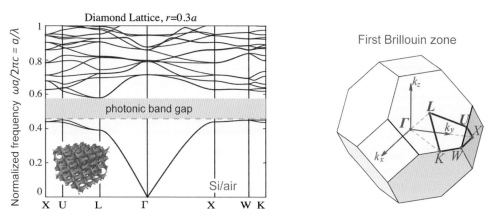

Figure 4.18 Photonic band diagrams of a Si–air holes-based diamond lattice photonic crystal. Reprinted from Chen and Bahl (2015), with the permission of OSA Publishing.

near the Γ-point (long wavelengths) the dispersion law is nearly linear, i.e., the effective medium approach can be applied to describe propagation of electromagnetic waves in periodic structures. Remarkably, it works till $\lambda / a > 3$ holds.

4.3.3 Experimental performance of 2D and 3D photonic crystals

Two-dimensional periodic structures can be developed by means of semiconductor anisotropic etching with a lithographically defined template on the surface. This is the most reliable and affordable approach, which is actually close to planar nanoelectronics components fabrication with the principal difference that etching depth should be bigger than the structure period. Many groups have demonstrated successful fabrication of Si 2D photonic crystals (Figure 4.19(a)). The high refractive index of silicon is beneficial for band gap formation, though its small band gap does not allow operation in the visible spectrum. Wide-band semiconductors and dielectrics should be used for the visible spectrum. Among them, electrochemically developed alumina has attracted strong attention. Alumina can be developed by means of anodization of aluminum. Remarkably, under certain anodization conditions aluminum etching gives rise to alumina arranged in regular pores whose diameter and spacing can be controlled to a certain extent by electrolyte concentration and temperature (Figure 4.19(b)).

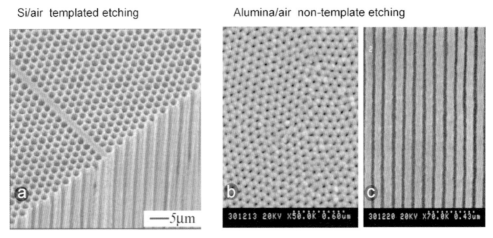

Figure 4.19 Two-dimensional photonic crystals fabricated by etching. (a) Electron micrograph of a 2D hexagonal porous silicon structure (lattice constant $a = 1.5$ µm and pore radius $r = 0.68$ µm), fabricated by anodization of p-type silicon with a lithographic template. Reprinted from Birner et al. (2001), with the permission of John Wiley and Sons. © 2001 WILEY-VCH Verlag GmbH, Weinheim, Fed. Rep. of Germany. (b,c) Scanning electron microscopy images of top and cross-sectional views of nanoporous alumina developed by means of self-organization without a template. Reprinted with permission from Lutich et al. (2004). Copyright 2004 American Chemical Society.

When self-organization is used, the porous sample thickness is determined by the processing time only. The thickness can be as large as hundreds of microns. Depending on the electrochemical treatment and preliminary surface treatment, the pore size can be from a few nanometers to hundreds of nanometers. The surface domains visible in the top view in Figure 4.19(b) are supposed to reproduce the polycrystalline domain structure of the raw aluminum foil used in the fabrication process.

Fabrication of 3D periodic structures remains a serious challenge for experimentalists. There are three principal approaches: (1) self-assembly and 3D-templating; (2) lithographically defined etching alignment; and (3) anisotropic etching in two dimensions.

Self-assembly is based on colloidal chemistry. Self-assembled structures are formed as colloidal crystals resembling naturally existing artificial *opals*. An unfortunate disadvantage of this technique is the rather limited set of materials for which 200–300 nm globules are available. Most experiments are restricted by either silica or polystyrene Figure 4.20(a). Because of the low refractive index, such structures cannot promise omnidirectional photonic band gap and feature angular-dependent reflection bands only (Figure 4.20(b)). The advantage is relative simplicity and feasibility of bulky size if necessary. It is important that, in close-packed spheres assemblies, pores form topologically continuous space, thus enabling impregnation with highly refractive materials. Polymers, TiO_2, and polycrystalline silicon have been successfully embedded. Selective etching can then be used to remove the silica core to arrive at the inverted opal structure shown in Figure 4.20(c). Such structures have approximately 30% volume-filling factor, which is favorable for band gap formation. Nevertheless, even when silicon is used, the photonic band gap becomes omnidirectional for higher frequencies only (Figure 4.20(d)).

Lithographically defined etching-alignment techniques for fabrication of 3D photonic crystals, though much more expensive and tricky, do offer a variety of topologies with highly refractive semiconductors. *Woodpile photonic crystals* can be developed by sequential multiple lithographical etching and alignment procedures (Figure 4.21). When made of highly refractive semiconductors like Si or GaAs, with air between rods, these structures feature photonic band gap with gap/midgap ratio about 20% when semiconductor volume fraction is about 25% (Ho et al. 1994).

Anisotropic etching in two dimensions with lithographically surface defined templates has been proposed lately for 3D photonic crystals. Combination of template etching in two orthogonal directions can generate the *inverse-woodpile* structure. It is very promising in terms of affordable bulky size as well as band gap formation. Calculations show that for Si–air photonic crystals of this topology, maximal relative band gap width $\Delta\omega/\omega_0$, so-called gap/midgap ratio, equals to 25% and occurs when pore radius is $0.3a$, and vertical period $c = \sqrt{a}$, where a is the period in the horizontal plane and c is the period in the vertical direction (Figure 4.22).

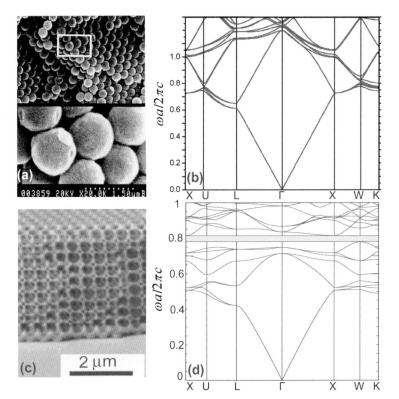

Figure 4.20 Opal-based 3D periodic structures. (a) Scanning electron microscopy image of an artificial silica colloidal crystal. The lower panel shows the magnified image of a portion of the crystal inside a white frame. Globule diameter is 250 nm. Reprinted figure with permission from Petrov et al. (1998). Copyright 1998 by the American Physical Society. (b) Photonic band structure diagram according to calculations by Reynolds et al. (1999). Reprinted figure with permission from Reynolds et al. (1999). Copyright 1999 by the American Physical Society. (c) Scanning electron microscopy image of a silicon replica of an artificial opal. Reprinted by permission from Macmillan Publishers Ltd.; Vlasov et al. (2001). (d) Computed photonic band structure diagram. Reprinted from Busch et al. (2007), with permission of Elsevier.

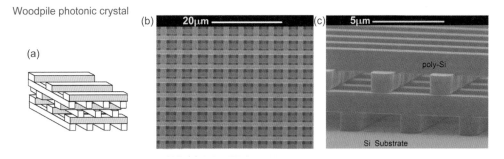

Figure 4.21 Woodpile photonic crystals. (a) Topology. (b,c) Electron micrographs of four-layer polycrystalline silicon woodpile periodic structure. Reprinted by permission from Macmillan Publishers Ltd.; Lin et al. (1998).

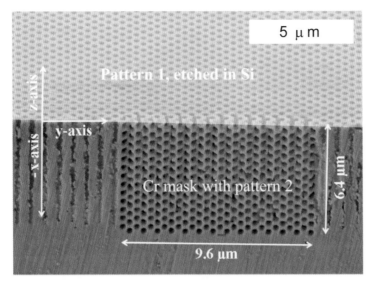

Figure 4.22 Inverse-woodpile structure developed by anisotropic etching of a silicon monocrystal. Reprinted with permission from Tjerkstra et al. (2011). Copyright American Vacuum Society.

4.3.4 Photonic crystal structures in nature

Periodic structures on a scale of optical wavelengths with pronounced interference-based colors can be found in nature (Vukusic and Sambles 2003). For example, 1D periodicity of coatings over wings can be found in several types of butterfly, in the feathers of pea-cocks, and on the shells of many beetles. Structures with 2D periodicity are present in eyes of a number of moths and other insects, and even more complex periodic structures are inherent in the corneas of human beings and mammalians. Interestingly, 2D periodicity is inherent in very ancient living subjects – diatom water plants that appeared about 500 million years ago, where it is believed to assist in light harvesting. Two-dimensional perio-dicity can be found in pearls that consist of a layered package of cylinders.

The functionality of creatures' structures has been optimized over hundreds of millions of years of evolution in nature. Presence of optical wavelength-scale periodicity is believed to be a result of such optimization. For example, the regular porous structure of insect eyes and mammalian corneas form antireflection interfaces and, probably, a cutoff filter for UV radiation. At the same time, porous topology assists in physico-chemical exchange processes.

The rationale for interference colors is not straightforward. Contrary to pigment-based colors, interferential (absorptionless) colors enable avoidance of heating and all photo-chemical processes, thus enabling durability of colors and resistance to overheating, e.g., for tropical butterflies.

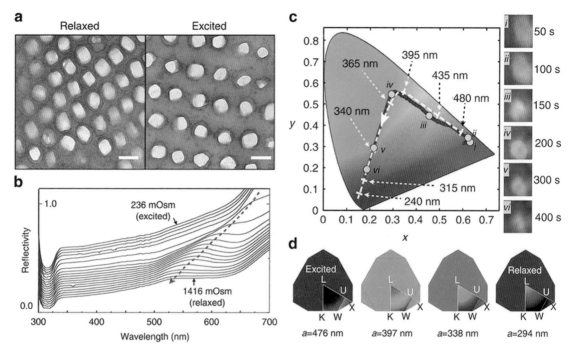

Figure 4.23 Chameleon dynamic colors in the context of tunable photonic crystals. (a) TEM images of the lattice of guanine nanocrystals in S-iridophores from the same individual in a relaxed and excited state (two biopsies separated by a distance <1 cm, scale bar, 200 nm). This transformation and corresponding optical response is recapitulated *ex vivo* by manipulation of white skin osmolarity (from 236 to 1416 mOsm). (b) Reflectivity of a skin sample and (c) time evolution (in the CIE chromaticity chart) of the color of a single cell (insets i–vi); both exhibit a strong blue shift (red dotted arrow in (b)) as observed *in vivo* during behavioral color change. Dashed white line: optical response in numerical simulations with lattice parameter indicated by dashed arrows. (d) Variation of simulated color photonic response for each vertex of the irreducible first Brillouin zone (color outside of the Brillouin zone indicates the average among all directions) shown for four lattice parameter values of the modeled photonic crystal. L-U-K-W-X are standard symmetry points. Reprinted from Teyssier et al. (2015).

Among all creatures featuring interference colors, chameleons are outstanding owing to the remarkable ability to perform dynamic color changes during social interactions, such as male contests or courtship. Recently it was found that chameleons shift color through active tuning of a lattice of guanine nanocrystals featuring tunable 2D biophotonic crystals (Figure 4.23).

Three-dimensionally periodic structures are present in nature as colloidal crystals. Biological colloidal crystals have been identified for the first time for viruses. When speaking about optical wavelength scale, opals are the familiar example. Opals are colloidal crystals with close-packed spherical silica globules and voids filled by another inorganic compound. The iridescent color of opals is because of lightwave interference in a three-dimensional lattice.

4.3.5   Photonic crystal waveguides

Optical fiber guides light owing to the total reflection phenomenon (Figure 4.24(a)). Modern technology makes possible light transfer over hundreds of kilometers through a silica core covered by a cladding with a lower refractive index. However, sharp bending of a fiber gives rise to leakage since the total reflection condition cannot be met (Figure 4.24(b)). This is the fundamental limit of downscaling existing fibers to microelectronic circuitry. Additionally, the fiber core should have a diameter greater than wavelength.

A linear defect in a photonic crystal offers a channel for light propagation assisted by multiple scattering and interference. This channel can have width smaller than wavelength and makes sharp bending possible without big losses (Figure 4.24(c)). This idea forms the basis of photonic crystal waveguides. Figure 4.24(d) shows an example of calculated light intensity distribution in straight and bent Z-shape waveguides. Photonic crystal waveguides, along with light tunneling, confinement in microcavities, and coupling between waveguides and cavities, form the principles of *photonic circuitry.*

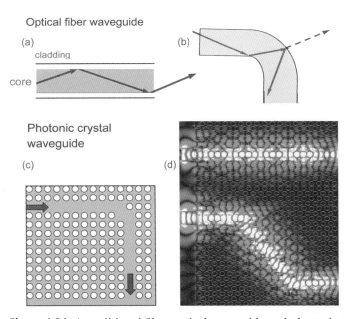

Figure 4.24 A traditional fiber optical waveguide and photonic crystal waveguide. (a) Total reflection in an optical fiber; (b) leakage in case of frustrated total reflection; (c) photonic crystal fiber; (d) electric field distribution in a linear and Z-shape defects (calculations by A. Lavrinenko).

4.4   METALLIC MIRRORS

The characteristic omnidirectional reflection inherent in polished metals results from inter-action of electromagnetic visible radiation with free electrons presenting in every metal. This interaction can be described without using quantum mechanics. The electric field $\mathbf{E}(\omega)$ gives rise to polarization described by the $\mathbf{P}$ vector, and the $\mathbf{D}$ vector in a medium can be written as

$$\mathbf{D}(\omega) = \varepsilon_0 \mathbf{E}(\omega) + \mathbf{P}(\omega) = \varepsilon_0 \varepsilon(\omega)\, \mathbf{E}(\omega) \tag{4.45}$$

where $\varepsilon(\omega)$ is relative dielectric permittivity of the medium. Consider now that the electric field oscillates with frequency ω, i.e.,

$$E(t) = E_0 \exp(i\omega t), \tag{4.46}$$

and derive the expression for $\varepsilon(\omega)$. The electric field makes the electron move owing to the force $\mathbf{F} = -e\mathbf{E}$, which in turn results in acceleration (consider the 1D case for simplicity):

$$\frac{d^2 x(t)}{dt^2} = -\frac{e}{m} E(t). \tag{4.47}$$

Displacement of a charge in space gives rise to elementary polarization $p = -ex$, which for particle density N results in polarization of the medium (per unit volume),

$$P = -exN. \tag{4.48}$$

Then, $\varepsilon(\omega)$ can be found as

$$\varepsilon = \frac{D}{\varepsilon_0 E} = 1 + \frac{P}{\varepsilon_0 E} = 1 - \frac{exN}{\varepsilon_0 E}. \tag{4.49}$$

We need to solve Eq. (4.47) and then substitute the x value into Eq. (4.49). The solution of Eq. (4.47) has the form of a function with the same time dependence as $E(t)$, i.e.,

$$x(t) = x_0 \exp(i\omega t). \tag{4.50}$$

Then

$$x = \frac{1}{\omega^2} \frac{e}{m} E(t) \tag{4.51}$$

and

$$\boxed{\varepsilon(\omega) = 1 - \frac{\omega_p^2}{\omega^2}, \quad \omega_p^2 = \frac{Ne^2}{m\varepsilon_0},} \tag{4.52}$$

where ω_p has the dimensionality of [time]$^{-1}$ and is called the *plasma frequency*. Eq. (4.52) is valid for all cases where electromagnetic waves propagate in a gas of charged particles, including, e.g., plasma in the upper atmospheric layers, with the only difference that charge value, mass, and concentration of particles should be properly modified.

Equation (4.52) is plotted in Figure 4.25. One can see $\varepsilon(\omega) < 1$ everywhere. Furthermore, $\varepsilon(\omega) < 0$ holds below ω_p. It tends to 1 asymptotically for $\omega \gg \omega_p$. Positive values of per-mittivity allow electromagnetic waves to propagate, whereas negative permittivity in the Helmholtz equation results in the evanescent electric field oscillating with frequency ω. The electromagnetic plane wave cannot exist in a medium with negative permittivity. *This is the reason for metallic reflectivity.* The incident wave when bordering a metal surface reflects back and a small portion of it penetrates into the metal, rapidly vanishing at a depth of the order of 100 nm. Considering the typical concentration of free electrons as $N = 10^{22}\,\mathrm{cm}^{-3}$, using electron charge and mass, one can see that plasma frequency for metals enters into the visible range (see Problem 4.4). Different crossover points on the frequency axis give rise to different colors of silver, gold, and copper.

The condition $\varepsilon(\omega) < 1$ allows electromagnetic waves to propagate but modifies the dispersion relation $\omega(k)$ considerably. Recalling relations for phase velocity $v = \omega/k$, $v = c/n$, and $n = \sqrt{\varepsilon}$, one may expect $\varepsilon(\omega) < 1$ to look as if superluminal speed $v = c/\sqrt{\varepsilon} > c$ becomes feasible. This is not the case. To reveal the properties of electromagnetic waves in plasma one should thoroughly account for the specific $\varepsilon(\omega)$ function in the range $0 < \varepsilon(\omega) < 1$. Indeed, the known relation

$$\omega = ck / \sqrt{\varepsilon} \tag{4.53}$$

with substitution of Eq. (4.52) for ε gives $\omega(k)$ in the form

$$\omega^2 = c^2 k^2 + \omega_p^2. \tag{4.54}$$

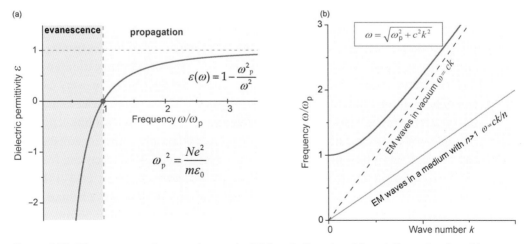

Figure 4.25 Electromagnetic waves in metals. Dielectric function (a) and dispersion law (b).

Table 4.4 **Optical parameters of common metals**

Metal	Skin depth at wavelength 600 nm (nm)	Reflection (%) at different wavelengths			
		400 nm	500 nm	600 nm	700 nm
Gold	31	38	40	90	97
Silver	24	80	91	93	95
Aluminum	13	85	88	89	87
Copper	30	30	44	72	83

This function is plotted in Figure 4.25. In the high-frequency limit $\omega \gg \omega_p$, the dispersion law (Eq. (4.54)) merges with the vacuum law, and phase velocity $v = \omega/k$ and group velocity $v_g = d\omega/dk$ remain less than c everywhere, in spite of $\sqrt{\varepsilon} < 1$.

The rapid exponential decay of electromagnetic field amplitude inside a conductive medium is referred to as *the skin effect.* The surface layer where field amplitude drops by the factor of $1/e$ is called the *skin layer.* Its thickness for the visible is given in Table 4.4, along with reflection coefficients for a few common metals. Note that even for frequencies lower than the plasma frequency, reflectance is not perfect since metals also absorb electromagnetic radiation (this results in heating), and their properties are not exhaustively described by the free electron gas model.

4.5   TUNNELING OF LIGHT

Interface of two media, one of which supports propagation of an electromagnetic field and another which does not, represents a barrier for lightwave propagation and gives rise to high reflection. However, *there is no infinite barrier for lightwaves.* In every case where light enters a medium which does not allow propagation of ordinary electromagnetic waves, the electromagnetic field penetrates inside and evanesces over a finite-length scale. One example was presented in Figure 4.12 for light inside a photonic crystal slab in the band gap range. Every photonic crystal slab whose length is not infinitely big features low but finite transmission defined by the amplitude of the electromagnetic field at the rear side of a slab. The evanescent wave gives rise to the transmitted plane wave behind the film. This phenomenon in many instances resembles the tunneling effect in quantum mechanics and is therefore referred to as *tunneling of light.*

Finite transparency of very thin metal films is another example of tunneling in optics (Figure 4.26). The evanescent wave with finite amplitude transforms into an ordinary plane

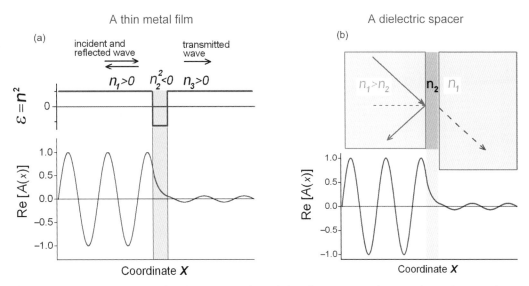

Figure 4.26 Tunneling in optics. (a) Propagation of the electromagnetic wave through a metal film. Dielectric function profile (top) and the real part of the electric field profile $A(x)$ (bottom). (b) Frustrated total reflection. Refractive index scheme (top) and the real part of the electric field profile (bottom).

wave behind a metal slab with finite thickness. The transmitted field amplitude, however, rapidly falls with thickness. Compare the electric field profile for light propagating through a metal slab with Figure 2.7, where the wave function for an electron in the case of a finite potential barrier is shown. One can see the real part of the electric field reproduces the wave function in quantum mechanics. This analogy is meaningful and important. It results from the mathematical similarity of the Schrödinger (quantum mechanics) and Helmholtz (optics) equations. Accordingly, the transmission coefficient for an electromagnetic wave propagating through a metal film with thickness a reads

$$T_{\text{metal}} = \left[1 - \frac{(\varepsilon_1 - \varepsilon_2)^2}{4\varepsilon_1\varepsilon_2} \sinh^2\left(\frac{\omega}{c}\sqrt{-\varepsilon_2}\,a \right) \right]^{-1} > 0. \tag{4.55}$$

In the limit of a thick, low-transparent barrier it reduces to an exponential law,

$$T_{\text{metal}} = \frac{(\varepsilon_1 - \varepsilon_2)^2}{4\varepsilon_1\varepsilon_2} \exp\left(-\frac{\omega}{c}\sqrt{-\varepsilon_2}\,a \right), \quad \text{when} \quad \frac{\omega}{c}\sqrt{-\varepsilon_2}\,a \gg 1. \tag{4.56}$$

This formula can be used for every common metal in the visible if its thickness exceeds 100 nm.

In Figure 4.27 the calculated transmittance spectra of silver films are presented as obtained with the real and imaginary parts of the refraction index taken into account. The figure represents the basic features evaluated on the basis of a very simple model of an electron gas in a metal. It is clearly seen that there are small but finite transmittance values for the films with thickness of the order of 100 nm, strong dependence of transmittance

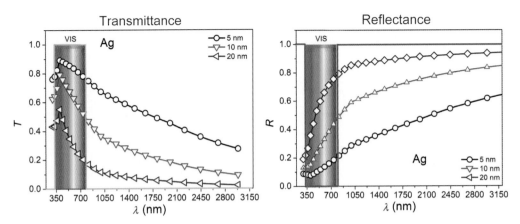

Figure 4.27 Spectral characteristics of transmission and reflection coefficients for Ag thin films with different thicknesses. Reprinted from Domaradzki et al. (2016).

on thickness, as well as a pronounced tendency for higher transparency for shorter wavelengths. Finite transparency of thin metal films is used, e.g., to make cheap semi-transparent conductive coatings in electro-optical experiments, to fabricate optical attenuating filters, and in commercial sunglasses. The latter then look like mirrors despite being partially transparent. This is a good commonly known example of tunneling in optics. Often, metal semi-transparent and reflective coatings are used in window glass in modern buildings.

An evanescent electromagnetic field develops also behind a highly refractive medium in the case of total reflection (Figure 4.26(b)). The evanescent wave generates the transmitted plane wave provided that a thin (as compared to wavelength) low-refractive layer is followed by the high-refractive medium. Note that the electric field profile looks similar to the case of light tunneling through a metal platelet and also similar to a wave function in the case of tunneling through a potential barrier in quantum mechanics. This phenomenon is used in scientific instruments known as *attenuated reflection spectrometers*.

4.6   MICROCAVITIES

A pair of parallel thin metal or dielectric mirrors form a Fabry–Pérot resonator. This resonator features high transparency for certain resonant wavelength and resembles an electromagnetic analog to quantum mechanical *resonant tunneling* (Section 2.2 and Figure 2.8). If losses in the mirrors can be neglected, transmission tends to 1 at wavelengths meeting the resonance condition, i.e., the integer number of half-waves should be equal to the inter-mirror spacing.

One can see in the case of Figure 4.28 that high transmission develops owing to "accumulation" of light in the cavity between mirrors. Electric field amplitude inside a cavity E in the course of multiple reflections and round trips rises iteratively and finally, in the steady-state regime, reaches the value $E_0/(1-r)$ where E_0 is the incident field amplitude and r is the mirror

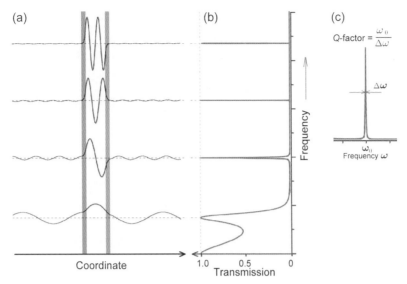

Figure 4.28 Resonant tunneling in optics: transparency of a plane cavity consisting of two metal plates in a dielectric ambient medium. (a) Accurate numerical solutions for electric field amplitude and (b) transmission spectrum. Two metal films with thickness d and refraction index $n^2 < 0$ are separated by a dielectric spacer of thickness d_0 and refraction index n_0, coinciding with that of ambient environment (courtesy of S. Zhukovsky). Parameters used in calculations are: $d_0 = 200$ nm, $n_0 = 1$, $d = 40$ nm , dielectric permittivity of the metal $\varepsilon = n^2 = -2$. (c) Q-factor evaluation.

reflection coefficient. Accordingly, for the steady-state regime to be established, the time interval is needed. The optical resonator or optical cavity is characterized by the Q-factor, which describes the capability for energy storage, and at the same time the finesse of the system in terms of the sharp transmission resonance. Q-factor can be defined as the ratio

$$Q\text{-factor} = \frac{\text{Energy stored in a system}}{\text{Energy lost per one period of oscillation}} \qquad (4.57)$$

For the first resonance, the round trip time over a cavity equals the oscillation period and therefore Q-factor defines the number of round trips necessary to arrive at the steady-state transmission value. Q-factor is the property inherent in every oscillator, including mechanical systems (pendulums, guitar strings, organ pipes, and loudspeakers) and electrical ones (LC-circuits). In mechanics, Q-factor has a straightforward intuitive meaning. It equals the number of periods an oscillating pendulum or a string performs after an external impact for oscillation amplitude to fall to zero. In the case of an optical cavity, Q-factor defines how many round trips will occur for the output intensity to reach zero after the moment when incident light has been switched off.

Q-factor defines the sharpness of transmission resonance and can be evaluated as the ratio of the resonance frequency to the resonance half-width (full-width at half-maximum), as is shown in Figure 4.28(c).

When spacing measures a single or a few wavelengths, the notation *microcavity* is used in optics. A Fabry–Pérot resonator can therefore be treated as a planar microcavity. Planar microcavities can be made using multilayer dielectric mirrors, often referred to as distributed Bragg reflectors (DBR). Though at first glance this is a more cumbersome fabrication method compared to metal mirrors, the results are much better because metal mirrors always feature undesirable absorptive losses that can be readily avoided in dielectric or semiconductor DBR mirrors.

Microcavities of high finesse can be developed as defects in photonic crystals. A representative example is given in Figure 4.29. The lower- or higher-refractive portion of space in a 2D or 3D periodic medium gives rise to standing waves and energy storage enabling big Q-factors. In this case, lightwave confinement is provided by reflecting periodic structures in two or three dimensions around a cavity.

A microdisk or a microsphere of a semiconductor or a dielectric material form microcavities as well. For these cavities, strong confinement occurs owing to the total reflection at the boundaries similarly to fiber waveguides. Surface-like standing waves in these cavities are referred to as *whispering gallery modes* since this type of resonance was first considered by Rayleigh when explaining the extraordinary acoustic audibility in the famous Whispering Gallery at St. Paul's Cathedral in London. Such cavities in optics are often referred to as *photonic dots* to emphasize an analogy to electron confinement in quantum dots (Figure 4.30). The electromagnetic mode in a spherical cavity can be excited by means of tunneling (frustrated total reflection) of waves from a fiber waveguide in the case of subwavelength proximity to a microsphere.

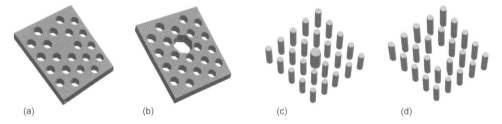

(a)   (b)   (c)   (d)

Figure 4.29 Microcavities inside a two-dimensional photonic crystal slab.

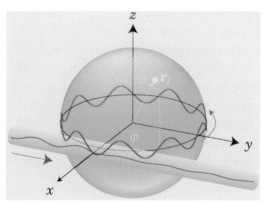

Figure 4.30 An example of a whispering gallery mode in a dielectric sphere and its coupling to a fiber mode. Adapted from Arnold et al. (2003), with permission from OSA Publishing.

4.7 LIGHT IN METAL–DIELECTRIC NANOSTRUCTURES: NANOPLASMONICS

Optical properties of tiny metal particles do remarkably differ from ordinary bulky metals. In 1857 Michael Faraday was the first to report on the reddish color of colloidal gold in a solution. The Lycurgus Cup, dating back to the fourth century, is the first identified artifact in which optical properties of nanometer-size metal (gold) particles have been purposefully used to obtain the marvelous pink–red transparent glass. Much later, gold and copper nanoparticles were widely used in amazing stained glasses. In the twentieth century copper and gold nanoparticles were industrially used to fabricate dark-reddish glasses. Optical properties of metal nanostructures and modification of light–matter interaction in metal nanostructures have led lately to the new field in nanoscience and nanotechnology, *nanoplasmonics*. Nanoplasmonics represents the dawn of nanoscience and nanotechnology.

Figure 4.31 presents the typical optical density spectra of metal nanoparticles dispersed in air or deposited on top of a dielectric substrate; the surface concentration of nanoparticles is low. One can see the optical density features a pronounced maximum in the visible spectrum, which lies in the green for gold and in the blue for silver. Optical density spectra correlate with the dielectric functions of these metals and can be well understood in terms of the effective medium approach.

BOX 4.1 NANOPLASMONICS, THE DAWN OF NANOSCIENCE AND NANOTECHNOLOGY

Lycurgus Cup

Mid XX cent.

The Lycurgus Cup from the fourth century, containing pink glass colored by gold nanoparticles, is the first known technological application of nanostructures in optics, though size effects were only unconsciously involved.

Other examples can be found in numerous masterpieces of antique stained glass. The first documented research on nano-optics is in the paper by Michael Faraday presented to the Royal Society in 1857. Since the middle of the twentieth century, routine commercial production of deep-red glass has been developed using copper dopants, in which copper nanoparticles are responsible for the color. Thus, plasmonics seems to be the oldest field of nanoscience and nanotechnology.

Michael Faraday
(1791–1867)

Experimental Relations of Gold (and Other Metals) to Light. *Philos. Trans. R. Soc. London*, **1857**, 147, 145

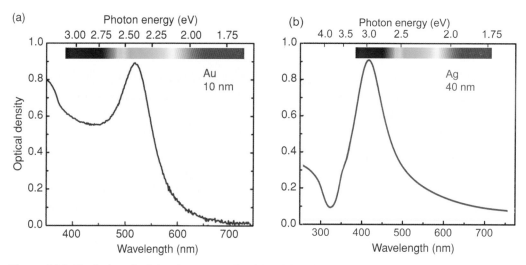

Figure 4.31 Typical optical density spectra for (a) gold and (b) silver nanoparticles dispersed on top of a dielectric substrate in air.

Consider the optical properties of a dielectric material – say, glass, water, or air – when nanometer-size metal nanoparticles are dispersed therein. If nanoparticle size is much less than optical wavelength, then an effective medium approach is applicable, i.e., the binary metal–dielectric mixture can be treated as a composite material to which the dielectric function can be ascribed. Recalling the Bruggeman formula (Eq. (4.44)), one can calculate the dielectric function of the composite medium, implying the actual dielectric function of the metal in question in the form

$$\varepsilon(\omega) = \varepsilon_\infty - \frac{\omega_p^2}{\omega^2 + i\omega\Gamma} \tag{4.58}$$

where ε_∞ accounts for the dielectric function in the limit of high frequencies, and Γ is the electron dephasing rate. It equals the time the electron spends between two successive scattering events. For example, for silver one has $\varepsilon_\infty = 5$, $\hbar\omega_p \approx 9$ eV, $\hbar\Gamma = 0.02$ eV.

The dielectric function becomes complex in every absorbing medium, not only in metals. It can be generally expressed as

$$\varepsilon(\omega) = \varepsilon_1 + i\varepsilon_2 \tag{4.59}$$

and gives rise to the *complex refractive index*

$$\tilde{n}(\omega) = \sqrt{\mu(\omega)\varepsilon(\omega)} = n + i\kappa. \tag{4.60}$$

In the optical range, $\mu(\omega) = 1$ for the known continuous media (it differs from 1 only for specially designed media referred to as metamaterials) and the real and the imaginary parts of the complex refractive index take the form

$$n = \sqrt{\frac{\varepsilon_1 + \sqrt{\varepsilon_1^2 + \varepsilon_2^2}}{2}} \quad , \quad \kappa = \frac{\varepsilon_2}{2n}. \tag{4.61}$$

Here, n is the familiar refractive index defining the speed of electromagnetic waves and their refraction, and κ is a dimensionless value referred to as the *extinction coefficient*. It allows for calculation of the *absorption coefficient*,

$$\alpha = \frac{4\pi\kappa}{\lambda} \qquad (4.62)$$

where λ is wavelength. The absorption coefficient, measured in [length]$^{-1}$ units, in turn, defines light intensity *transmission coefficient* for a material layer thickness L as

$$T = \exp(-\alpha L). \qquad (4.63)$$

Notably, optical spectra of metal suspensions are strongly dependent on dielectric permittivity of the ambient environment. This phenomenon can be purposefully used to monitor or measure small deviations in the refractive index of liquids or gases resulting, e.g., from chemical reactions or pollution. This approach enables development of *plasmonic sensors* for biomedicine and ecology, areas in which monitoring of liquid and gas purity is important.

When metal nanoparticle size rises and cannot be treated as negligibly small on the optical wavelength scale, the effective medium approach should be replaced by the accurate numerical calculations based on scattering theory. In this case, light attenuation in the composite medium is defined not only by energy dissipation in metal but also by light scattering. Accordingly, the notion of *extinction* should be introduced to indicate the sum of dissipative and scattering contributions to light attenuation.

In Figure 4.32, calculated optical extinction spectra are presented for very small (a few nanometers) and larger silver nanoparticles in air. Lightwave scattering can be neglected for particle diameters less than 20 nm. In this case, light attenuation comes from dissipation of energy through electron scattering. Remarkably, metal nanoparticles do not exhibit quantum size effects like semiconductor ones, referred to as quantum dots. Because of the high Fermi energy, electrons in metal have very small de Broglie wavelengths, so that strong confinement of electrons cannot be performed (see Section 3.2 and Table 3.1 for more detail). However, an optical size effect for smaller particles does exist. When nanoparticle size becomes smaller than the electron mean free path, it is the nanoparticle diameter that defines dephasing time because of the electron scattering by its surface. Therefore, the absorption spectrum of smaller particles broadens (Figure 4.32(a)). For particles larger than 20 nm, a size-dependent scattering component presents in the extinction spectrum which shifts to the longwave side with size (Figure 4.32(b)).

A similar size-dependent contribution from scattering governs optical absorbance of rod-like metal nanoparticles (Figure 4.33). Intrinsic resonant absorbance near 520 nm (as in Figure 4.31(a)) is overtaken by morphological resonance, i.e., size- and shape-dependent resonance resulting from light scattering. To summarize, optical properties of metal nanoparticles feature pronounced size-dependent optical properties that are well understood and accurately described theoretically.

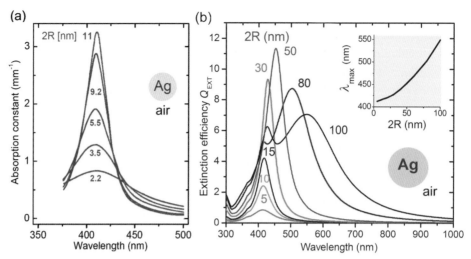

Figure 4.32 Calculated size-dependent optical properties of Ag nanoparticles. (a) Lightwave scattering can be neglected; light attenuation comes from dissipation of energy through electron scattering. (b) Light attenuation comes from both energy dissipation and size-dependent scattering for nanoparticles larger than 30 nm. Two extinction maxima are clearly resolved, the longwave one being size-dependent and shifting to the red with size. The insert shows the position of the main maximum versus size. Ambient medium refractive index is 1.5. Adapted from Kreibig and Vollmer (1995), Copyright Springer. Courtesy of S. M. Kachan.

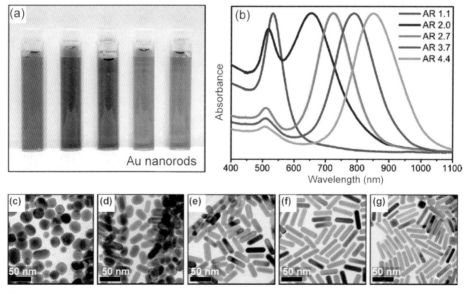

Figure 4.33 Size-dependent optical absorption by Au nanorods. (a) Photograph of gold nanorod solutions with increasing aspect ratio (AR) left to right. (b) Absorbance spectra of gold nanorods with different ARs. (c–g) Electron microscopy images of gold nanorods: (c) AR = 1.1, (d) AR = 2.0, (e) AR = 2.7, (f) AR = 3.7, and (g) AR = 4.4. Adapted with permission from Abadeer et al. (2014), copyright 2014 American Chemical Society.

When electromagnetic radiation shines onto metal nanostructures, electrons in metal oscillate because of plasmon resonance if radiation frequency falls into the range of noticeable extinction. This results in enhancement of electromagnetic energy near nanotextured metal surfaces with the texture scale comparable to or less than the wavelength. Light intensity, or more generally speaking, electromagnetic radiation intensity, is always higher near a curved metal surface than otherwise. Figure 4.34 shows a number of model metal nanostructures examined in the context of possible electric field/light intensity enhancements to promote stronger light–matter interaction, e.g., enhanced photoluminescence of molecules, atoms, or quantum dots located near a metal nanotextured surface. One spherical particle offers just a few times enhancement in the case of gold (Figure 4.34(a)) and about ten or slightly more in the case of silver. Two particles offer several times higher enhancement than one in the space between the particles (Figure 4.34(b)). A number of spherical particles properly scaled can give up to six orders of the magnitude for light intensity ($|\mathbf{E}|^2$) in the small area between the particles. Elliptical particles, both single and coupled, offer much greater enhancement than spherical ones. One can see 10^6 enhancement factors for a couple of silver spheroids. Generally, silver enables higher enhancement than gold, and for every metal the wavelength range should fall into the range of noticeable extinction – e.g., for larger particles the enhancement spectrum shifts to the longer wavelength in accordance with their extinction spectra.

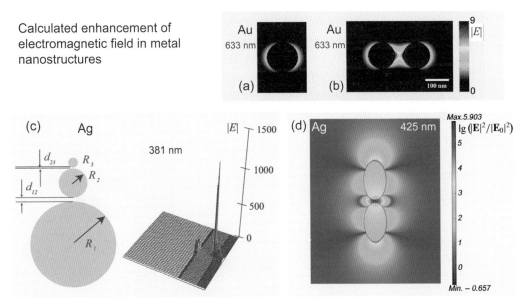

Figure 4.34 Examples of calculated local electric field enhancement for a number of model metal nanostructures: (a) a single sphere, (b) two spheres, (c) three spheres, (d) two spheroids. Numbers (nm) indicate wavelength for which calculations were made. These data provide an intuitive introduction to the optical antenna notion. Adapted from (a,b) Chung et al. (2011) under the Creative Commons License, (c) with permission from Li et al. (2003), copyright American Physical Society, and (d) Guzatov and Klimov (2011) under the Creative Commons License.

4.8 OPTICAL ANTENNAS

An antenna is a geometrical construction that promotes receiving and transmitting of waves by means of modification of the waveform to adjust the transceiver or emitter parameters to the properties of the ambient environment. It is not a mirror or a lens, but rather a device that modifies the properties of waves on a wavelength or even subwavelength scale. Naturally existing receiving antennas are the aural cavities of humans and animals that are designed through evolution to efficiently deliver acoustic waves to the receptor in ears. An acoustical transmitting antenna is a well-known horn widely used in loudspeakers and musical instruments.

In radiophysics, antennas are commonly used to promote transmitting and receiving radio- and microwaves. These are familiar pieces of metal, single or coupled rods, various frames, and more complicated constructions. One can see from Figure 4.34 and the relevant discussion in the previous section that single metal nanoparticles and their assemblies feature the remarkable property of locally enhancing the incident electromagnetic field in the optical range. This property leads to the notion of *optical antenna* – the emerging concept in photonics of bridging optical technologies with radio- and microwave engineering. Notably, localization of electromagnetic energy occurs at the subwavelength scale, where ray optics vanishes and light beams cannot be drawn. In radiophysics, the antenna matches impedance of a receiver or a transmitter with ambient space. In acoustics, impedance matching is provided by horns. In optics, one can also speak about impedance matching, recalling that, for a continuous medium, impedance is $Z = \sqrt{\mu / \varepsilon}$ and μ in the optical range is 1 for the wealth of existing materials except for the special case of artificially fabricated nanostructures referred to as metamaterials.

In addition to the examples given in Figure 4.34 with nanoantennas using spherical nanoparticles, Figure 4.35 shows further designs that can be implemented by submicron

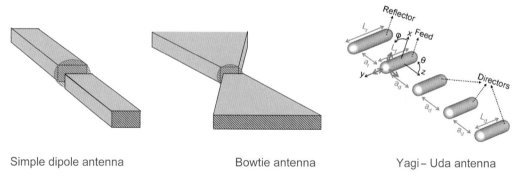

Simple dipole antenna | Bowtie antenna | Yagi – Uda antenna

Figure 4.35 Examples of optical antenna design using metal components. In a simple dipole and bowtie antenna the incident field concentration occurs in the central area showed by red dashes. In the case of a Yagi–Uda antenna, field concentration occurs at the feed element position. Design of the Yagi–Uda antenna is reprinted from Taminiau et al. (2008), with permission from OSA Publishing.

lithography. For a simple dipole antenna and a bowtie antenna the incident field concentrates in the middle, between antenna parts. A more sophisticated design is presented by the Yagi–Uda construction in which a feed element is surrounded by a longer reflector and a set of shorter directors. Here, field enhancement occurs at the feed element position.

How can experimenters measure the local electromagnetic radiation intensity in an optical antenna with subwavelength lateral resolution? This is a tricky but feasible task. Gold nanoparticles feature weak intrinsic luminescence resulting from electron transitions between subbands within the conduction band. This luminescence can be excited by two-photon absorption. Since probability of two-photon processes is very low, high optical power is necessary for their study. It can be provided by using pico- or subpicosecond laser pulses. Unlike the one-photon process, the absorbed power in two-photon processes is proportional to I^2, i.e., to E^4 rather than I and E^2 inherent in one-photon absorption. Since a real light beam spatial profile is typically Gaussian, departing from E^2 to E^4 reduces the effective cross-section of the incident beam and allows for mapping of the optical events with subwavelength resolution beyond the diffraction limit. Figure 4.36 shows measured two-photon-induced gold luminescence and the relevant theoretical modeling for an antenna consisting of a couple of metal sticks. One can see that field enhancement in the experiment well agrees with the theory, it occurs between the sticks and shows wavelength dependence defined by the material and the gap between sticks. Figure 4.37 shows a similar experiment for a gold *bowtie antenna*. One can see that field concentration agrees with theory, and features gap dependence approaching the result for a single metal triangle at spacing comparable with optical wavelengths.

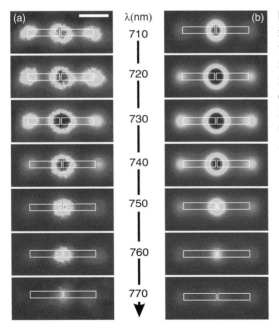

Figure 4.36 Mapping of (a) the two-photon intrinsic gold luminescence intensity and (b) the theoretically predicted E^4 distribution for a number of wavelengths, demonstrating resonant local enhancement of the incident electromagnetic field. The scale bar is 500 nm. Reprinted figure with permission from Ghenuche et al. (2008). Copyright 2008 by the American Physical Society.

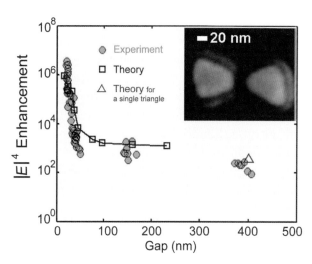

Figure 4.37 Measured and calculated local incident electric enhancement for gold bowtie antennas with various gaps. Wavelength is 830 nm. Adapted figure with permission from Schuck et al. (2005). Copyright 2005 by the American Physical Society.

4.9   NON-PERIODIC STRUCTURES AND MULTIPLE LIGHT SCATTERING

Most information about the world around us is gained through *vision*. It is light scattering that makes human vision possible. We sense light coming from an external source like the sun or a lamp, and then scattered by things around us to our eyes. Transparent things that do not scatter are not visible. We can only guess about them using partial light reflection.

What happens if light waves or any other electromagnetic waves propagate in an inhomogeneous non-absorbing medium with imperfections whose size is not negligible compared to wavelength? Imagine randomly distributed nanoparticles of different sizes and shapes made of material with $\varepsilon_1 > 1$ dispersed in the ambient material with $\varepsilon_2 \neq \varepsilon_1$. The representative examples are water droplets or ice microcrystals in air, solid powders in solutions, and polymers with microcrystalline impurities like crystalline luminophore grains dispersed in silicone in white LEDs. Waves experience scattering at every obstacle. The important parameter is the *mean free path*, which is the average distance the wave travels between scattering events.

There are three distinct cases of electromagnetic wave propagation in scattering media, depending on the relations between wavelength λ, slab thickness L, and the mean free path ℓ. These cases correspond to the *single scattering* regime, the *multiple scattering* regime, and the *localization* limit (Figure 4.38). The single scattering regime occurs if the following condition is met:

$$\lambda < L < \ell. \tag{4.64}$$

In this case, every scattering event removes irreversibly a portion of energy and the output intensity detected behind a slab in the direction of the incident beam propagation decreases. The elementary intensity change can be described as

$$dI = -\alpha I dx \tag{4.65}$$

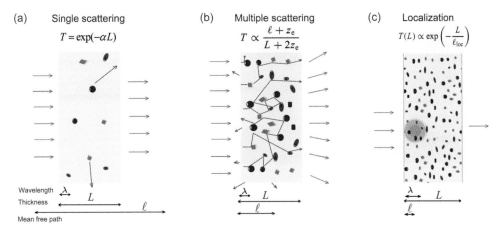

Figure 4.38 Different propagation regimes of light through scattering media. See text for detail.

with α describing scattering rate per unit length. Integration of this equation over a sample length gives rise to the familiar expression

$$T = \exp(-\alpha L). \tag{4.66}$$

One can see that Eq. (4.66) coincides with the Bouguer law for absorptive media. Therefore, *propagation through a slightly scattering medium in the regime of single scattering features the same transmission versus thickness dependence as does propagation in the case of absorption.* However, it is possible to distinguish scattering from absorption by observing the intensity scattered outside the original beam propagation direction.

The multiple scattering regime occurs when a slab thickness is bigger than the mean free path, i.e.,

$$\lambda < \ell < L. \tag{4.67}$$

Now, waves experience complicated motion, the output field features complex angular dependence, a portion of incident intensity scatters even in the backward direction, and inside a slab loop-like paths exist (Figure 4.38(b)). The analytical solution is not possible and numerical analysis of arising integro-differential equations is necessary. Notably, propagation of waves in this case follows to a large extent the laws of diffusion and is described by similar equations. Light spends much more time in a sample compared to the single scattering case, and a short input pulse becomes many times longer and its shape changes accordingly. Owing to complicated paths, the probability exists that scattered light will propagate in the direction coinciding with the incident flux. The theory in this case gives the following relation for transmitted intensity versus thickness:

$$T \propto \frac{\ell + z_e}{L + 2z_e}, \tag{4.68}$$

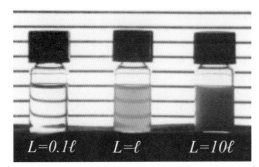

Figure 4.39 Three vials filled with turbid liquid corresponding to different scattering regimes: left, weak scattering ($L / \ell = 0.1$); middle, intermediate regime ($L / \ell = 1$); right, opaqueness because of multiple scattering ($L / \ell = 10$). Illumination comes from the back. Courtesy of F. Poelwijk.

where the z_e parameter defines the boundary properties. Eq. (4.68) describes the basic trend of transmission versus thickness, whereas the exact solution should explicitly account for boundary events in every case under consideration.

Figure 4.39 shows that a vial filled with liquid is transparent for single scattering $L / \ell = 0.1$ and fully opaque for the case of multiple scattering ($L / \ell = 10$). Multiple scattering completely obscures the image behind a scattering layer. Precise experimental manifestation of the law described by Eq. (4.68) is presented in Figure 4.40 using nanoporous semiconductor material. The linear dependence is apparent and the slope depends on the sample structure, which defines the mean free path value.

In the limit of a very long scattering layer, multiple scattering gives rise to the inverse dependence of transmission versus length, $T \propto \ell / L$. This property remarkably resembles Ohm's law in electricity – electric current is inversely proportional to a conductor length or, in other words, electrical resistance of a conductor is proportional to its length. This analogy is essentially based on similarity of the phenomena in question. Both lightwaves and electrons (also waves) experience multiple scattering. In the case of electromagnetic waves, scattering results from refraction index inhomogeneity, whereas in the case of electrons scattering results from potential inhomogeneities which in turn comes from material imperfections (impurities, disorder in atoms displacement and atoms vibrations, grain boundaries in polycrystalline materials).

What happens if small but very strong scatterers (i.e., consisting of highly refractive material) densely dispersed in an ambient environment lower the mean free path to values considerably smaller than the wavelength? Then neither beam can be drawn inside a medium. The theory considers that the case $\ell < \lambda / 2\pi$ gives rise to *localization* of waves. This condition has been first suggested for electron localization in disordered solids and is referred to as the *Ioffe–Regel criterion*. Localization of electrons in disordered solids was thoroughly analyzed by Philip Anderson in 1958 and is extensively used in the electron theory of solids (Nobel Prize in 1977). Much later, in 1984 Sajeev John highlighted that *Anderson localization* is the general phenomenon that occurs for every type of wave in disordered media, provided the mean free path falls to $\ell < \lambda / 2\pi$.

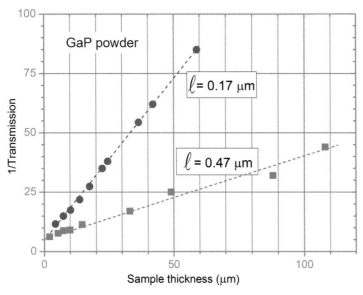

Figure 4.40 Thickness dependence of optical transmission for porous GaP samples (data by Schuurmans et al. 1999) for a wavelength of 685 nm.

For an electron, every ion is a very strong scatterer because of Coulomb interaction. Therefore, Anderson localization of electrons is the common phenomenon in disordered solids. For electromagnetic waves, highly refractive materials are necessary to perform the localization regime. For radio- and microwaves, high refraction is feasible but for the visible range even $n > 3$ is hard to perform and in the near-IR $n > 4$ is not possible (see Table 4.1). It is important that highly refractive materials should necessarily be non-absorptive. Experimental observation of Anderson localization in photonics remains a challenge, made difficult by the facts that minor absorption is hard to avoid, and conditions of multiple scattering – because of the long time spent by lightwaves – give rise to drastic falls in transmission, which is hard to discriminate against a localization regime. If a lossless medium features the localization regime, it can be used for light storage elements.

While light localization is an issue and has no straightforward applications to date, disordered materials with multiple scattering are commonly used in daily life. Every color *ceramic* or *pigment* on walls, cars, or artworks actually features multiple scattering, otherwise it would look transparent. In color pigments, multiple scattering occurs selectively for wavelengths where absorption is negligible. Otherwise, light energy dissipates in the matter after a few scattering events. White *porcelain* in tableware, white *paper*, white powders, and white screens are another set of multiple scattering materials. White coloring develops by equal scattering of lightwaves within the visible spectrum. Therefore, to get white matter one needs to make a random powder or other type of submicron disorder using non-absorbing materials.

Dielectric or semiconductor materials with disorder on the light wavelength scale can be used to make selective filters, so-called Christiansen's filters named after C. Christiansen, a Danish physicist who in 1884 discovered selective transmission of light by a mixture of two powders in the spectral range, where both components have the same refractive index n. Since n is typically wavelength-dependent, for certain pairs of materials intersection of $n(\lambda)$ curves is possible. Then a dense mixture of these materials will look homogeneous at the intersection points, with efficient light scattering otherwise.

Multiple scattering in a medium with optical gain may result in local loop-like paths. If gain exceeds losses along a single loop, then local random ring microlasers develop. This phenomenon is referred to as *random lasing* and was predicted for the first time in 1968 by V. Letokhov, a Russian physicist. It often occurs in ceramic laser materials.

4.10 USEFUL ANALOGIES OF ELECTRONIC AND OPTICAL PHENOMENA

The Helmholtz equation used in optics and the Schrödinger time-independent equation used in quantum mechanics look alike. It is reasonable to recall their simplest 1D forms (Eqs. (4.21) and (2.17)) to highlight the similarity. This similarity reflects the wave properties of light and electrons. The only difference is that the Helmholtz equation describes the real electric or magnetic field that can be sensed by humans, animals, or artificial detectors, whereas the Schrödinger equation describes the special function that gives us the probability of an electron being found at a given point in space. Both equations can be written as

$$\frac{d^2\Phi(x)}{dx^2} + k^2\Phi(x) = 0 \ , \quad \Phi_{QM} \equiv \psi, \ k_{QM}^2 = \frac{2m(E-U)}{\hbar^2} \ ; \quad \Phi_{EM} \equiv A, \ k_{EM}^2 = \varepsilon\frac{\omega^2}{c^2}. \tag{4.69}$$

Here, A is the electric field amplitude. Material or space parameters are defined by $U(x)$ in quantum mechanics and by $\varepsilon(x)$ in optics. Comparing quantum mechanical (subscript QM) and electromagnetic (subscript EM) counterparts, one can see that changes in the refractive index will result in optics in the same effect as change in potential energy U in quantum mechanics. Let us recall the properties of electrons discussed in Chapters 2 and 3 and optical phenomena resulting from confined electromagnetic waves considered in this chapter to reveal the analogies of electrons and electromagnetic waves to their full extent. These are summarized in Table 4.5.

Table 4.5 **A set of quantum mechanical phenomena and their optical counterparts**

Medium inhomogeneity	Phenomena	
	Quantum mechanics	Optics
Refraction step, potential step	Reflection/transmission	Reflection/transmission (Figure 4.3)
Refraction barrier (well), potential barrier (well)	Resonant propagation, reflection/transmission	Reflection/transmission and Fabry–Pérot modes in thin films
Potential well/optical cavity	Discrete energy spectrum (Figure 2.3)	Resonant modes (Figure 2.2)
Drop from positive to negative permittivity, high potential step ($U > E$)	Extension of wave function under barrier	Evanescence of electromagnetic field in the area with negative permittivity
A slab with negative permittivity, potential barrier with finite length	Reflection/tunneling (Figure 2.7)	Reflection/tunneling, transparency of thin metal films (Figure 4.26)
Two parallel mirrors, double potential barrier ($U > E$)	Resonant tunneling (Figure 2.8)	Resonant transmission of optical interferometers (Figure 4.28)
Periodic refractive index change in space, periodic potential in space	Energy bands formation separated by forbidden energy gaps in crystals (Figure 2.11)	Development of reflection and transmission bands and band gaps in photonic crystals (Figures 4.9–4.11)
Material with weak disorder of refraction index, crystalline materials with defects	Ohm's law, $1/L$ current dependence	$1/L$ transmission of opaque materials (Figure 4.40)
Impurity (defect) in a crystal, a cavity or defect in a photonic crystal	Localized electron state	High-Q microcavity
Strong disorder of refraction index, highly disordered solids	Anderson localization of electrons	Anderson localization of electromagnetic waves (hard to observe in the optical range)

Conclusion

- Every step of refraction index n (which for a non-absorbing, non-magnetic medium equals the square root of its dielectric permittivity ε) gives rise to electromagnetic wave reflection which rises for oblique as compared to normal incidence.

- Periodic alteration of n in space leads to the notion of photonic crystal and results in development of transmission/reflection bands whose spectral position depend on parameters/sizes of materials/components involved. Reflection in this case correlates with evanescent waves that can give rise to tunneling for finite-length structures.

- Negative permittivity ε inherent in metals below plasma frequency gives rise to characteristic wide-band metal reflection and evanescent waves inside, resulting in light tunneling and finite transparency of thin metal layers.

- Cavities made as a couple of mirrors with a dielectric spacer or a defect in a photonic crystal feature the ability to store energy and this property is expressed in terms of the Q-factor.

- Metal nanoparticles and nanostructures with subwavelength sizes offer many options to control transmission of light as well as local concentration of incident electromagnetic field, which has led to the development of nanoplasmonics and elaboration of the new optical conception – optical antenna.

- Multiple scattering results in universal $1/L$ transmission behavior and is responsible for coloring ceramics and pigments as well as for whiteness of paper, porcelain, and screens. Light scattering enables human vision.

- Generally, electromagnetic waves in optics and electrons in complex media feature a number of remarkable analogies coming from the harmony of nature embodied in the notable mathematical analogy of the Helmholtz equation in optics and the time-independent Schrödinger equation in quantum mechanics. These analogies, when applied to complex dielectric media and nanostructures consisting of non-absorbing materials, allow for tracing a number of instructive phenomena:

  - light reflection at every refraction step;

  - development of transmission/reflection bands in periodic structures;

  - light tunneling through a periodic multilayer or a metal film;

  - resonant tunneling of light through an interferometer or a planar cavity;

  - development of high-Q defects (microcavities) in photonic crystals;

  - universal $1/L$ transmission dependence for disordered media with multiple scattering;

  - possible light localization in strongly disordered nanostructures made of highly refractive materials.

Problems

4.1 Recall Snell's law and explain why n is called *refractive* index.

4.2 Calculate reflection and transmission for GaN/SiC, GaN/Al$_2$O$_3$, SiC/air, and Al$_2$O$_3$/air interfaces. These pairs of semiconductors and dielectrics are used in LEDs.

4.3 Using quarter-wave stacks, design dielectric mirrors for 400, 500, and 600 nm, choosing materials from Table 4.1. Keep in mind transparency of materials at the desirable wavelength.

4.4 Calculate wavelength where dielectric permittivity crosses zero for metals, assuming $N = 10^{22}\,\text{cm}^{-3}$. Predict optical properties of a sheet and a thin film made of such a metal.

4.5 Provide examples from different fields of physics and technology where Q-factor matters.

4.6 Recall the principal phenomena in the context of plasmonics.

4.7 Using the value $\hbar\Gamma = 0.02$ eV, calculate how much time an electron on average moves without scattering in metal.

4.8 Explain size-dependent properties of metal nanoparticles dispersed in a dielectric medium.

4.9 Explain how plasmonic sensors can be used in biomedicine or ecology.

4.10 Explain why multiple scattering gives rise to transmission versus length in the different form as compared to absorption and single scattering.

4.11 Explain why single scattering cannot be distinguished from weak absorption by means of optical transmission measurements. Suggest (a) possible experimental way(s) to reveal single scattering versus weak absorption. Hint: consider spatial patterns and energy dissipation.

4.12 Find multiple scattering phenomena in daily life. Explain the difference in optical absorption of materials used to fabricate white and red bricks.

4.13 Based on band gap data for a number of common semiconductors (Table 2.3), predict the color of their disordered microstructures resulting from multiple scattering. Suggest materials for red, orange, yellow, and white powders/pigments. Explain why blue or green pigment cannot be designed based on dielectric or semiconductor materials without impurities.

4.14 Human vision occurs via scattering of light coming from an external source. Design a technical vision system that is not based on light scattering.

4.15 Recall the analogies between electrons and electromagnetic waves. What is the physical origin of these?

4.16 Try to add your own finding(s) to extend the list of analogies in Table 4.5.

Further reading

Barber, E. M. (2008). *Aperiodic Structures in Condensed Matter: Fundamentals and Applications.* CRC Press.

Bharadwaj, P., Deutsch, B., and Novotny, L. (2009). Optical antennas. *Adv Opt Photonics*, **1**(3), 438–483.

Born, M., and Wolf, E. (1999). *Principles of Optics: Electromagnetic Theory of Propagation, Interference and Diffraction of Light.* Cambridge University Press.

Dal Negro, L. (ed.) (2013). *Optics of Aperiodic Structures: Fundamentals and Device Applications.* CRC Press.

Joannopoulos, J. D., Johnson, S. G., Winn, J. N., and Meade, R. D. (2011). *Photonic Crystals: Molding the Flow of Light.* Princeton University Press.

Kavokin, A., Baumberg, J. J., Malpuech, G., and Laussy, F. P. (2007). *Microcavities.* Oxford University Press.

Krauss, T. F., and De La Rue, R. M. (1999). Photonic crystals in the optical regime: past, present and future. *Progr Quant Electron*, **23**, 51–96.

Lagendijk, A., van Tiggelen, B., and Wiersma, D. S. (2009). Fifty years of Anderson localization. *Phys Today*, **62**(8), 24–29.

Lekner, J. (2016). *Theory of Reflection: Reflection and Transmission of Electromagnetic, Particle and Acoustic Waves.* Springer.

Li, J., Slandrino, A., and Engheta, N. (2007). Shaping light beams in the nanometer scale: a Yagi–Uda nanoantenna in the optical domain. *Phys Rev B*, **76**, 25403.

Limonov, M. F., and De La Rue, R. M. (2012). *Optical Properties of Photonic Structures: Interplay of Order and Disorder.* CRC Press.

Lu, T., Peng, W., Zhu, S., and Zhang, D. (2016). Bio-inspired fabrication of stimuli-responsive photonic crystals with hierarchical structures and their applications. *Nanotechnology*, **27**(12), 122001.

Maciá, E., (2005). The role of aperiodic order in science and technology. *Rep Prog Phys*, **69**(2), 397.

Novotny, L., and van Hulst, N. (2011). Antennas for light. *Nat Photonics*, **5**(2), 83–90.

Park, Q. H., (2009). Optical antennas and plasmonics. *Contemp Phys*, **50**(2), 407–423.

Parker, A. R. (2000). 515 million years of structural color. *J Optics A*, **2**, R15–R28.

Poelwijk, F. J. (2000). *Interference in Random Lasers.* PhD thesis, University of Amsterdam.

Thompson, D. (2007). Michael Faraday's recognition of ruby gold: the birth of modern nanotechnology. *Gold Bulletin*, **40**(4), 267–269.

Van den Broek, J. M., Woldering, L. A., Tjerkstra, R. W., et al. (2012). Inverse-woodpile photonic band gap crystals with a cubic diamond-like structure made from single-crystalline silicon. *Adv Funct Mater*, **22**(1), 25–31.

Vukusic, P., and Sambles, J. (2003). Photonic structures in biology. *Nature*, **424**, 852–855.

Vukusic, P., Hallam, B., and Noyes, J. (2007). Brilliant whiteness in ultrathin beetle scales. *Science*, **315**(5810), 348.

Wilts, B. D., Michielsen, K., De Raedt, H., and Stavenga, D. G. (2012). Hemispherical Brillouin zone imaging of a diamond-type biological photonic crystal. *J R Soc Interface*, **9**(72), 1609–1614.

References

Abadeer, N. S., Brennan, M. R., Wilson, W. L., and Murphy, C. J. (2014). Distance and plasmon wavelength dependent fluorescence of molecules bound to silica-coated gold nanorods. *ACS Nano*, **8**(8), 8392–8406.

Arnold, S., Khoshsima, M., Teraoka, I., Holler, S., and Vollmer, F. (2003). Shift of whispering-gallery modes in microspheres by protein adsorption. *Opt Lett*, **28**(4), 272–274.

Baldycheva, A., Tolmachev, V., Perova, T., et al. (2011). Silicon photonic crystal filter with ultrawide passband characteristics. *Opt Lett*, **36**, 1854–1856.

Bendickson, J. M., Dowling, J. P., and Scalora, M. (1996). Analytic expressions for the electromagnetic mode density in finite, one-dimensional, photonic band-gap structures. *Phys Rev E*, **53**, 4107–4121.

Birner, A., Wehrspohn, R. B., Gösele, U. M., and Busch, K. (2001). Silicon-based photonic crystals. *Adv Mat*, **13**, 377–382.

Bruggeman, D. A. G. (1935). Berechnung verschiedener physikalischer Konstanten von heterogenen Substanzen. I. Dielektrizitetskonstanten und Leitfehigkeiten der Mischkorper aus isotropen Substanzen. *Ann Phys*, **416**, 636–664.

Busch, K., von Freymann, G., Linden, S., et al. (2007). Periodic nanostructures for photonics. *Phys Rep*, **444**, 101–202.

Chen, Y. C., and Bahl, G. (2015). Raman cooling of solids through photonic density of states engineering. *Optica*, **2**(10), 893–899.

Chung, T., Lee, S. Y., Song, E. Y., Chun, H., and Lee, B. (2011). Plasmonic nanostructures for nanoscale bio-sensing. *Sensors*, **11**(11), 10907–10929.

Domaradzki, J., Kaczmarek, D., Mazur, M., et al. (2016). Investigations of optical and surface properties of Ag single thin film coating as semitransparent heat reflective mirror. *Mater Sci Poland*, **34**(4), 747–753.

Gaponenko, S. V. (2010). *Introduction to Nanophotonics*. Cambridge University Press.

Ghenuche, P., Cherukulappurath, S., Taminiau, T. H., van Hulst, N. F., and Quidant, R. (2008). Spectroscopic mode mapping of resonant plasmon nanoantennas. *Phys Rev Lett*, **101**(11), 116805.

Guzatov, D. V., and Klimov, V. V. (2011). Optical properties of a plasmonic nano-antenna: an analytical approach. *New J Phys*, **13**(5), 053034.

Ho, K. M., Chan, C. T., Soukoulis, C. M., Biswas, R., and Sigalas, M. (1994). Photonic band gaps in three dimensions: new layer-by-layer periodic structures. *Solid State Commun*, **89**(5), 413–416.

Joannopoulos, J. D., Johnson, S. G., Winn, J. N., and Meade, R. D. (2011). *Photonic Crystals: Molding the Flow of Light*. Princeton University Press.

Kreibig, U., and Vollmer, M. (1995). *Optical Properties of Metal Clusters*. Springer.

Li, K., Stockman, M. I., and Bergman, D. J. (2003). Self-similar chain of metal nanospheres as an efficient nanolens. *Phys Rev Lett*, **91**(22), 227402.

Lin, S.-Y., Fleming, J. G., Hetherington, D. L., et al. (1998). A three-dimensional photonic crystal operating at infrared wavelengths. *Nature*, **394**(6690), 251–253.

Lutich, A. A., Gaponenko, S. V., Gaponenko, N. V., et al. (2004). Anisotropic light scattering in nanoporous materials: a photon density of states effect. *Nano Lett*, **4**, 1755–1758.

Madelung, O. (2012). *Semiconductors: Data Handbook*. Springer Science & Business Media.

Pellegrini, V., Tredicucci, A., Mazzoleni, C., and Pavesi, L. (1995). Enhanced optical properties in porous silicon microcavities. *Phys Rev B*, **52**, R14328–R14331.

Petrov, E. P., Bogomolov, V. N., Kalosha, I. I., and Gaponenko, S. V. (1998). Spontaneous emission of organic molecules in a photonic crystal. *Phys Rev Lett*, **81**, 77–80.

Reynolds, A., Lopez-Tejeira, F., Cassagne, D., et al. (1999). Spectral properties of opal-based photonic crystals having a SiO_2 matrix. *Phys Rev B*, **60**, 11422–11426.

Sakoda, K. (2004). *Optical Properties of Photonic Crystals*. Springer.

Schuck, P. J., Fromm, D. P., Sundaramurthy, A., Kino, G. S., and Moerner, W. E. (2005). Improving the mismatch between light and nanoscale objects with gold bowtie nanoantennas. *Phys Rev Lett*, **94**(1), 017402.

Schuurmans, F. J. P., Vanmaekelbergh, D., van de Lagemaat, J., and Lagendijk, A. (1999). Strongly photonic macroporous gallium phosphide networks. *Science*, **284**, 141–143.

Strutt, J. W. (Lord Rayleigh) (1887). On the maintenance of vibrations by forces of double frequency, and on the propagation of waves through a medium endowed with a periodic structure. *Phil Mag S*, **24**, 145–159.

Taminiau, T. H., Stefani, F. D., and van Hulst, N. F. (2008). Enhanced directional excitation and emission of single emitters by a nano-optical Yagi–Uda antenna. *Opt Expr*, **16** (14), 10858–10866.

Teyssier, J., Saenko, S. V., Van Der Marel, D., and Milinkovitch, M. C. (2015). Photonic crystals cause active colour change in chameleons. *Nat Commun*, **6**, 6368.

Tjerkstra, R. W., Woldering, L. A., van den Broek, J. M., et al. (2011). Method to pattern etch masks in two inclined planes for three-dimensional nano-and microfabrication. *J Vac Sci Technol*, **29**(6), 061604.

Vlasov, Y. A., Bo, X. Z., Sturm, J. C., and Norris, D. J. (2001). On-chip natural assembly of silicon photonic band gap crystals. *Nature*, **414**, 289–293.

Voitovich, A. P. (2006). Spectral properties of films. In: B. Di Bartolo and O. Forte (eds.), *Advance in Spectroscopy for Lasers and Sensing*. Springer, 351–353.

Vukusic, P., and Sambles, J. (2003). Photonic structures in biology. *Nature*, **424**, 852–855.

Wang, R., Wang, X., -H., Gu, B., -Y., and Yang, G., -Z. (2001). Effects of shapes and orientations of scatterers and lattice symmetries on the photonic band gap in two-dimensional photonic crystals. *J Appl Phys*, **90**, 4307–4312.

Yariv, A., and Yeh, P. (1984). *Optical Waves in Crystals*. Wiley & Sons.

5 Spontaneous emission of photons and lifetime engineering

Excited atoms, molecules, and solids emit spontaneously photons, with the rate of emission being dependent both on their intrinsic properties and the properties of ambient space. The latter is capable or not capable of carrying on certain electromagnetic modes. This property of space is described by means of the density of electromagnetic modes or, in quantum language, photon density of states. Photon density of states can be engineered in a desirable way using spatially arranged components with different dielectric permittivity on the wavelength scale. This chapter provides a brief introduction to spontaneous photon emission control in nanostructures based on confinement of electromagnetic waves to get photon density of states' enhancement or inhibition.

5.1 EMISSION OF LIGHT BY MATTER

5.1.1 Spontaneous and stimulated transitions

Electrons in atoms, molecules, and solids experience upward and downward transitions between states with different energies, E. In the course of these transitions, matter and radiation perform continuous energy exchange, with photons being the energy quanta meeting the condition

$$E_i - E_j = \hbar\omega_{ij} \ , \ i, j = 1, 2, 3, ..., \ i > j. \tag{5.1}$$

There are two types of transitions, namely stimulated and spontaneous ones (Figure 5.1). The rate of stimulated transitions is proportional to the electron concentration in the given state (n_1 or n_2) and to radiation density u (joules per cubic meter). The rate of spontaneous emission is independent of radiation intensity. It is simply proportional to the excited state population, n_2. This scheme of light–matter interaction was introduced by Niels Bohr (in 1913) and Albert Einstein (in 1916). The B and A factors are referred to as Einstein coefficients for stimulated and spontaneous transitions, respectively.

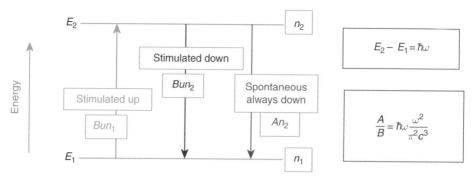

Figure 5.1 Upward and downward transitions in a simple two-level system.

5.1.2 Thermal equilibrium: the Boltzmann relation and Planck formula

In thermal equilibrium, for every pair of energy levels the rate of all upward transitions equals the rate of all downward transitions. In this case, populations obey the *Boltzmann relation*,

$$n_2 = n_1 e^{-\left(\frac{E_2-E_1}{k_B T}\right)}, \quad k_B = 1.380662 \ldots \cdot 10^{-23} \text{ J/K} \tag{5.2}$$

and the emitted energy spectrum obeys the *Planck formula*,

$$u(\omega) = \hbar\omega \frac{\omega^2}{\pi^2 c^3} \frac{1}{e^{\frac{\hbar\omega}{k_B T}} - 1}. \tag{5.3}$$

This relation was suggested by Max Planck in 1900 to describe experimentally observed spectra of hot bodies. It contained for the first time a hypothesis on light quanta whose energy equals $\hbar\omega$. It gives spectral energy density per unit volume on a frequency scale (joules per cubic meter per hertz).

Relations (5.2) and (5.3) describe the population of states and radiation intensity when the finite temperature is the only physical reason for nonzero population of upper levels. The radiation emitted in this case is referred to as *thermal radiation*. Its spectra for various temperatures are presented in Figure 5.2. For experimental and engineering purposes, $u(\lambda)$ rather than $u(\omega)$ is used (see Problem 5.1),

$$u(\lambda) = u(\omega) \frac{d\omega}{d\lambda} = \frac{8\pi hc}{\lambda^5} \frac{1}{e^{\frac{hc}{\lambda k_B T}} - 1}. \tag{5.4}$$

The function $u(\lambda)$ gives spectral energy density per unit volume (joules per cubic meter per meter or micrometer as wavelength unit). To calculate how much energy is contained in 1 m³ in a certain spectral range, one needs to integrate $u(\lambda)$ over λ within the range of interest. Consideration in terms of emitted intensity $I(\lambda)$ in W/m² per unit solid angle looks much more reasonable for many practical purposes.

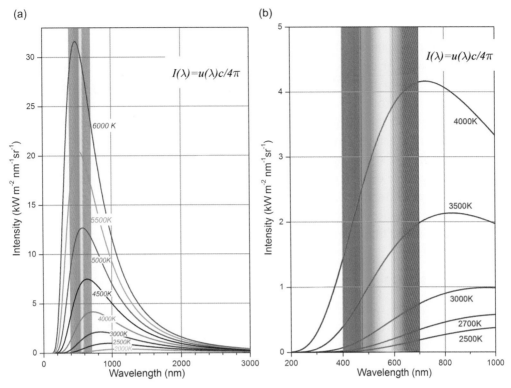

Figure 5.2 Spectral density of black body radiation for different temperatures. (a) Wider spectral and temperature ranges; (b) the visible portion of radiation and the temperature range that is typical for lighting devices.

To get power per unit square we need to calculate the product of $u(\lambda)$ and light speed, c. To calculate power per unit solid angle, we need to add the factor $1/4\pi$. Therefore, one has

$$I(\lambda) = \frac{1}{4\pi} cu(\lambda). \tag{5.5}$$

It is this function that is presented in Figure 5.2. Notably, intensity $I(\lambda)$ has the same shape as energy density $u(\lambda)$, since both functions differ by the constant factor only.

Equations (5.3) and (5.4) are often referred to as the *black body radiation spectrum* since a black body is an ideal system that has continuous emission and absorption spectra and exhibits no reflection. Hot bodies emit electromagnetic radiation quanta in the form of the continuous spectrum, owing to a multitude of transitions involving oscillations of atoms in molecules and solids. Atoms and also quantum dots dispersed in gases, liquids, or solid matrices emit narrow spectral lines with relative intensity meeting the Planck formula.

Higher temperatures give rise to higher intensities of the black body radiation spectrum for all wavelengths, and the maximum emission spectrum shifts to shorter wavelengths (Figure 5.2; see also the Problem 5.2). Therefore, when the temperature rises, the matter at first emits only invisible infrared (IR) radiation; then, it becomes red for $T > 1000$ K, and at temperatures of several thousand kelvin it changes color to yellowish or even white. Notably, solar radiation has a spectrum close to a black body spectrum at 5250 K. Incandescent lamps emit light owing to exclusively thermal emission (in proportion with the familiar I^2R relation from electroengineering, where I is current and R is resistance) and have the spectrum close to the Planck formula, their temperature being in the range 2500–3000 K, depending on power. It is important that their emission spectrum is continuous like the solar one, but the spectral shape is different because of lower temperature. Regretfully, only a very small portion of emitted energy falls into the visible spectrum, typically just a few percent. The visible portion could be enhanced by raising the temperature, but this is not possible because of melting of a metal filament.

5.1.3 Luminescence

Matter can be excited beyond simple heating. For example, one can apply an external light source or electric power to get the number of electrons in the excited states higher than that defined by the Boltzmann relation (Eq. (5.2)). This situation is referred to as non-equilibrium conditions. The parity of upward and downward transitions for every pair of energy levels vanishes, and the matter will now emit extra photons as compared to the thermal radiation spectrum defined by Eqs. (5.3)–(5.5). This extra emission and the variety of processes resulting in extra emission are referred to as *luminescence*. Owing to luminescence, a piece of matter shines brighter than it should at a given temperature. If extra excitation comes from an optical source, then the term *photoluminescence* is used. In the case of electric pumping, one speaks about *electroluminescence*. These two types of excitation are the most common in photonics. Less common are *chemiluminescence* (excitation comes as a result of a chemical reaction) and *piezoluminescence* (excitation is produced mechanically).

Commercial *luminescent lamps* and semiconductor *LED*-based luminaries make use of both electro- and photoluminescence. Bulb-type luminescent lamps all use the electric discharge in *mercury* gas, which efficiently converts electric power into electromagnetic radiation – regrettably mainly in the near-ultraviolet (UV) range. To convert UV radiation further into the visible spectrum, various luminophores are used. In common tubular lamps widely used in offices, workshops, and foundries, and in residential compact fluorescent lamps (CFL), a solid powder is used as a luminophore, whereas in yellowish street lamps sodium vapor is used to convert mercury UV light into a yellow light. White LEDs and LED-based luminaries typically use blue LEDs and a yellow–red solid luminophore to get white light. Notably, spectral shapes of all luminescent light sources differ from the solar spectrum and incandescent lamp spectrum because of pronounced discrete emission lines or bands (Figure 5.3).

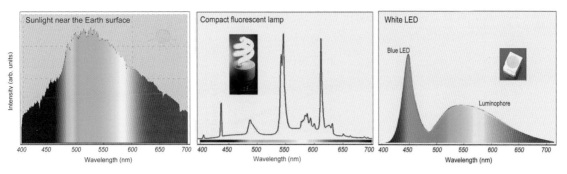

Figure 5.3 Emission spectra of the sun, a compact fluorescent lamp, and a white LED.

5.1.4 Lifetime and quantum yield

Every excited quantum system (an atom, a molecule, a quantum dot, or just a piece of solid matter) after a certain time will appear in the ground, unexcited state. In the simplest case of a two-level system, either induced downward transitions or spontaneous decay results in a system transition from the upper state E_2 to the ground state E_1 (Figure 5.1). Notably, when considering the optical range ($E_2 - E_1 > 1$ eV), one can see that at room temperature and at typical experimental or operation conditions for optical devices, say about 100 °C, the role of stimulated transitions is negligible since power density u is very low to enable efficient downward stimulated transitions whose rate is Bun_2. In optics, under the above conditions the contribution of thermal transitions can be neglected and the dominant way of excited matter relaxation is represented by spontaneous downward transitions. Then the lifetime τ of an excited two-level system will be equal to $1/A$, and the time-dependent population of the upper state $n_2(t)$ will decay exponentially with the factor $-t/\tau$, i.e.,

$$n_2(t) = n_2(0)e^{-t/\tau}, \quad \tau = \frac{1}{A}. \tag{5.6}$$

Rare earth ions used as luminophores and active laser media represent a good example of a real system that corresponds to the simplest two-level presentation with exponential decay of excited states. Since luminescence intensity is proportional to the transition rate, i.e., to the product of $n_2(\tau)A$, their luminescence intensity decays exponentially after excitation is over in accordance with Eq. (5.6). Figure 5.4(a,b) shows the experimental example of Eu^{3+}. These ions emit a narrow line peaking near 620 nm. The decay exhibits with high accuracy a single exponential behavior with lifetime $\tau = 716.3$ μs. Intensity versus time dependence in a semi-logarithmic presentation obeys a straight line with high precision.

Many real systems cannot be represented by a simple two-level scheme because typically a number of paths do exist for a system to come from an excited to a ground state. Spontaneous decay from a given excited state (E_2 in Figure 5.5) is often bypassed by competitive *radiative* transitions to possible intermediate states (E_3 and E_4) as well as by non-radiative transitions. The latter can originate from many events – e.g., in semiconductors

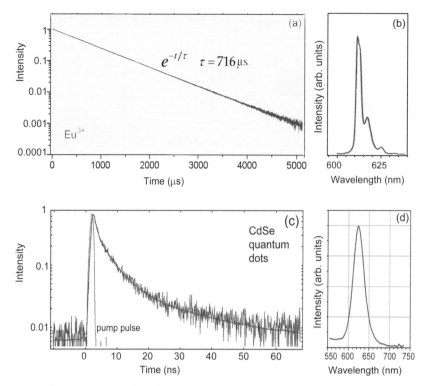

Figure 5.4 Representative photoluminescence kinetics and emission spectra. (a,b) Eu^{3+} ions in a complex with an organic ligand in toluene solution; (c,d) CdSe quantum dots in a solution.

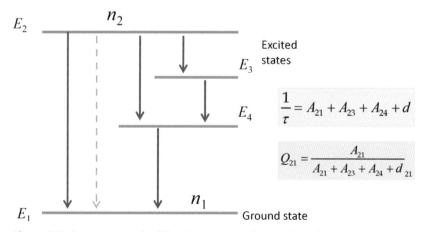

Figure 5.5 Spontaneous (red lines) and nonradiative (gray line) transitions in a multilevel system. Thermal transitions are neglected.

and semiconductor nanostructures *phonon* emission is possible, i.e., transfer of the energy stored in the electron subsystem to the energy of atomic vibrations. A similar phenomenon occurs in molecules, but phonon notion is not applicable. In dense or closely packed molecular aggregates or quantum dot assemblies, various energy transfer processes are possible that can form another nonradiative bypass with respect to $E_2 \rightarrow E_1$ transition.

Then the total probability for excited state decay is the sum of all possible downward transition probabilities and the following relations can be written:

$$\frac{1}{\tau} = A_{21} + A_{23} + A_{24} + d, \tag{5.7}$$

$$\frac{1}{\tau} = \sum_i \frac{1}{\tau_i}, \tag{5.8}$$

where d implies all possible nonradiative processes, and τ_i equals the particular decay time in every decay channel.

Various bypasses in the decay process competing with the transition of interest, $E_2 \rightarrow E_1$ in the case under consideration, give rise to the fact that the number of photons emitted in the channel $E_2 \rightarrow E_1$ will always be lower than the number of photons absorbed in the same channel (transitions $E_1 \rightarrow E_2$). This gives rise to the notion of *quantum yield*,

$$\text{Quantum yield} \quad Q_{21} = \frac{A_{21}}{A_{21} + A_{23} + A_{24} + d_{21}} \tag{5.9}$$

which ranges from 0 to 1 or from 0% to 100%. Quantum yield shows what portion of absorbed photons is transformed into emitted photons for a given emission channel.

For many real systems, decay dynamics of excited states is essentially non-exponential, which means the decay process cannot be described in terms of decay rate or decay probability. This is the case for energy transfer effects depending on distance variations within an ensemble of emitters, and for local environment fluctuations affecting transition rates. In semiconductor quantum dot ensembles, in addition to the above mechanisms, intrinsic decay rate is size-dependent because of size-dependent overlap of electron and hole wave functions. Therefore, photoluminescence decay is typically non-exponential, as shown in Figure 5.4(c,d).

5.1.5 Fluorescence and phosphorescence

Originally these terms were coined in molecular physics to distinguish between photon emission in the course of allowed (short lifetime, high quantum yield) and forbidden (long lifetime, low quantum yield) transitions. In science, it is still the case. Fluorescence usually means the emission of light by molecules owing to the radiative decay process of excited singlet states with the typical intrinsic lifetimes in the range 10^{-8}–10^{-9} s, whereas phosphorescence is used for low-rate emission, typically from triplet states with lifetimes of the order of 10^{-3} s. However, in technology nowadays both notations are widely used as synonyms to luminescence – e.g., electroluminescence in mercury gas combined with phosphorescence in luminophores gives rise to the name of the "compact *fluorescent* lamp." In white LEDs, a luminescent compound adding green, yellow, and red light emission to get white light is typically referred to as a *phosphor*.

5.1.6 Fundamental relation between Einstein coefficients for spontaneous and stimulated transitions

Presentation in terms of discrete upward and downward transitions together with the Planck formula (Eq. (5.3)) gives rise to the basic relation between Einstein coefficients for spontaneous and stimulated transitions,

$$\frac{A}{B} = \hbar\omega\,\frac{\omega^2}{\pi^2 c^3}. \tag{5.10}$$

This relation means that relative "weights" of downward spontaneous transitions versus stimulated ones essentially depend on frequency in the third power. Since stimulated transitions rate Bu depends on external radiation intensity u, the latter should also be taken into consideration. At low radiation intensities, when the population of an excited state is much lower than the population of the ground state, only frequency matters. For low frequencies of electromagnetic radiation, i.e., in the radiofrequency range, stimulated downward transitions typically dominate over spontaneous ones. In the optical range high frequencies enable spontaneous emission to dominate over stimulated downward transitions. High radiation intensity can change these relations. At high intensities (as will be considered in Chapter 6), stimulated downward transitions in optics can dominate, resulting in optical gain.

5.2 PHOTON DENSITY OF STATES

Let us extend the notion of the density of states (see Section 3.1) to photons. Considering standing waves in a cavity, we arrived at a conclusion that the number of modes in the range of wave numbers $[k, k + dk]$ takes the following form, depending on the dimensionality of space indicated by subscripts (see Eq. (3.7)):

$$dN_3 = D_3(k)dk = \frac{k^2}{2\pi^2}\,dk, \;\; dN_2 = D_2(k)dk = \frac{k}{2\pi}\,dk, \;\; dN_1 = D_1(k)dk = \frac{1}{\pi}\,dk. \tag{5.11}$$

These relations are independent of the wave type under consideration. They give rise to the notion of mode density, which is a universal function for all waves for the wave number scale,

$$D_3(k) = \frac{k^2}{2\pi^2}, \;\; D_2(k) = \frac{k}{2\pi}, \;\; D_1(k) = \frac{1}{\pi}. \tag{5.12}$$

The density of modes depends also on every value related to k, i.e., on wave frequency ω and wavelength λ. Based on the fact that density of modes is the derivative, dN/dk, and recalling the mathematical rule for derivative of a complicated function, one has for other scales the simple rules

$$D(\lambda) = D(k)\frac{dk}{d\lambda}, \quad D(\omega) = D(k)\frac{dk}{d\omega}. \tag{5.13}$$

For electromagnetic radiation $D(\omega)$ is important. One can see that it essentially depends on the dispersion law $\omega(k)$. Recalling $\omega = ck$ for a vacuum (Eq. (4.8)), one arrives at the relation

$$D_3(\omega) = \frac{\omega^2}{2\pi^2 c^3}. \tag{5.14}$$

Taking into account that electromagnetic waves are transversal, the factor of two should be added to account for the two orthogonal polarizations possible. Then, we finally have

$$D_3(\omega) = \frac{\omega^2}{\pi^2 c^3} \quad \text{for EM-waves in a vacuum}. \tag{5.15}$$

Now we can see that this term directly enters the Planck formula (Eq. (5.3)) and Einstein relation for spontaneous emission rate (Eq. (5.10)). Actually, it was first suggested by Rayleigh in 1900 that counting modes can help to describe the experimental black body radiation spectrum.

Implying photons as light quanta and recalling the basic expression for photon momentum, $p = \hbar k$, one can immediately derive the photon density of states over momentum from $D_3(k)$ in Eq. (5.12) by applying, $D(p) = D(k)\frac{dk}{dp}$:

$$D_3(p) = \frac{4\pi p^2}{h^3} \quad \text{for photons in a vacuum}. \tag{5.16}$$

As was discussed in Section 3.1, the same relation can be derived without counting modes by implying that phase space consists of elementary h^3 cells. The idea of elementary cells conforms to the quantum mechanical uncertainty relation and actually was suggested by Max Planck well prior to the uncertainty relation principle. We can see that the consideration of modes in cavities directly results in Eq. (5.16) without any additional suppositions.

An important element is the conceptional transition from electromagnetic modes to light quanta. It is possible to perform a consistent consideration of *black body radiation as a gas of photons* in equilibrium (Bose 1924). First, we have to count the number of states per unit volume per unit frequency interval, i.e., $D(\omega)$. Next, we need to consider the occupation function

$$F(\hbar\omega) = \frac{1}{e^{\frac{\hbar\omega}{k_B T}} - 1} \tag{5.17}$$

derived by Bose as a common property of quantum particles for which occupation number of a given state can be more than 1 (these are called *bosons*). Then one can see that the product

$$F(\hbar\omega)D_3(\omega) = \frac{1}{e^{\frac{\hbar\omega}{k_B T}} - 1} \; \frac{\omega^2}{\pi^2 c^3} \qquad (5.18)$$

gives the number of photons in the range of $d\omega$. Then by accounting for energy $\hbar\omega$ carried by every photon we immediately arrive at the Planck formula (Eq. (5.3)).

BOX 5.1 BOHR, EINSTEIN, DIRAC, AND PURCELL

Niels Bohr (1913): An atom emits light with frequency ω when coming from one steady state to another, and frequency meets the condition $\omega = (E_2 - E_1)/\hbar$.

Albert Einstein (1916): Spontaneous emission rate is proportional to $\hbar\omega \cdot \omega^2/(\pi^2 c^3)$.

Paul Dirac (1927): $\omega^2/(\pi^2 c^3)$ is the density of electromagnetic modes.

Edward Purcell (1946): $\omega^2/(\pi^2 c^3)$ should be replaced by explicit calculation for every case other than a vacuum.

Niels Bohr	Albert Einstein	Paul Dirac	Edward Purcell

Let us now come back to the Einstein relation for spontaneous emission probability (Eq. (5.10)). Comparing it with Eq. (5.15), one can see that the *spontaneous emission probability of a photon with energy $\hbar\omega$ by an excited quantum system is directly proportional to the density of photon states for the given energy or frequency*. This is in agreement with the more general consideration in quantum mechanics which states that *the probability of a quantum transition is proportional to the number of ways this transition can be performed*. More accurately, a transition of a given electron from the valence band (v-band) to the conduction band (c-band) in semiconductors for a given energy is proportional to the density of final states in the c-band. *Vice versa*, a transition probability of an electron from the c-band to the v-band is proportional to the density of electron states in the v-band. Accordingly electron scattering within the c-band will always be proportional to the density of final electron states as well. This important property was highlighted by E. Fermi in 1950 as the golden rule and is often referred to as the *Fermi golden rule*. The statement "Spontaneous photon emission rate is proportional to photon density of states" nowadays is often referred to as the Fermi golden rule as well, though Enrico Fermi never applied this notation to spontaneous emission of photons.

Quantum consideration in terms of photons as light quanta, the conception of photon density of states, and photon creation in the course of spontaneous downward transitions of atoms, molecules, and solids constitute the basic principles of quantum electrodynamics. However, in engineering, many prefer to consider emission of light in terms of mode density. Then, one can say that *spontaneous emission of light has the rate which is proportional to the mode density for the given frequency*.

The concept of mode density is possibly more instructive and apparent than photon density of states. Energy can be taken by an electromagnetic field from a piece of matter by putting it into a certain mode. Space should enable radiation in this mode to exist. Otherwise, energy cannot go from the matter to the field.

5.3 THE PURCELL EFFECT

Edward Purcell, an American physicist, was the first to realize (in 1946) that spontaneous emission probability can be controlled by means of space geometry engineering to enhance electromagnetic mode density. He was challenged by the problem that in radiophysics spontaneous transitions have very, very low probability and stimulated transitions typically dominate, in accordance with Eq. (5.10). He was the first who understood that spontaneous emission is governed by the properties of space. Changing geometry of space around an emitter on a wavelength scale will result in changing the mode density (photon density of states), enabling control over spontaneous emission rate for electromagnetic radiation and simultaneously control over lifetimes of excited atoms, molecules, and solids. Therefore, *the Einstein coefficient A for spontaneous decay rate is not the entirely intrinsic property of a quantum system in question (an atom, a molecule, or a piece of solid)* but essentially it is the parameter defined by the space property in the sense of mode propagation enabling/inhibition. Purcell suggested that for an atom in a cavity, the cavity resonant mode will dominate in the mode spectrum, and therefore, to account for the modified density of modes one has to consider that spontaneously emitted power W (joules per second per cubic meter) will be enhanced by the factor

$$\frac{W_{\text{cavity}}}{W_{\text{vacuum}}} = \frac{3}{4\pi^2} \frac{\lambda^3}{V} Q, \tag{5.19}$$

which is now referred to as the *Purcell factor*. Here, λ is emission wavelength, V is the actual volume the mode occupies, and Q is the quality factor of the cavity for this mode. Enhancement of spontaneous decay rate by the Purcell factor will occur provided the frequency of atomic transition belongs to the set of cavity modes.

The general phenomenon of spontaneous emission rate control by means of electromagnetic waves confinement is now referred to as the *Purcell effect*. In fact, because of the confined space volume of a cavity, the Purcell factor gives an immediate hint to the notion of the photon *local density of states* (LDOS) in the case of discontinuous space with various topological singularities. For photon LDOS, Q-factor is critical. Furthermore, as Purcell outlined, the ratio of λ^3 versus actual volume of a localized mode V contributes as well. One can see this is a straightforward consequence of electromagnetic radiation confinement. The λ^3/V factor is essential for surface modes in microspheres. Recalling that Q-factor describes the ability of a cavity to store energy, one can see that enabling electromagnetic energy storage and electromagnetic wave confinement in a certain portion of space gives rise to enhancement of spontaneously emitted electromagnetic radiation (and accordingly to decrease of radiative lifetime).

What will happen in the case of a cavity resonant frequency ω_c detuning with respect to the atomic transition frequency ω_a? F. Bunkin and A. Oraevsky (1959) extended the Purcell formula to the case of arbitrary atomic transition frequency and arrived at the conclusion that *strong detuning will result in Q-times inhibited spontaneous decay.* Their factor reads

$$\frac{W_{\text{cavity}}}{W_{\text{vacuum}}} \propto \frac{Q}{\dfrac{(\omega_a - \omega_c)^2}{\omega_c^2}Q^2 + 1}. \tag{5.20}$$

One can see at resonance ($\omega_a \rightarrow \omega_c$) Eq. (5.20) gives a Q-fold increase of spontaneous decay rate, whereas detuning results in a steady decrease of W_{cavity}. For example, for $(\omega_a - \omega_c) = 0.5\omega_c$ and $Q \gg 1$, one has $W_{\text{cavity}}/W_{\text{vacuum}} = 4/Q$. Figure 5.6 shows the frequency dependence of a cavity-enhanced spontaneous emission rate. Note that enhancement coexists with inhibition (the right-hand panel).

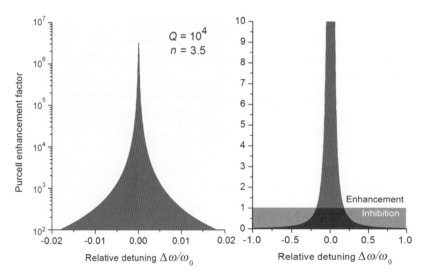

Figure 5.6 Purcell enhancement factor for spontaneous emission probability of an excited atom or molecule in a cavity. Note logarithmic and linear scales for left and right panels.

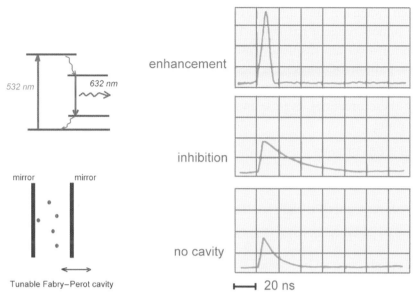

Figure 5.7 First experimental observation of the Purcell effect for organic molecules in a tunable Fabry–Pérot cavity. Adapted figure with permission from De Martini et al. (1987). Copyright 1987 by the American Physical Society.

Experimental evidence for the Purcell effect in the optical range was for the first time observed in Roma University by De Martini and co-workers in 1987 (Figure 5.7). Molecules were excited by a green laser pulse with a duration of 2 ns and exhibited luminescence in the red range (632 nm) with either shorter (upper oscilloscope trace) or longer (middle trace) decay with respect to the free space (lower trace). One of the mirrors was movable by means of a piezo-positioner to make the cavity tunable. Remarkable is that both enhancement (acceleration) and inhibition (slowing down) of the decay rate have been observed. Note that the apparent decay rate in the upper pulse shape is most probably limited by the oscilloscope time resolution, which is evident from longer buildup times as compared to the actual excitation pulse duration.

To summarize, once confinement of electromagnetic radiation is possible in a certain portion of space (a cavity), then spontaneous emission of photons by excited matter can be either enhanced or inhibited, depending on perfect tuning or detuning of the transition frequency versus cavity resonance. The quick estimate of enhancement/inhibition factor is the cavity's ability to store electromagnetic energy expressed by the Q-factor.

5.4 PHOTON DENSITY OF STATE EFFECTS IN OPTICS

Modification of electromagnetic mode density (photon density of states; DOS) occurs not only in cavities but in every medium or nanostructure that has spatially dependent dielectric permittivity. In other words, modification of photon DOS is the case for every medium differing from a vacuum. Only air possesses the unperturbed DOS expressed by Eq. (5.15) since in the optical range it has $\varepsilon = 1$. In other words, every time electromagnetic wave propagation for a certain frequency (wavelength) experiences perturbation with respect to a vacuum, photon DOS has to be calculated explicitly to examine modification of the optical process related to photon emission.

What are these processes, events, and phenomena that are modified owing to DOS effects? They are:

1 thermal emission of hot bodies (intensity and directionality);

2 spontaneous emission of photons (both rate and directionality);

3 spontaneous resonant (Rayleigh) scattering of photons (rate and directionality);

4 spontaneous non-resonant (Raman) scattering of photons (rate and directionality).

The general conclusion can be reached that *every time a photon is emitted or scattered, the probability of this event will be proportional to the density of photon states.*

When considering photon DOS effects, one has to distinguish between local and integral DOS. A cavity represents a typical example of *local* DOS modification. A homogeneous medium with permittivity differing from $\varepsilon = 1$ represents the case of modified *integral* DOS. The same is valid for inhomogeneous but evenly modified permittivity inherent in photonic crystals. Photonic crystals feature zero photon DOS in the gap, enhanced DOS near the gap, and a DOS modification factor tending to constant value with respect to Eq. (5.15) for longer wavelengths when the effective medium approach is applicable.

Modification of thermal emission is the primary effect in complex media and nanostructures and can be purposefully used to improve obsolete incandescent light sources, though it is not so easy to integrate a tungsten filament into a microcavity or a photonic crystal medium. Spontaneous emission of photons is the second case of DOS effects which can be purposefully used in all luminescent light sources, first of all in solid-state lighting since, again, it is not easy to add cavities or photonic crystals to mercury or sodium bulbs.

Further cases are resonant (elastic) and non-resonant (inelastic) photon scattering. The former is a familiar phenomenon and is discussed in detail in Section 4.9. It occurs every time a wave meets an obstacle in its propagation. The latter is more complicated and results essentially from light interaction with molecules and solids. This interaction gives rise to a shift of light frequency up or down by the frequency of intrinsic vibrations of atoms in molecules and solids. It is said that the matter is virtually excited while interacting with the electromagnetic field and then emits a photon that can have the same energy (elastic, Rayleigh) scattering or up- or downshifted by a certain value in the case of the inelastic process.

Notably, resonant scattering can be fully described by means of classical wave physics and does not need the notion of light quanta (photons) to be involved. However, the generalized conception of photon scattering including both elastic and inelastic events is notable for the presentation of light scattering being consistent. Rayleigh was the first to explain the regularities in lightwave scattering in the air in terms of ω^4-like dependence, which actually has a contribution from DOS in the form of the ω^2-like dependence. It is this dependence that defines bluish sky (we see the high-frequency portion of sunlight scattered by air inhomogeneities) and reddish sunrise and sunset (we observe the transmitted low-frequency portion of sunlight that has been less scattered by air).

Inelastic scattering of light is essentially a quantum phenomenon since neither classical process can change energy (frequency) of a photon. As we discussed in Chapter 4 (Figure 4.1 and comments on it), we cannot move a photon along its dispersion curve. In quantum electrodynamics, photon scattering is seen as emission of a new photon whose energy differs from that of the incident one by the portion of energy subtracted or added from the matter (more accurately, from atomic vibrations). Inelastic light scattering was discovered in 1928 by C. V. Raman and K. S. Krishnan in India for molecules, and G. S. Landsberg and L. I. Mandelstam in the USSR for crystals. It is referred to as *Raman scattering.*

Continuous and regular *media* – e.g., photonic crystals – can be described by the integral density of states. However, local discontinuities like microdroplets in a continuous medium, cavities in a photonic crystal, or single or assembled dielectric or metal nanobodies that disturb local dielectric permittivity are to be characterized by means of LDOS. There is a reasonable approach to definition for photon LDOS suggested by D'Aguano et al. (2004). It is considered that a probe classical dipole emitter of electromagnetic radiation should be examined at a given point of interest whose radius vector is $\mathbf{r}$ at a given frequency ω. Then, it is anticipated that the photon LDOS will be modified at that point and frequency in accordance with calculated or measured modification of electromagnetic radiation emission rate for this probe dipole with respect to a vacuum.

There exists a special type of functions, *Green's functions*, that are used in calculations. Green's functions play an important role in electrodynamics. In a general case, $\mathbf{G}(\mathbf{r}, \mathbf{r}')$ is a tensor defining the field $\mathbf{E}(\mathbf{r})$ at a point $\mathbf{r}$ from a radiating dipole $\boldsymbol{\mu}$ located at the point $\mathbf{r}'$,

$$\mathbf{E}(\mathbf{r}) = \omega^2 \mu_0 \mu \mathbf{G}(\mathbf{r}, \mathbf{r}')\boldsymbol{\mu}. \tag{5.21}$$

For free space it takes the scalar form

$$G_0(\mathbf{r}, \mathbf{r}') = \frac{\exp(\pm ik|\mathbf{r} - \mathbf{r}'|)}{4\pi|\mathbf{r} - \mathbf{r}'|}. \tag{5.22}$$

In terms of Green's function, the emission rate of a classical dipole can be calculated as

$$W(\mathbf{r}, \omega) = \frac{2\omega^2}{\hbar} \mu_i \mu_j \, \mathrm{Im}\, G_{ij}^T(\mathbf{r}, \mathbf{r}, \omega) \tag{5.23}$$

where $G_{ij}^T(\mathbf{r},\mathbf{r},\omega_0)$ is the transverse Green's function defined by the solution of the partial differential equation,

$$-\left(\nabla^2 + \frac{\omega^2}{c^2}\varepsilon(\mathbf{r},\omega)\right)G_{ij}^T(\mathbf{r},\mathbf{r}',\omega) = \frac{1}{\varepsilon_0 c^2}\delta_{ij}^T(\mathbf{r}-\mathbf{r}') \qquad (5.24)$$

and $\varepsilon(\mathbf{r},\omega)$ is the medium complex dielectric function at point $\mathbf{r}$, $\delta_{ij}^T(\mathbf{r}-\mathbf{r}')$ being the transverse part of the delta function. The reader is referred to the work of Novotny and Hecht (2012) for more detail.

Based on the above operational definition for LDOS and the recipe for a classical dipole emission rate (Eq. (5.21)), we can write the general expression for the photon LDOS calculation as

$$D(\mathbf{r},\omega) \equiv D_{\text{vacuum}}(\omega)\frac{W(\mathbf{r},\omega)}{W_{\text{vacuum}}(\omega)} = D_{\text{vacuum}}(\omega)\frac{6\pi c}{\omega}\operatorname{Im}G(\mathbf{r},\mathbf{r},\omega),\ D_{\text{vacuum}}(\omega) = \frac{\omega^2}{\pi^2 c^3}. \qquad (5.25)$$

To summarize, modification of the photon LDOS, which allows for calculation of *photon emission rate by a quantum system* at a given point in any type of inhomogeneous space, is taken to be equal to the modification of the electromagnetic emission rate by a *classical* oscillator placed at the point of interest with respect to the rate in a vacuum. This notable circumstance is worthy of a closer discussion. Taking *de fide* the concept of photon emission in the course of *quantum jumps*, we use *classical* equations to calculate modification of transition rate (probability) in a space with complicated topology. In other words, while the *classical* electrodynamics cannot *explain* the event of a photon emission in the course of atomic or molecular transition, it is capable of offering the *computational technique* for its rate calculation. This convergence of quantum and classical electrodynamics is discussed in more detail in Section 5.10 when speaking about nanoantennas.

5.5 THE BARNETT–LOUDON SUM RULE

One can see from Eq. (5.20) that when an atom or another quantum emitter is placed in a cavity, its spontaneous decay rate (and radiative lifetime) can be either enhanced or inhibited depending on the spectral position of atom transition frequency versus cavity resonance frequency. Therefore, enhancement of decay with respect to a vacuum in certain spectral ranges is accompanied by inhibition otherwise. This is a manifestation of the very general property. Departing from a vacuum to inhomogeneous space will generally result in mode density spectral redistribution in a way that all positive changes will be fully compensated by all negative ones provided the whole spectral range from $\omega = 0$ to $\omega \to \infty$ is considered. S. Barnett and R. Loudon (1996) proved the general theorem that formulates the sum rule for modified spontaneous emission probabilities. It states that in a given point of space, modification of spontaneous emission rate of a dipole emitter in a certain frequency range will be necessarily compensated by the opposite modification otherwise. Precisely, modified rates are constrained by an integral relation for the relative emission rate modification

$$\int_0^\infty \frac{W(\mathbf{r},\omega) - W_{\text{vacuum}}(\omega)}{W_{\text{vacuum}}(\omega)} d\omega = 0. \tag{5.26}$$

Here, $W(\mathbf{r},\omega)$ is the emission rate of an atom or molecule at position $\mathbf{r}$ as modified by the environment, whose effect is assumed to vary by a negligible amount across the extent of the emitting object.

The Barnett–Loudon sum rule means the photon density of states (and electromagnetic mode density) should meet the same conservation law. Namely, departure from a vacuum to structured space will result in deviation of density of states (modes) from the vacuum value defined by Eq. (5.15) in a way that overall changes integrated over $0 < \omega < \infty$ will give zero. This important property is seen in Figure 5.6 for a cavity and is additionally illustrated in Figure 5.8 for a photonic crystal. Not only dips occur in the density of states versus vacuum curve $D(\omega) \propto \omega^2$, but also multiple frequency intervals

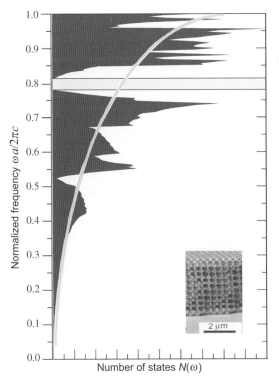

Figure 5.8 Calculated photon density of states for a 3D photonic crystal whose design and band structure were presented in Figure 4.20(c,d). The pink band indicates the band gap. The yellow line is ω^2-like dependence inherent in a vacuum. Adapted from Busch et al. (2007), with permission from Elsevier.

with enhancement of the density of states are apparent. In the low-frequency limit one has $D(\omega) \propto \omega^2$, i.e., $D(\omega)$ reduces to a vacuum since this corresponds to the effective medium case because a photonic crystal for waves much longer than its period looks like a homogeneous medium.

5.6 MIRRORS AND INTERFACES

In front of a mirror, an atom, a molecule, or a tiny piece of solid (whose size should be much smaller than the light wavelength), e.g., a quantum dot, a nanorod, or a nanoplatelet, will necessarily change spontaneous emission rate, lifetime, and angular distribution of emitted radiation in accordance with electromagnetic mode density modification. This phenomenon is shown in Figure 5.9. Note the oscillating character of lifetime deviation from its vacuum value, which is reached at a distance larger than the wavelength of emitted light (620 nm). The theory reasonably describes the observed behavior. Proximity of a metal surface to the experiment shown in Figure 5.9 gives rise to luminescence quenching at distances shorter than the electron mean free path in the metal. This is the reason for rapid lifetime decrease for distances shorter than 40 nm. In the case of a dielectric multi-layered periodic mirror, this quenching will not occur.

Every interface of two dielectric media with different permittivities represents a reflecting border. This border also changes mode density and lifetimes. Figure 5.9 shows the calculated radiative rate near the border of dielectrics, one of which is a vacuum or air with $\varepsilon = 1$, and another features higher permittivity up to 10 (e.g., a wide-band semiconductor like ZnO or GaN). One can see that at distances comparable to or shorter than

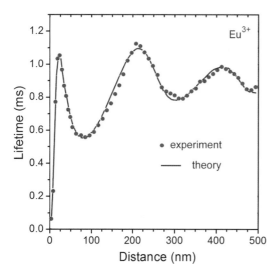

Figure 5.9 Radiative decay of an emitter in front of a mirror. Lifetime of Eu^{3+} ions in front of an Ag mirror as a function of separation between the Eu^{3+} ions and the mirror and the theoretical dependence obtained within the classical electrodynamics. Reprinted figure with permission from Amos and Barnes (1997). Copyright 1997 by the American Physical Society.

the wavelength, the emission rate oscillates similarly to the data presented for a mirror (Figure 5.9). The period of oscillations equals half of the wavelength in the medium under consideration, the amplitude of oscillations rising up for higher permittivity. The effect vanishes at distances considerably exceeding the wavelength value. Note that oscillations occur with respect to radiation rate (lifetime) in a medium, not in a vacuum. In every medium with refractive index $n > 1$, mode density rises since wavelength falls down in proportion with $n = \varepsilon^{1/2}$.

At first glance, when comparing density of states in a dielectric with refractive index n versus vacuum, simple substitution of c/n instead of c in Eq. (5.15) leads to n^3 enhancement for radiative decay rate, i.e., $W_{rad}(n) = n^3 W_{rad}^{vacuum}$. However, to account for the LDOS effect properly, one needs to consider the local field correction factor that accounts for the difference between the field at the emitter position and the field in the ambient medium. This factor arises owing to the presence of a small but finite-size emitter that changes the local electric field of an electromagnetic wave. Then, one has to deal with the following expression:

$$W_{rad}(n) = \left(\frac{F_{loc}^2(n)}{n^2} \right) n^3 W_{rad}^{vacuum} = F_{loc}^2(n) n W_{rad}^{vacuum}. \qquad (5.27)$$

The simplest way to account for the local field factor is to use the known expression from classical electrodynamics. Under the impact of the external field, the internal field inside a particle reads

$$E_{int} = E_0 \frac{3\varepsilon_{out}}{\varepsilon_{in} + 2\varepsilon_{out}}, \qquad (5.28)$$

where ε_{in} is the particle dielectric permittivity and ε_{out} is the host medium dielectric permittivity. Therefore, combining Eqs. (5.27) and (5.28) and assuming a purely dielectric, lossless case ($\varepsilon_{out} = n^2$, $\varepsilon_{in} = n_{in}^2$), one can write

$$W_{rad}(n) = \left(\frac{3n^2}{n_{in}^2 + 2n^2} \right)^2 n W_{rad}^{vacuum}. \qquad (5.29)$$

An immediate example is the modified lifetime of semiconductor quantum dots in a dielectric matrix. Semiconductor quantum dots represent an interesting example of artificial atom-like objects with size-dependent spectra of light absorption and emission as well as decay rates. It is possible to further control their luminescent properties by means of environmental dependence of the intrinsic radiative lifetime. In Chapter 4 (Table 4.1), the general trend was outlined for dielectric properties of various materials. Namely, higher refractive indices are inherent in narrow-band materials that are transparent only well into the IR range. However, the same materials feature very small effective mass of electrons, thus resulting in big shortwave shifts of the absorption edge. Therefore it is possible, starting from narrow-band original crystals, by means of nanometer-size restrictions to

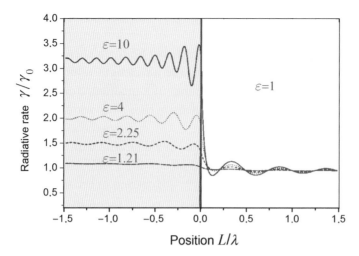

Figure 5.10 Calculated decay rate of an atom near a dielectric vacuum interface for the four different values of dielectric permittivity. The dielectric area is shaded pink. Reprinted from Cho (2003) with permission of Springer.

move their transparency range significantly toward the visible (defined by electron and hole confinement) while having a refractive index close to that of the bulk parent crystal. It becomes possible as, far enough from the sharp absorption in the spectrum, the refraction index is defined mainly by the crystal lattice – e.g., narrow-band semiconducting crystal PbSe possesses rather high permittivity, $\varepsilon = 23$, which gives the modified decay rate.

Assuming $\varepsilon_{\text{out}} = 1$ and using Eq. (5.29), one has the decay rate $W_{\text{rad}}(\varepsilon) = 0.014 W_{\text{rad}}^{\text{vacuum}}$ without any purposeful structuring of the matter on the light wavelength scale! Assuming $\varepsilon_{\text{out}} \rightarrow \varepsilon$ (quantum dot solids), one has the upper limit $W_{\text{rad}}(\varepsilon) \rightarrow W_{\text{rad}}^{\text{vacuum}}$. One can see the dielectric environment changes lifetime about 70 times! And it actually manifests itself experimentally (Allan and Delerue 2004).

It is well known from numerous experiments that lifetimes of atoms and molecules also exhibit radiative lifetime dependence on solvent or matrix refractive indices. In many cases the approximate effect of the surrounding matrix can be estimated using Eq. (5.29), with $n_{\text{in}} = 1$.

At the interface of the two dielectric media, lifetime oscillates with distance from the interface. A representative example is shown in Figure 5.10. From both sides of the interface at great enough distance the rate tends to the constant value $W_{\text{rad}}(n) = n W_{\text{rad}}^{\text{vacuum}}$, whereas in close vicinity of the interface the rate oscillates around those values. Here, the local field correction factor was assumed to be equal to unity since the purpose of the calculations was to reveal the interface effects on decay rate.

Note that Figures 5.9 and 5.10 present the data on lifetime modification, and the oscillating character of the observed effects should not be treated as interference. *Neither interference phenomenon can change the excited state lifetime of an atom or a molecule.* However, interference phenomena in front of a mirror or at an interface change the electromagnetic wave propagation conditions and therefore modify the density of modes (photon states),

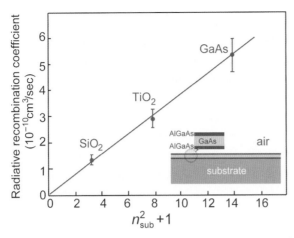

Figure 5.11 Modification of spontaneous decay rate for a 500 nm-thick GaAs slab depending on substrate material refractive index. Adapted figure from Yablonovitch et al. (1988). Copyright 1988 by the American Physical Society.

providing a hint of lifetime modification. Therefore one can say that there is correlation between various interference phenomena in light propagation and lifetime modification. Modification of lifetimes at the interface or in a thin dielectric film has been observed in numerous experiments and had even been applied to monitor the depth profile of fluorescent molecule displacement in certain biosystems.

Figure 5.11 presents an instructive experimental example from semiconductor photonics. A double heterostructure consisting of a 500 nm-thick GaAs active layer and two AlGaAs barriers has been examined on substrates with different refractive indices. A special technique has been elaborated to transfer the heterostructure from the original substrate used in epitaxy to another one. The radiative recombination rate has been monitored versus substrate type, $B_r np$ (number of events per cubic centimeter per second), where B_r is the radiative recombination coefficient, and n (p) is electron (hole) concentration.

Yablonovitch et al. (1988) considered the case of radiative decay rate modification for a thin (thickness much less than emission wavelength) and thick (thickness exceeds wavelength) layer with refractive index n_{int} placed inside a medium with refractive index n. These authors found that for the thin layer the spontaneous emission rate should scale as

$$W_{rad}(n) \propto \frac{n}{n_{int}} W_{rad}^{vacuum}, \tag{5.30}$$

and for the case of a thick layer,

$$W_{rad}(n) \propto \frac{n^2}{n_{int}^2} W_{rad}^{vacuum}. \tag{5.31}$$

In the experiment shown in Figure 5.10 there is a substrate ($n = n_{sub}$) from one side of a thick film and air ($n = 1$) from another side. So, considering the average of two contributions, one has

$$W_{rad}(n) \propto \frac{1}{2}\left(\frac{n_{sub}^2}{n_{int}^2} + \frac{1}{n_{int}^2}\right)W_{rad}^{vacuum}. \quad (5.32)$$

The latter has been remarkably pronounced in the experiment, with three different substrates, as is seen in Figure 5.11.

5.7 MICROCAVITIES

There are numerous experimental evidences for enhancement of spontaneous emission in planar microcavities, spherical microcavities – often referred to as photonic dots – and also in cavities developed inside a photonic crystal. The pioneering example of lifetime modification was shown in Figure 5.7. Further to that example, it is important to highlight the following:

1 Not only does lifetime go down, but experimentally measured luminescence intensity rises accordingly.

2 The emission spectrum is squeezed into the available cavity modes.

3 The radiation pattern and spectrum changes in accordance with the angular-dependent properties of a cavity in the case of a planar microcavity.

Experimentally documented 50-fold enhancement in photoluminescence intensity is presented in Figure 5.12, along with the apparent spectrum squeezing. The example is from erbium in silica, which is actually the active material of optical communication amplifiers.

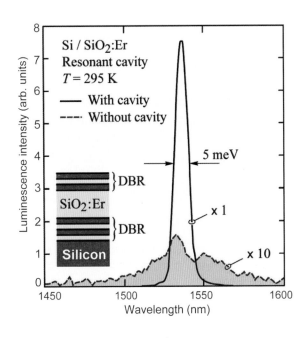

Figure 5.12 Photoluminescence spectra of Er ions in silica in and out of a planar cavity composed by two multilayer mirrors, so-called distributed Bragg reflectors (DBRs). Adapted from Schubert (2006).

Therefore, one can see that the basic phenomenon of spontaneous decay modification of matter by a cavity allows for straightforward experimental implementation in terms of luminescence enhancement.

Notably, real cavities, even planar ones, are essentially 3D structures. Therefore to calculate modification of spontaneous emission one has to involve not only cavity resonant modes that are normal to the planar mirrors, but oblique ones as well. A portion of oblique modes may offer efficient additional emission channels by means of waveguiding. Therefore, experimental results are not as impressive as the simple Eqs. (5.19) and (5.20) predict, but nevertheless do demonstrate the solid base the theory suggests for photonic engineering.

5.8 PHOTONIC CRYSTALS

In a photonic crystal, there are no propagating modes within the band gap and therefore photon density of states rapidly drops to zero at the band edges. However, near the band edges photon DOS noticeably exceeds the vacuum value (see Figure 5.7) so that the Barnett–Loudon sum rule is met. Therefore, the spontaneous emission rate in a photonic crystal will be either inhibited or enhanced depending on spectral position with respect to the band edge(s). This is illustrated in Figure 5.13, where calculations made for an atom in an infinite photonic crystal are shown. One can see that far from the band edge the decay rate tends toward its vacuum value.

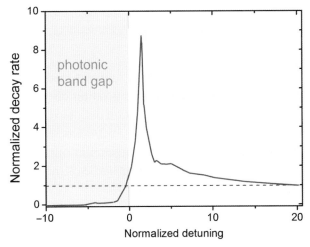

Figure 5.13 Calculated decay rate within the assumption of the perturbative approach of a two-level quantum system in a photonic crystal normalized with respect to vacuum rate versus normalized dimensionless detuning of the transition frequency with respect to photonic band gap edge. The dashed line shows the vacuum rate. Adapted from Lambropoulos et al. (2000). © IOP Publishing. Reproduced with permission. All rights reserved.

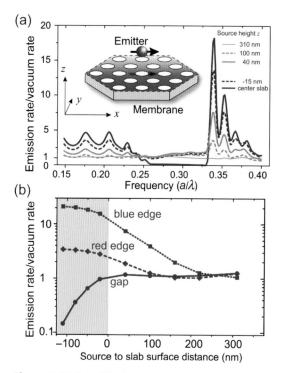

Figure 5.14 Modified spontaneous emission rate of a model dipole emitter near the surface of a 2D photonic crystal membrane. (a) Emission rate normalized to the vacuum rate versus normalized frequency (a is the crystal period) for an x-oriented dipole at the central hole of a photonic crystal membrane. (b) Emission rate modification as a function of the height of a dipole above the PC membrane. Diamonds, circles, and squares show data for frequencies below, in, and above the gap, respectively. The shaded region shows the range of positions in the membrane. Adapted from Koenderink et al. (2005), with the permission of OSA Publishing.

In the case of a photonic crystal bordering a continuous ambient medium, the effect of the photonic crystal on spontaneous emission rate versus position of a probe emitter with respect to photonic crystal surface is similar to the situation discussed in the previous paragraph. Calculations for this case are presented in Figure 5.14. One can see that, by and large, the situation presented in Figure 5.14 is valid for an emitter outside a photonic crystal slab vanishing at a distance of about 300 nm. The bluish side of the gap features a stronger effect on lifetime than the reddish side.

The experimental demonstration of the photonic band gap effect on spontaneous emission rate is presented in Figure 5.15. Semiconductor crystalline material GaInAsP exhibits luminescence near 1.5 μm and is used to fabricate photonic crystal slabs with various periods from 350 to 500 nm to develop photonic band gaps in various spectral ranges with respect to the intrinsic luminescence spectrum. One can see that every case in which the emission spectrum falls inside the band gap, the spontaneous decay curve shows pronounced slow-down, which is evidence of inhibition of spontaneous emission. In Figure 5.15(b), emission spectra of slabs are shown measured in the vertical direction.

Notably, when the decay exhibits slowing down, luminescence intensity detected in the vertical direction rises. This is an important result indicating modification of the angular pattern of emitted radiation. It reflects an angular redistribution of photon density of states. Density of states for frequency range within the band gap tends to zero in the sample plane, but at the same time enhances otherwise. This phenomenon correlates with angular-dependent transmission. Transmission is extremely low within the sample plane because of high reflection (stop band) and is high along the direction normal to the sample surface. This is illustrated in Figure 5.15(d).

Another issue worthy of a closer look is luminescence kinetics. Since luminescence intensity exhibits angular redistribution following photon density of states' redistribution, will kinetics be different for different detection angles? The answer is "no." Luminescence

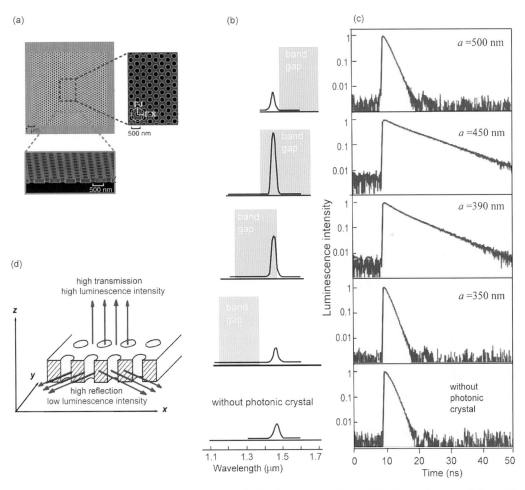

Figure 5.15 Modification of spontaneous emission in semiconductor 2D photonic crystal slab. (a) Image of photonic crystal slab; (b) luminescence spectrum in the vertical direction superimposed over the band gap position; (c) luminescence decays; (d) anisotropic emission patterning. Reprinted with permission from Macmillan Publishers Ltd.; Noda et al. (2007).

kinetics monitors the number of excited emitters, which reduces with time owing to photon emission into all possible modes. Therefore, kinetics do not depend upon the emission channel chosen for monitoring the decay process.

5.9 NANOPLASMONICS

In Chapter 4 we saw multiple examples of local concentration of the incident electromagnetic field near metal nanobodies. Spatial concentration of electromagnetic energy is a hint toward enhanced density of photon states. Therefore, the spontaneous emission rate should also rise for an emitter near a metal nanobody. Analysis and experimental implementation of this phenomenon for sensors and light-emitting devices constitutes an important part of nanoplasmonics.

The modified radiative rate of a dipole near a metal nanobody can be calculated as

$$\frac{\gamma_{rad}}{\gamma_0} = \frac{|\mathbf{d}_0 + \delta\mathbf{d}|^2}{|\mathbf{d}_0|}, \tag{5.33}$$

where $\mathbf{d}_0$ is the emitter dipole moment in vacuum, and $\delta\mathbf{d}$ is the induced dipole moment that a nanoparticle acquires in the presence of an emitter. For a spherical nanoparticle whose size is small compared to the emission wavelength (i.e., scattering contribution to the extinction is ignored), analytical expressions can be derived (Klimov 2009):

$$\left(\frac{\gamma_{rad}}{\gamma_0}\right)_{norm} = \left|1 + 2\left(\frac{\varepsilon-1}{\varepsilon+2}\right)\left(\frac{a}{r}\right)^3\right|^2,$$

$$\left(\frac{\gamma_{rad}}{\gamma_0}\right)_{tang} = \left|1 - \left(\frac{\varepsilon-1}{\varepsilon+2}\right)\left(\frac{a}{r}\right)^3\right|^2. \tag{5.34}$$

Here, the "norm" subscript indicates the radial orientation of a molecule dipole moment with respect to a spherical nanoparticle surface (see insert in Figure 5.15), whereas the "tang" subscript implies its tangential orientation with respect to a spherical nanoparticle surface. Numerical calculations with the correct account of scattering contribution show that these formulas are relevant for metal nanospheres with diameter up to 20 nm. For bigger nanospheres only numerical modeling becomes appropriate. Extensive modeling shows that normal orientation of a dipole with respect to a metal nanoparticle surface is more favorable than the tangential ones in terms of plasmonic enhancement of luminescence. For normal orientation, explicit relations that allow calculation of radiative and nonradiative transition rates with the finite size of a metal nanoparticle taken into account read (Klimov and Letokhov 2005):

$$\frac{\gamma_{rad}}{\gamma_0} = \frac{3}{2}\sum_{n=1}^{\infty} n(n+1)(2n+1)\left|\frac{\psi_n(k_0 r_0)}{(k_0 r_0)^2} + A_n\frac{\zeta_n(k_0 r_0)}{(k_0 r_0)^2}\right|^2, \tag{5.35}$$

$$\frac{\gamma_{rad}+\gamma_{nr}}{\gamma_0}=1+\frac{3}{2}\sum_{n=1}^{\infty}n(n+1)(2n+1)\,\mathrm{Re}\left\{A_n\left(\frac{\zeta_n(k_0r_0)}{(k_0r_0)^2}\right)^2\right\},\qquad(5.36)$$

where $\psi_n(x)=xj_n(x)$, $\zeta_n(x)=xh_n^{(1)}(x)$, $j_n(x)$, and $h_n^{(1)}(x)$ are the spherical Bessel functions; k_0 is wave number in a vacuum; $r_0=a+\Delta r$ is the distance from a metal nanoparticle center to an emitter, and

$$A_n=-\left(\frac{\sqrt{\varepsilon}\psi_n\left(k_0a\sqrt{\varepsilon}\right)\psi_n'(k_0a)-\psi_n'\left(k_0a\sqrt{\varepsilon}\right)\psi_n(k_0a)}{\sqrt{\varepsilon}\psi_n\left(k_0a\sqrt{\varepsilon}\right)\zeta_n'(k_0a)-\psi_n'\left(k_0a\sqrt{\varepsilon}\right)\zeta_n(k_0a)}\right)$$

is one of the Mie coefficients for the field reflected from a metal nanoparticle surface, in which primes denote derivatives, a is metal nanoparticle radius, and ε is the metal complex dielectric permittivity. When an emitter is moved far from the metal nanoparticle, i.e., when $k_0r_0\to\infty$ holds, one has $\gamma_{rad}/\gamma_0\to1$ and $(\gamma_{rad}+\gamma_{nr})/\gamma_0\to1$, and then $Q\to Q_0$ is the case.

Figure 5.16 represents sample calculations for the radiative decay rate modification by a silver nanoparticle in air for the most favorable, normal, orientation of an emitter dipole moment. Radiative decay rate γ_{rad} is calculated with respect to its value for the same emitter in a vacuum γ_0. One can see that the effect can reach two orders of magnitude and depends on emission wavelength, nanoparticle size, and emitter–nanoparticle spacing Δr. Enhancement vanishes for $\Delta r>100$ nm. The spectral shape of the enhancement graphs remarkably correlates with the size-dependent extinction spectra (Figure 5.17).

Unfortunately, metal proximity promotes fast nonradiative decay of an excited emitter. In this process, energy stored in an excited emitter is nonradiatively transferred to metal, resulting in metal heating instead of photon emission. This effect is often referred to as luminescence quenching. Enhancement of the nonradiative decay rate is very big and typically dominates enhancement of the radiative decay rate at close distances, then rapidly

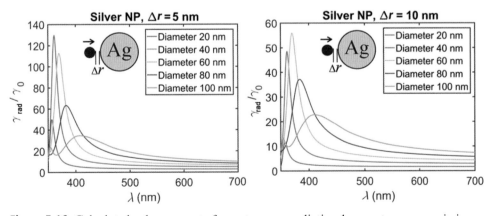

Figure 5.16 Calculated enhancement of spontaneous radiative decay rate versus emission wavelength for a dipole near a spherical silver nanoparticle (diameter ranging from 20 to 100 nm) in air at distance $\Delta r=5$ nm and 10 nm. The dipole moment is normal to a particle surface (insert). Courtesy of D. V. Guzatov.

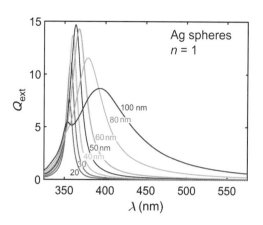

Figure 5.17 Calculated extinction spectra of silver spherical nanoparticles in air. Numbers indicate the diameter. Reprinted from Guzatov et al. (2018a) under the Creative Commons License.

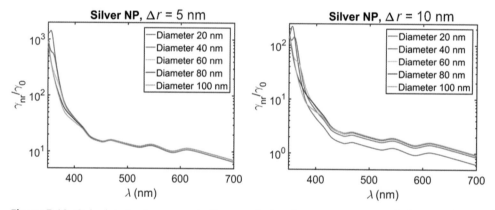

Figure 5.18 Calculated enhancement of nonradiative decay rate versus emission wavelength for a dipole near a spherical silver nanoparticle (diameter ranging from 20 to 100 nm) in air at distance $\Delta r = 5$ nm and 10 nm. Courtesy of D. V. Guzatov.

falls with distance, vanishing at $\Delta r > 50$ nm. In Figure 5.18, nonradiative decay rate γ_{nr} is calculated with respect to the radiative decay rate for the same emitter in vacuum γ_0. Different distance and size dependencies of radiative and nonradiative rates offer a possibility to choose optimal metal–emitter spacing where radiative rate enhancement can be bigger than nonradiative enhancement. Notably, size-dependent radiative rate enhancement follows size-dependent extinction spectra and contains a size-dependent peak moving to longer wavelengths for bigger metal particles. However, nonradiative decay enhancement is nearly size-independent and is defined mainly by the intrinsic silver dielectric function rather than by size-dependent extinction. Therefore, bigger metal nanoparticles in many cases appear to be more efficient for luminescence enhancement since these allow for radiative rate enhancement to overtake nonradiative rate enhancement.

When speaking about plasmonic effect on luminescence, the important figure of merit is quantum yield modification, Q/Q_0 where Q and Q_0 read

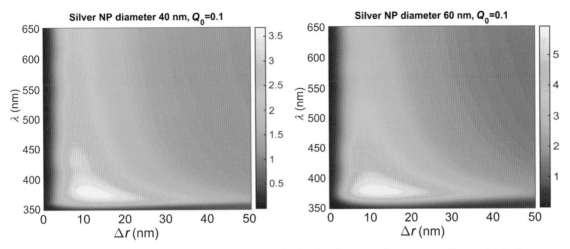

Figure 5.19 Modification of the quantum yield Q/Q_0 for an emitter near a silver spherical nanoparticle in air versus emission wavelength λ and metal–emitter spacing Δr. Reprinted from Guzatov et al. (2018a) under the Creative Commons License.

$$Q = \frac{\gamma_{\text{rad}}}{\gamma_{\text{rad}} + \gamma_{\text{nr}} + \gamma_{\text{int}}}, \quad Q_0 = \frac{\gamma_0}{\gamma_0 + \gamma_{\text{int}}}, \tag{5.37}$$

with γ_{int} being the internal intrinsic nonradiative decay rate of an emitter in a vacuum. This internal intrinsic nonradiative path may arise, e.g., from singlet-to-triplet state transition in a molecule, energy transfer processes in molecular or quantum dot ensembles, or non-desirable impurity or surface states promoting recombination in semiconductor nanostructures. Because of this internal nonradiative decay channel, quantum yield of many emitters is less than 1. If this is the case, quantum yield can be enhanced with metal nanoparticles. The representative calculations are shown in Figure 5.19 for a hypothetical emitter with $Q_0 = 0.1$ and silver nanoparticles. One can see that for distances of about 10 nm the original quantum yield $Q_0 = 0.1$ can be enhanced by a factor of 3–5. In all cases, however, even being enhanced, quantum yield is still noticeably less than 1. Higher original quantum yield can experience lower enhancement, and for $0.5 < Q_0 < 1$, metal nanoparticle(s) typically cannot increase quantum yield.

The results presented in Figure 5.19 for quantum yield enhancement can be directly applied to electroluminescent devices, including semiconductor LEDs and OLEDs (organic LEDs). Metal nanoparticles of the proper size and appropriately displaced at the optimal distance (5–10 nm) from the interface where electron–hole pairs recombine can raise internal quantum efficiency in accordance with Eq. (5.35), with corresponding increase in the output electroluminescence intensity,

$$\text{Electroluminescence} \quad \frac{I}{I_0} = \frac{Q}{Q_0} = \frac{(\gamma_{\text{rad}} / \gamma_0)/Q_0}{\gamma_{\text{rad}} / \gamma_0 + \gamma_{\text{nr}} / \gamma_0 + (1 - Q_0)/Q_0}. \tag{5.38}$$

Here, internal nonradiative rate γ_{int} is written as $\gamma_{\text{int}} = (1 - Q_0)/Q_0$ to emphasize that enhancement of electroluminescence intensity is fully defined by the intrinsic value Q_0 (the emitter property) and the radiative and nonradiative decay enhancement factors, γ_{rad}/γ_0 and γ_{nr}/γ_0, respectively (the plasmonic effects).

Photoluminescence has an additional option to experience plasmonic enhancement, namely incident electromagnetic field enhancement $|E|^2/|E_0|^2$, which has been discussed in detail in Section 4.7 (Figure 4.34). Local increase in incident light intensity gives rise to higher excitation (absorption) rate, and photoluminescence intensity rises accordingly. The overall plasmonic effect on photoluminescence intensity is expressed as

$$\text{Photoluminescence} \quad \frac{I}{I_0} = \frac{|E|^2}{|E_0|^2}\frac{Q}{Q_0} = \frac{|E|^2}{|E_0|^2}\frac{(\gamma_{rad}/\gamma_0)/Q_0}{\gamma_{rad}/\gamma_0 + \gamma_{nr}/\gamma_0 + (1-Q_0)/Q_0}. \quad (5.39)$$

Now the problem actually breaks down into two subtasks. First, one needs to calculate the local field enhancement factor at a given point near a metal nanobody for excitation wavelength. Then, one needs to calculate radiative and nonradiative decay rates, and enhancement versus spontaneous decay rates in a vacuum. In the first subtask, polarization of incident light is important. In the second subtask dipole moment orientation of an emitter is essential. Calculations show that the optimal alignment is when the incident light E vector and the emitter dipole moment are aligned along the line connecting the given point with the nanoparticle center (see the insert in Figure 5.20(c)). The problem appears to become a complicated multiparametric task: (1) incident field polarization and its wavelength should fit conditions for strong incident intensity enhancement and at the same time its wavelength should fall into the absorption maximum; (2) dipole orientation should fit the above alignment and at the same time correspond to strong enhancement of radiative decay rate; (3) spacing should be adjusted to perform the most favorable combination of intensity rise and quantum yield minimal loss – or better, quantum yield gain – whenever possible. Further parameters are metal type, nanoparticle shape, and nanoparticle mutual displacement/distance. The general recipe is to put excitation wavelength (absorption maximum) close to extinction resonance, whereas emission wavelength should be kept somewhat at longer wavelengths, which is typical for photoluminescence (an emitted quantum is typically smaller than the absorbed one).

Figures 5.20 and 5.21 give the two representative examples of modeling attempted to evaluate the most favorable conditions for experimental implementation of photoluminescence enhancement of a dipole emitter with a spherical silver nanoparticle. In Figure 5.20 the fixed parameter is the emission wavelength – 530 nm (green light) is chosen. Then, excitation wavelength and metal–emitter spacing are treated as arguments and nanoparticle diameter is an adjustable parameter. Incident field and dipole moment versus nanoparticle alignment are used in the most favorable configuration, as is shown in the insert. An ideal emitter with $Q_0 = 1$ is considered. 50 nm particles were found to promise the maximal enhancement of photoluminescence intensity (50 times); however, it occurs when excitation is performed near 400 nm (extinction maximum), which often does not correspond with the absorption maximum for a typical 530 nm emitter. For example, fluorescein, an

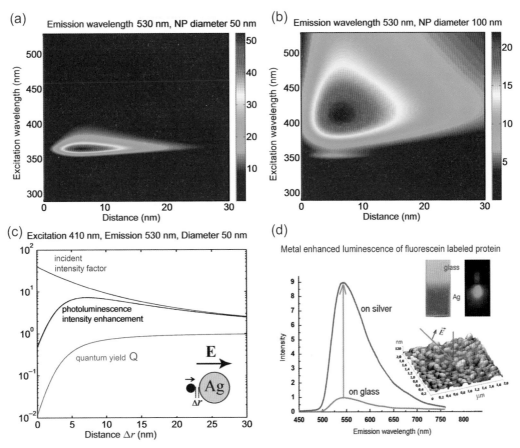

Figure 5.20 Plasmonic enhancement of molecular luminescence near a spherical silver particle. (a–c) Calculations; (d) experiment. See text for details. Adapted with permission from Guzatov et al. (2012). Copyright 2012 American Chemical Society.

organic commercial fluorophore for 530 nm, has its absorption peak near 500 nm. In this case, bigger particles (e.g., 100 nm diameter) appear to be more efficient since their extinction maximum shifts to a longer wavelength and in the spectral ranges of interest (excitation at 500 nm, emission at 530 nm) offers better combination for overall photoluminescence enhancement factor. The optimal distance typically ranges from 5–6 nm to 10–12 nm. Theoretical predictions have been reasonably confirmed experimentally (Figure 5.20(d)).

Figure 5.21 shows optimization results for the case in which the excitation wavelength is fixed at 450 nm and the emission wavelength scans to longer wavelengths within the visible. This task directly corresponds to a white LED, in which a blue semiconductor LED excites color-converting phosphor to get white light for lighting (see Figure 5.3, right-hand panel). One can see that manifold photoluminescence intensity enhancement becomes feasible, essentially owing to the fact that for the ambient medium with $n = 1.5$ (a typical

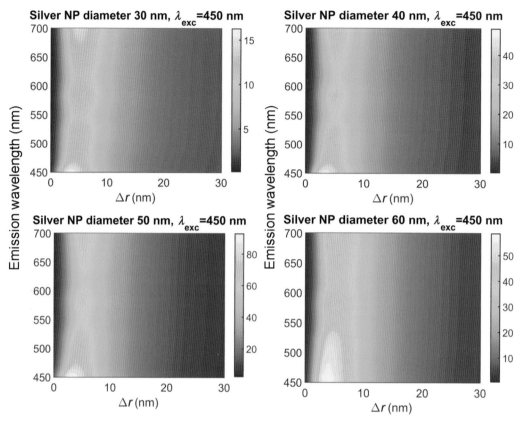

Figure 5.21 Calculated enhancement factor for intensity of a phosphor photoluminescence near an Ag nanoparticle with diameter 30 to 6 nm as a function of emission wavelength and spacing; excitation wavelength is 450 nm. Intrinsic quantum yield $Q_0 = 1$, and ambient medium refractive index $n_m = 1.5$. Reprinted from Guzatov et al. (2018b).

polymer), extinction maximum is close to 450 nm (see Figure 4.34), the prerequisite excitation wavelength.

When comparing theoretical results in Figures 5.20 and 5.21, a possibility exists to evaluate the role of the ambient refraction index in nanoplasmonic luminescence enhancement. One can see that for the same excitation wavelength (450 nm) $n_m = 1.5$ suggests much higher enhancement than $n_m = 1$. At the same time, the optimal distance falls from 5–7 nm to 3–4 nm. This is not surprising and can be explained in terms of the size/wavelength ratio, where wavelength should be considered in the ambient medium. Then, changing refractive index gives rise to the above ratio change with the corresponding shift of extinction spectra. This shift changes radiative decay graphs similarly to the effect of size change, but does not affect nonradiative graphs since the latter are independent of size. Therefore, using a highly refractive ambient medium can considerably increase positive impact of plasmonics on luminescence efficiency.

Modification of spontaneous emission rate near a 3-axial silver ellipsoid

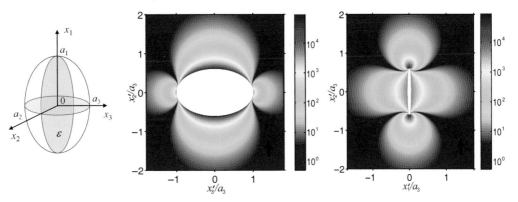

Figure 5.22 Modification of the radiative decay rate of a probe dipole near a silver nanoellipsoid with respect to a vacuum. The black arrow shows the dipole moment orientation. Calculations are made for emission wavelength 632 nm. $a_1/a_2/a_3 = 0.043/0.6/1$. Adapted from Guzatov and Klimov (2005), with permission from Elsevier.

Modification of radiative decay rate near two spherical Ag nanoparticles

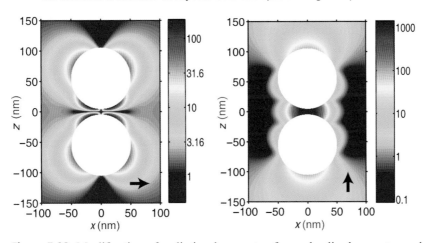

Figure 5.23 Modification of radiative decay rate of a probe dipole near two spherical silver nanoparticles with respect to a vacuum. The black arrow shows the dipole moment orientation. Calculations are made for emission wavelength 400 nm. Courtesy of D. Guzatov.

Radiative lifetime experiences even more dramatic enhancement in more complicated metal nanostructures as compared to a single solid sphere. Figure 5.22 shows the case of a three-axial silver ellipsoid, and Figure 5.23 presents results for a couple of spherical particles. One can see that up to 10^3-fold acceleration of radiative decay rate is possible in both cases. To evaluate the overall effect on electroluminescence intensity, nonradiative decay rates should be examined as well. To evaluate photoluminescence intensity enhancement, incident light intensity enhancement analysis should be involved as well (see Figure 4.34). Such general extensive studies involving interplay of all the above factors for complicated metal systems have not been performed to date.

5.10 NANOANTENNAS REVISITED

In Chapter 4 we saw that the radiophysical notion of *antenna* can be extended to the optical range of the electromagnetic spectrum, and multiple examples of incident light concentration have been provided to demonstrate that optical antennas enable higher light absorption similar to radio or TV antennas, enabling more efficient detection of radiowaves. There is an important principle in classical electrodynamics, the *reciprocity principle*, which states that once a system enhances detection (i.e., absorption) of electromagnetic radiation it should necessarily promote (enhance) its emission at the same wavelength. Because of this principle, radio antennas can equally improve both detection of radiation by a radio or TV and its emission by a transmitter.

There is a definite convergence of classical and quantum electrodynamics that manifests itself in the optical nanoantenna notion. Every system promoting incident light concentration will necessarily enhance light emission at the same wavelength (frequency). Multiple examples of spontaneous radiative decay rate enhancement by a single metal nanoparticle or by a more complicated nanoparticle arrangement given in the previous section can all be interpreted in terms of the nanoantenna notion.

However, classical electrodynamics cannot explain how atom(s) or any other quantum system, e.g., an electron–hole pair (exciton) in a semiconductor emits light. There are no discrete transitions in atoms and there are no photons in classical electrodynamics. Nevertheless, taking *de fide* the concept of photon emission in the course of *quantum jumps*, we can efficiently use *classical* equations to calculate modification of transition rate in various nano-environments. In other words, while the *classical* electrodynamics cannot *explain* the event of photon emission in the course of downward transition of an excited atom or a molecule, it is capable of offering the *computational technique* for its rate calculation. Therefore, the nanoantenna-based approach should be treated as a useful *computational procedure* provided that we understand that photon creation in the course of a quantum system downward relaxation by no means can be derived or explained by classical radiophysical techniques.

This remarkable convergence does manifest itself in the definition of LDOS for photons, which is based on the statement that modification of photon LDOS with respect to a vacuum can be calculated as a modification of the emission rate for a classical dipole placed in the same position, where the photon LDOS value is sought (see discussion in Section 5.4, especially related to Eq. (5.25)).

The nanoantenna notion can even shed more light on understanding of the essence of local DOS in optics. Once we accept that high LDOS objects in optics mimic radiophysical antennas, we understand that high LDOS values mean the capability of arranged material objects to concentrate energy contained in electromagnetic waves. Therefore, once there is evidence or expectation for local concentration of lightwave energy in space, one can speak about a similar concentration of LDOS in the same place (point) of space. The difference is that for incident electromagnetic radiation, concentration effects occur with respect to real lightwaves, whereas when speaking about LDOS we consider imaginary probe waves as if concentration of these waves is analyzed. Note that when considering photoluminescence, the incident field enhancement and LDOS are to be examined for different frequencies; therefore, the incident field calculation to be absorbed by an emitter will not necessarily ensure similar enhancement of LDOS. This becomes clear if one recalls the Barnett–Loudon theorem (Section 5.5) stating that DOS redistributes over the frequency scale and therefore favorable light concentration at incident light frequency may not be followed by similar DOS enhancement.

An antenna modifies power radiated by a classical emitter by a factor F_{antenna}, which reads (Bharadwaj et al. 2009)

$$F_{\text{antenna}} = \frac{(P_{\text{rad}}/P_0)}{P_{\text{rad}}/P_0 + P_{\text{antenna loss}}/P_0 + (1-\eta_i)/\eta_i}. \qquad (5.40)$$

Here, P_0 is the emitter power in free space, P_{rad} is the power emitted in the presence of an antenna, $P_{\text{antenna loss}}$ describes undesirable losses because of non-ideality of an antenna, and η_i stands for the intrinsic internal losses in the emitter itself. Comparing this formula with Eq. (5.38), one can see it has the same structure as the modified quantum yield of an emitter near a metal nanostructure. Moreover, all rates in Eq. (5.38) transform into appropriate powers by multiplying the energy carried by a photon, $\hbar\omega$. Thus, consideration of radiophysical (classical) terms merges with the quantum one.

Consideration of the antenna effect on a classical emitter described by Eq. (5.40) shows that only in the case $\eta_i < 1$ can the antenna enhance the emitted power, whereas for an ideal emitter without internal losses (i.e., $\eta_i = 1$) the antenna will always reduce the emitted power because of unavoidable losses of an antenna itself. This fully coincides with plasmonically enhanced spontaneous emission – namely, only intensity of an imperfect emitter with $Q_0 < 1$ can be enhanced, whereas for a perfect emitter ($Q_0 = 1$) intensity will always be reduced because of nonradiative losses promoted by proximity of a metal nanobody. The analogy of optics with radiophysics is complete for the case of electroluminescence; however, for the case of *photo*luminescence, metal-induced losses of quantum yield are often superimposed and overtaken by incident field enhancement (see Figure 5.20(c)).

Does the above link between classical and quantum electrodynamics provided by the optical antenna notion exhaust its usefulness for nanophotonics? Surely, it does not. Not only the elegant bridge between radiophysics and optics is established, but also the antenna design can be borrowed from radioengineering and used in photonic devices. A simple dipole antenna, bowtie antenna, and Yagi–Uda antenna shown in Chapter 4 (Figure 4.35) can be used in photonics both for incident intensity concentration in photoluminescence and for emissivity enhancement in photo- and electroluminescent devices. One should bear in mind that electroluminescence enhancement is possible for emitters with $Q_0 < 1$, whereas for optical excitation, luminescence enhancement can be observed even for perfect emitters with $Q_0 = 1$. Though for optical excitation in accordance with Eq. (5.39), lower Q_0 enables huge enhancement factors since the metal-induced nonradiative path is not so sensible because it is added to already existing internal losses. For example, for molecules with low intrinsic quantum yield, experiments with a gold bowtie antenna made it possible to obtain photoluminescence enhancement of the order of 10^3 (Kinkhabwala et al. 2009). It is also important that optical antennas can help to enhance emission directivity, which can be useful in certain applications.

5.11 CONTROLLING EMISSION PATTERNS

A plane light source which features equal brightness over the surface in the far field obeys Lambert's law, known since 1760. It states that intensity I has a peak for the normal direction with respect to the surface and follows cosine dependence on the observation angle Θ,

$$I(\Theta) = I_{max} \cos \Theta. \tag{5.41}$$

The relevant radiation pattern is referred to as *Lambertian*. It is presented in Figure 5.24 in Cartesian and polar coordinate systems. Many films and crystals with plane surfaces exhibit radiation patterns close to the Lambertian one.

Complicated topology of a light-emitting system gives rise to non-Lambertian patterns – e.g., if a phosphor is embedded inside a 2D photonic crystal slab, the radiation pattern is redistributed, featuring enhancement of emissivity along the directions where the medium is homogeneous and inhibition in the directions (planes) where the medium

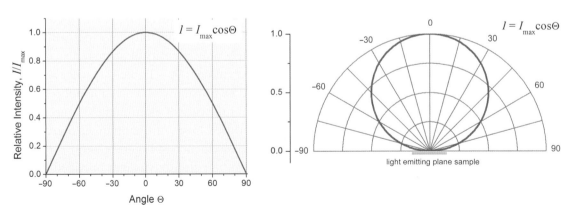

Figure 5.24 Lambert's law in (a) Cartesian and (b) polar systems.

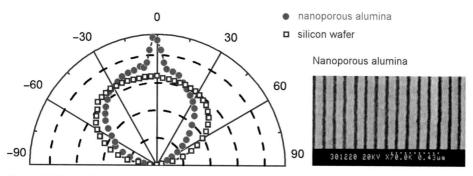

Figure 5.25 Radiation patterns of Eu-based phosphor deposited on a silicon wafer (blue squares) and embedded in nanoporous alumina (red circles).

possesses a periodic refractive index (Figure 5.25). This phenomenon can be attributed to photon density of states effect and can be purposefully used to increase light extraction efficiency in LEDs.

Optical antennas offer another way to modify the emission pattern of a point-like light source, similar to radiowaves patterning in radio transmitters. Several groups have reported on directional emission for quantum dots coupled to lithographically fabricated Yagi–Uda antennas (Maksymov et al. 2012). The same effect should also be observed for other possible dielectric and metal–dielectric nanostructures featuring light confinement and photon density of states modification with respect to vacuum or air.

It should be emphasized that neither of the above examples should be treated as a filtering phenomenon. A filter alters light intensity from a detector side, but cannot modify an emitter's properties. A photonic crystal and a nanoantenna modify angular emission diagrams simultaneously with modification of the decay rates (lifetimes) of emitters owing to modification of space properties on the subwavelength scale, i.e., it is not a kind of small pocket flash effect. No light beam can be drawn within each of the above schemes. Both above examples mean that space arrangement near an emitter changes photon density of states.

Conclusion

- Excited matter spontaneously emits light in portions called light quanta, or photons.
- Photons are emitted in the course of quantum downward transitions of electrons in atoms, molecules, and solids.
- The probability of spontaneous downward transitions of every quantum system is inversely proportional to its lifetime and is defined by the intrinsic properties of a system in question and the capability of space to maintain electromagnetic wave existence and propagation at the frequency $\omega = E / \hbar$, with E being the energy difference between the upper and lower states of the system.

- Space capability to maintain electromagnetic wave existence and propagation at ω frequency is described by means of electromagnetic mode density (density of modes, DOM) which reads

$$D(\omega) = \omega^2 / \pi^2 c^3$$

for a vacuum or another continuous medium with refractive index $n = 1$. DOM depends on dielectric properties of space, including topology of space inhomogeneities, and can be calculated for a given point using Green's functions. Spontaneous emission rate (number of transitions per second) is proportional to DOM. DOM is the common and basic notion of all waves, e.g., acoustic waves, not necessarily electromagnetic ones, and does not directly relate to light emission.

- When speaking about photons, the density of modes notion evolves to the photon density of states (DOS).

- Areas of space with pronounced singularities, e.g., a microcavity or a tiny piece of metal, are characterized by the local density of states (LDOS), which to a large extent describes the ability of space to accumulate electromagnetic energy and can be characterized in terms of the Q-factor by analogy with other systems related to oscillations and energy storage/release (LC-circuits in electric engineering and pendulums in mechanics).

- Modification of DOM (photon DOS and LDOS) by means of space organization on the optical wavelength scale offers a way to control spontaneous emission rate and directionality. The actual scaling factor is half- or quarter-wave length divided by the refraction index of inhomogeneity(ies) so that for semiconductor nanostructures in the visible spectrum it measures about 50–100 nm. For every given point of space the overall change in DOS features redistribution over frequency scale, and all cases of DOS enhancement in certain frequency range(s) are necessarily compensated by the opposite DOS change otherwise, so that the total DOS at every given point calculated over a wide frequency range remains the same as in a vacuum.

- Microcavities, photonic crystals, interfaces, and metal–dielectric nanostructures change photon density of states significantly.

- Lifetime engineering using metal–dielectric nanostructures is an essential part of nanoplasmonics, along with using metal–dielectric nanostructures to perform concentration of light. Plasmonic nanostructures simultaneously offer incident light concentration and radiative decay enhancement, but also enhance nonradiative losses. In the case of a perfect emitter with quantum yield = 1 (internal quantum efficiency), photoluminescence can be enhanced by means of absorption enhancement overtaking quantum yield losses. Electroluminescence can be enhanced with plasmonic nanostructures for emitters with quantum yield < 1 only. Plasmonic nanostructures reduce lifetime for all emitters, and if this effect is not accompanied by high quantum yield losses, it can be used to accelerate modulation speed in light-emitting systems, including semiconductor LEDs.

• Many issues related to lifetime engineering by means of lightwave confinement can be treated in terms of nanoantenna effects bridging radiophysics and optics and providing an efficient route to mimic certain radiophysical devices in optics. This analogy reflects a definite convergence of classical and quantum electrodynamics and offers an efficient computational approach for lifetime engineering in nanophotonics. However, since there are no photons in classical electrodynamics, we still need quantum physics to explain light emission by excited matter.

Problems

5.1 Derive Eq. (5.4), starting from Eq. (5.3).

5.2 Based on Eq. (5.3) or (5.4), derive and analyze the value of photon energy or wavelength in the emission maximum of a black body depending on temperature.

5.3 Explain physical limits for efficiency of incandescent lamps.

5.4 Explain differences in emission spectra for different light sources (incandescent lamps, luminescent gas mercury and sodium lamps, semiconductor LEDs).

5.5 Compare the relation expressed by Eq. (5.8) with the formula for resistance of a number of parallel resistors. Explain the physical reason for the apparent similarity.

5.6 Derive electromagnetic mode density $D(\lambda)$ in a vacuum. Highlight this term in Eq. (5.4).

5.7 Recall a number of typical cases in which structuring of space will result in modified light emission.

5.8 Consider Figure 5.8 and compare it to Figure 4.9 for a dispersion curve in the long-wave limit.

5.9 Explain different factors for photo- and electroluminescence in plasmonic nanostructures.

5.10 Compare extinction spectra for metal nanoparticles in media with $n = 1$ (Figure 5.17) and 1.5 (Figure 4.32) and explain the difference.

5.11 Explain why plasmonic enhancement of luminescence rises similarly for higher refraction of an ambient medium and for bigger metal nanoparticle sizes. Hint: consider nanoparticle size versus wavelength.

5.12 Explain similarities of spontaneous emission enhancement in a complex medium versus the antenna effect in radiophysics.

Further reading

Andrew, P., and Barnes, W. L. (2001). Molecular fluorescence above metallic gratings. *Phys Rev B*, **64** (12), 125405.

Barnes, W. L. (1998). Fluorescence near interfaces: the role of photonic mode density. *J Mod Optics*, **45**, 661–699.

Bharadwaj, P., Deutsch, B., and Novotny, L. (2009). Optical antennas. *Adv Opt Photonics*, **1**, 438–483.

Biagioni, P., Huang, J. S., and Hecht, B. (2012). Nanoantennas for visible and infrared radiation. *Rep Prog Phys*, **75**, 024402.

Bykov, V. P. (1993). *Radiation of Atoms in a Resonant Environment*. World Scientific.

Cho, K. (2003). *Optical Response of Nanostructures: Nonlocal Microscopic Theory*. Springer.

De Martini, F., Marrocco, M., Mataloni, P., Crescentini, L., and Loudon, R. (1991). Spontaneous emission in the optical microscopic cavity. *Phys Rev A*, **43**, 2480.

Drexhage, K. H. (1970). Influence of a dielectric interface on fluorescence decay time. *J Luminescence*, **1–2**, 693–701.

Fujita, M., Takahashi, S., Tanaka, Y., Asano, T., and Noda, S. (2005). Simultaneous inhibition and redistribution of spontaneous light emission in photonic crystals. *Science*, **308**, 1296–1298.

Gaponenko, S. V. (2010). *Introduction to Nanophotonics*. Cambridge University Press, chs. 13 and 14.

Gaponenko, S. V. (2014). Satyendra Nath Bose and nanophotonics. *J Nanophotonics*, **8**, 087599.

Geddes, C.D., and Lakowicz, J.R. (eds.) (2007). *Radiative Decay Engineering*. Springer Science & Business Media.

Klimov, V. (2014). *Nanoplasmonics*. CRC Press.

Klimov, V. V., and Ducloy, M. (2004). Spontaneous emission rate of an excited atom placed near a nanofiber. *Phys Rev A*, **69**, 013812.

Lee, K. G., Eghlidi, H., Chen, X. W., et al. (2012). Spontaneous emission enhancement of a single molecule by a double-sphere nanoantenna across an interface. *Opt Express*, **20**(21), 23331–23338.

Oraevskii, A. N. (1994). Spontaneous emission in a cavity. *Physics – Uspekhi*, **37**, 393–405.

Parker, G. J. (2010). Biomimetically-inspired photonic nanomaterials. *J Mater Sci Mater Electron*, **21**, 965–979.

Törmä, P., and Barnes, W. L. (2015). Strong coupling between surface plasmon polaritons and emitters. *Rep Prog Phys*, **78**, 013901.

References

Allan, G., and Delerue, C. (2004). Confinement effects in PbSe quantum wells and nanocrystals. *Phys Rev B*, **70**, 245321.

Amos, R. M., and Barnes, W. L. (1997). Modification of the spontaneous emission rate of Eu^{3+} ions close to a thin metal mirror. *Phys Rev B*, **55**, 7249.

Barnett, S. M., and Loudon, R. (1996). Sum rule for modified spontaneous emission rates. *Phys Rev Lett*, **77**, 2444–2448.

Bharadwaj, P., Deutsch, B., and Novotny, L. (2009). Optical antennas. *Adv Opt Photonics*, **1**, 438–483.

Bose, S. N. (1924). Planck's Gesetz und Lichtquantenhypothese. *Zs. Physik*, **26**, 178–181.

Bunkin, F. V., and Oraevskii, A. N. (1959).) Spontaneous emission in a cavity. *Izvestia Vuzov, Radiophysics*, **2**, 181–188. (In Russian).

Busch, K., von Freymann, G., Linden, S., et al. (2007). Periodic nanostructures for photonics. *Phys Rep*, **444**, 101–202.

D'Aguanno, G., Mattiucci, N., Centini, M., Scalora, M., and Bloemer, M. J. (2004). Electromagnetic density of modes for a finite-size three-dimensional structure. *Phys Rev E*, **69**, 057601.

De Martini, F., Innocenti, G., Jacobowitz, G. R., and Mataloni, P. (1987). Anomalous spontaneous emission time in a microscopic optical cavity. *Phys Rev Lett*, **59**, 2955–2958.

Guzatov, D. V., and Klimov, V. V. (2005). Radiative decay engineering by triaxial nanoellipsoids. *Chem Phys Lett*, **412**, 341–346.

Guzatov, D. V., Gaponenko, S. V., and Demir, H. V. (2018a). Plasmonic enhancement of electroluminescence. *AIP Advances*, **8**, 015324.

Guzatov, D. V., Gaponenko, S. V., and Demir, H. V. (2018b). Possible plasmonic acceleration of LED modulation for Li-Fi applications. *Plasmonics*. DOI 10.1007/s11468-018-0730-6.

Guzatov, D. V., Vaschenko, S. V., Stankevich, V. V., et al. (2012). Plasmonic enhancement of molecular fluorescence near silver nanoparticles: theory, modeling, and experiment. *J Phys Chem C*, **116** (19), 10723–10733.

Kinkhabwala, A., Yu, Z., Fan, Sh., et al. (2009). Large single-molecule fluorescence enhancements produced by a bowtie nanoantenna, *Nature Phot*, **3**, 654–657.

Klimov, V. V. (2009). *Nanoplasmonics*. Fizmatlit. (In Russian)

Klimov, V. V., and Letokhov, V. S. (2005). Electric and magnetic dipole transitions of an atom in the presence of spherical dielectric interface. *Laser Phys*, **15**, 61–73.

Koenderink, A. F., Kafesaki, M., Soukolis, C. M., and Sandoghdar, V. (2005). Spontaneous emission in the near field of two-dimensional photonic crystals. *Opt Lett*, **30**, 3210–3212.

Lambropoulos, P., Nikolopoulos, G. M., Nielsen, T. R., and Bay, S. (2000). Fundamental quantum optics in structured reservoirs. *Rep Prog Phys*, **63**, 455–503.

Maksymov, I. S., Staude, I., Miroshnichenko, A. E., and Kivshar, Y. S. (2012). Optical Yagi–Uda nanoantennas. *Nanophotonics*, **1**(1), 65–81.

Noda, S., Fujita, M., and Asano, T. (2007). Spontaneous-emission control by photonic crystals and nanocavities. *Nat Photonics*, **1**(8), 449–458.

Novotny, L., and Hecht, B. (2012). *Principles of Nano-Optics*. Cambridge University Press.

Purcell, E. M. (1946). Spontaneous emission probabilities at radio frequencies. *Phys Rev*, **69**, 681.

Schubert, E. F. (2006). *Light-Emitting Diodes*. Cambridge University Press.

Yablonovitch, E., Gmitter, T. J., and Bhat, R. (1988). Inhibited and enhanced spontaneous emission from optically thin AlGaAs/GaAs double heterostructures. *Phys Rev Lett*, **61**, 2546–2549.

6 Stimulated emission and lasing

Stimulated transitions in quantum systems represent the principal background for laser operation. Lasers play important roles in daily life and form a valuable part of photonics. Though the first lasers were supposed to have an impact on material processing owing to extreme energy concentration, currently the main applications of lasers are not based on high-energy abilities, but essentially on their compactness, cheapness, and efficiency. Optical communication, laser printers, and DVD players are the most common laser applications, and here semiconductor lasers are definitely unsurpassed. In this chapter we provide a brief introduction to the principles of laser operation and laser device engineering to emphasize where nanostructures can be useful and which components of lasers can be improved by means of nanophotonic solutions. Because of the scope of this book, a brief explanatory style is used which may confuse laser experts but at the same time may still appear complicated for newcomers. This chapter may appear complicated for readers not at all familiar with lasers, and in this case appropriate textbooks are suggested for further reading.

6.1 STIMULATED EMISSION, ABSORPTION SATURATION, AND OPTICAL GAIN

6.1.1 A two-level system at high intensities

Absorption and emission of light occurs by means of the balance between upward and downward transitions in atoms, molecules, and solids. Upward transitions are always stimulated, i.e., their rate is proportional to the density of energy in the input radiation interacting with a quantum system. Downward transitions can occur in two ways. The first way is spontaneous downward transitions; the second is stimulated transitions. At low-intensity input radiation, population of upper states is typically much lower than population of the ground state. In this case, downward stimulated transitions only slightly contribute to the overall balance and the matter emits light in the form of luminescence, with luminescence intensity being in proportion to input radiation (see Chapter 5). The portion of absorbed energy rises directly proportional with input intensity. This is called the *linear optics* regime, to emphasize that absorbed and emitted light intensity grows in a linear fashion with input intensity.

The situation changes dramatically when input intensity rises, so that upper state population grows and can become comparable to the ground state population. Neglecting the background thermal radiation, we can write the simple and instructive rate equation for a two-level system,

$$Bun_1 = Bun_2 + An_2, \qquad (6.1)$$

which is to be examined together with the conservation condition

$$n_1 + n_2 = N. \qquad (6.2)$$

Here, n_1, n_2, and N are the ground state, excited state populations, and total number of atoms or molecules in the system in question; u is the radiation density, and A and B are Einstein coefficients for spontaneous and stimulated transitions, respectively. We will examine what happens with populations n_1 and n_2 upon unlimited growth of radiation energy density u. An elementary calculation gives rise to the relations

$$n_1 = N \frac{Bu + A}{2Bu + A} \xrightarrow{u \to \infty} \frac{N}{2} \; ; \; \xrightarrow{u \to 0} N$$

$$n_2 = N \frac{Bu}{2Bu + A} \xrightarrow{u \to \infty} \frac{N}{2} \; ; \; \xrightarrow{u \to 0} 0 \qquad (6.3)$$

One can see that at low u ground state, population is close to N, and excited state population is close to zero. Then, with growing u infinitely, both populations tend to $N/2$. In this case, the rate of stimulated upward transitions equals the rate of simulated downward transitions, i.e., in Eq. (6.1) $Bun_1 = Bun_2$ holds, whereas the intensity-independent An_2 term can be neglected.

The absorption coefficient α, defining intensity attenuation by a slab of matter (see Eq. (4.63)) is related to the difference between rates of upward and downward stimulated transitions, i.e., $\alpha = \alpha(u)$ and reads

$$\alpha(u) = \alpha_0 \frac{n_1(u) - n_2(u)}{N} = \frac{\alpha_0}{1 + \frac{2B}{A} u}, \quad \alpha_0 = \alpha(0). \qquad (6.4)$$

For infinitely growing u, absorption coefficient tends to zero. Then, transmission coefficient T of a slab of matter with thickness L which is defined by the familiar Bouguer law,

$$T = I(L) / I_0 = e^{-\alpha L}, \qquad (6.5)$$

tends to 1, i.e., initially absorbing matter looks absolutely transparent. Does this mean the matter does not absorb electromagnetic energy? Not the case. The matter absorbs as much energy per unit time as it is capable of doing. Therefore, this phenomenon is called *absorption saturation*. Rewriting Eq. (6.4) in terms of intensity (W/cm^2) instead of u, one has

$$\alpha(I) = \frac{\alpha_0}{1 + I / I_{\text{sat}}}, \qquad (6.6)$$

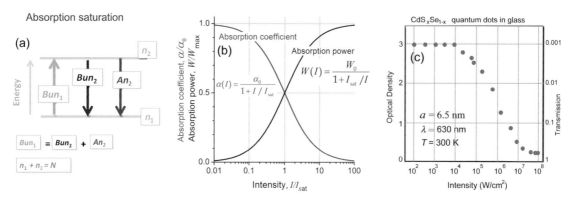

Figure 6.1 Absorption saturation. (a) Optical transitions and rate equations for a two-level system; (b) absorption coefficient and absorption power versus intensity; (c) experiment for quantum dots in glass – dependence of optical density and transmission on light intensity.

where I_{sat} is termed *saturation intensity* and depends on material parameters B, A, and N. I_{sat} defines intensities where absorption becomes nonlinear. The absorbed power W (W/cm^3) is the product of intensity and absorption coefficient,

$$W(I) = I\alpha(I) = I \frac{\alpha_0}{1 + I / I_{sat}} \xrightarrow{I \to \infty} \alpha_0 I_{sat},$$ (6.7)

and tends to its maximal possible value, which reads $W_{max} = \alpha_0 I_{sat}$ and is fully defined by the material parameters (Figure 6.1(b)).

Semiconductor quantum dots represent a discrete-level system in which absorption saturation is strongly pronounced. A typical example is given in Figure 6.1(c), where intensity-dependent optical density $D = -\lg T$ is shown for a commercial cutoff filter based on silicate glass containing CdS$_x$Se$_{1-x}$ quantum dots upon excitation by 10 ns laser pulses. One can see that transmission rises from about 0.001 to nearly 1, i.e., the material becomes transparent.

6.1.2 Multilevel system at high intensities

Absorption coefficient resulting from transitions between any pair of levels in a quantum system depends on population difference between upper and lower states, similar to Eq. (6.4). In a system with more than two levels, upper state population can become higher than the lower state one. This situation is called *population inversion*. In this case, the absorption coefficient acquires a negative value and absorption changes to *optical gain*. Propagation of radiation through a slab of matter still follows the Bouguer law (Eq. (6.5)), but since α is negative, attenuation changes to amplification. **L**ight **A**mplification by **S**timulated **E**mission *of* **R**adiation gave rise to the *laser* notation. Figure 6.2 shows how population inversion can be obtained.

Optical gain in a three-level system (Cr^{3+}, Nd^{3+}, Yb^{3+}, Er^{3+}, Tm^{3+}, Ho^{3+})

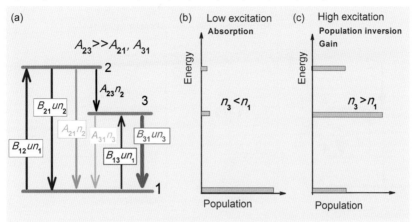

Figure 6.2 Optical gain in a three-level system. (a) Optical transitions; (b) level populations at low excitation in channel $1 \to 2$; (c) level populations at high excitation in channel $1 \to 2$. Population inversion results in optical gain in channel $3 \to 1$ provided that level 3 is metastable, i.e., spontaneous decay rate in channel $2 \to 3$ should be much faster than in channels $2 \to 1$ and $3 \to 1$.

At low incident light intensity with frequency corresponding to $1 \to 2$ transitions, populations of levels 2 and 3 are much less than that of level 1. Atoms or molecules excited to level 2 experience fast relaxation to level 3, which dominates direct spontaneous transitions to level 1, i.e., the system in question meets the condition $A_{23} \gg A_{21}$. Then, under condition that $A_{23} \gg A_{31}$ after the time period about $1/A_{23}$, the population of level 3 will become higher than that of level 1. Therefore, for radiation whose frequency corresponds to $E_3 - E_1$ energy separation, stimulated downward transitions rate (shown by a wide red arrow in Figure 6.2) will always dominate over the stimulated upward transitions rate. Intensity of radiation meeting the resonance condition between third and second levels will be amplified in the course of propagation through such a medium. When the condition $A_{23} \gg A_{21}$, A_{31} is met, level 3 is referred to as the metastable one. Several ions in transparent dielectric crystalline or glass matrices feature a three-level system with a metastable intermediate level: Cr^{3+}, Nd^{3+}, Yb^{3+}, Er^{3+}, Tm^{3+}, and Ho^{3+}.

Optical gain can readily develop also in a four-level system (Figure 6.3). The four-level system has an important advantage as compared to the three-level one. Here, population inversion is to be developed for level 3 versus level 4 rather than the ground level 1. Since initial population of level 4 is significantly lower than that of the ground (level 1) state, population inversion and optical gain develops at lower pump level without significant depletion of the ground state, as is mandatory in the case of every three-level system. The four-level scheme is inherent in the following ions embedded in transparent crystalline or glass materials: Ce^{3+}, Ti^{3+}, Cr^{2+}, Nd^{3+}, Cr^{4+}, and Er^{3+}.

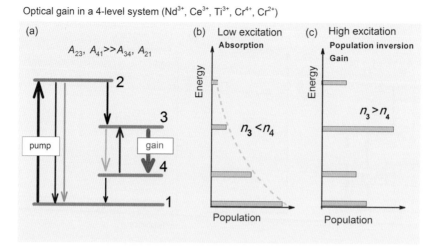

Figure 6.3 Optical gain in a four-level system. (a) Optical transitions; (b) populations at low excitation; (c) populations at high excitation. Since initial population of level 4 is lower than that of level 1, population inversion for level 3 with respect to level 4 occurs at lower intensity as compared to population inversion in the channel $3 \rightarrow 1$. Spontaneous relaxation rates should be fast in channels $2 \rightarrow 3$, $4 \rightarrow 1$, and slow in channels $3 \rightarrow 4$ and $2 \rightarrow 1$.

One can see that stimulated downward transitions, first discovered by Albert Einstein in 1916, play a decisive role in development of optical gain. Interestingly, the idea of optical gain possibility was first suggested in 1934 by V. A. Fabrikant, a Russian physicist. Experimental implementation happened only a few decades later. A three-level system for optical gain was proposed by N. G. Basov and A. M. Prokhorov and implemented by T. H. Maiman in the first laser in 1960. It was based on ruby, a sapphire crystal containing Cr^{3+} ions, and generated red light at 694.3 nm.

6.2 LASERS

6.2.1 Optical quantum generators

From radioengineering it is known that *an amplifier with positive feedback becomes a generator*. Generation occurs owing to multi-pass amplification of inevitable noise present in every system at a frequency at which positive feedback features the highest value. This radiophysical principle has been used together with the idea of optical gain to develop lasers. In optics, positive feedback can be organized using a resonator (a cavity), consisting of a pair of mirrors, one being partially transparent to let light leave a cavity (Figure 6.4). Initial radiation from luminescence after multiple passes between cavity mirrors with optical gain per pass exceeding losses in a cavity (scattering, absorption, partial leakage through the mirrors) will give an optical quantum generator.

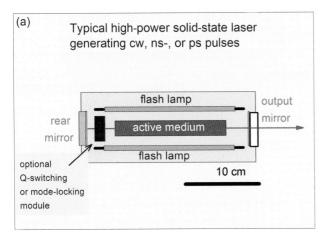

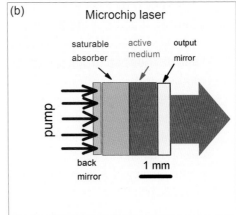

Figure 6.4 Laser design. (a) Traditional laser design with transverse pumping of elongated active element. Nowadays lamps are being replaced by LED arrays. (b) The modern trend in solid-state laser design, a microchip laser, containing a thin slab of highly concentrated active medium pumped longitudinally by a semiconductor laser diode or an LED.

Lasers are essentially nonlinear systems, and a transition from stochastically emitted spontaneous radiation to generation of coherent radiation can be treated in terms of non-equilibrium second-order phase transition. Lasing becomes possible if optical gain per pass exceeds optical losses. Therefore, a specific threshold pump energy should be surpassed to get to the generation regime. Upon increasing the external pump level, generation will start for that mode in which gain exceeds loss. Therefore, typical laser radiation is polarized and monochromatic. Generation of wide-band laser radiation is a serious challenge.

For several decades, solid-state lasers have been pumped by flash lamps with rather wide emission spectrum, whereas the absorption spectrum of ions in the active media is typically very narrow. Most of the optical pump energy has been wasted, resulting in very low overall efficiency – 1% or less, depending on the operation mode (continuous or pulsed). During recent decades, great progress in development of semiconductor laser diodes and light-emitting diodes (LEDs) has enabled application of properly tuned monochromatic or narrow-band radiation for pumping to replace flash lamps and to give much higher efficiency, measuring in a number of cases double-digit percentages. Lasers with flash lamp pumping are still present in the market at the time of writing, but a strong tendency to replace lamps with LEDs is well pronounced. Not only do LEDs promise lower energy consumption owing to higher electric to optical power conversion, but the narrower and adjustable emission spectrum allows more efficient use of optical energy for absorption by an active medium, thus allowing reduction in waste of unused optical pump energy and undesirable heating of laser components. Additionally, in the case of pulse lasers, LED pulses can be reasonably controlled by the power supply driver to eliminate wasted pump energy in the time domain.

Recent progress in fabrication of highly concentrated active media, with nearly 10% content of active centers (ions), enables high-gain coefficients at rather thin (1 mm or less) thickness. These achievements resulted in microchip lasers pumped through a rear Bragg mirror transparent for pump source (a powerful LED or a semiconductor laser diode) spectrum.

Figure 6.5 summarizes the currently available active media for optically pumped solid-state lasers. One can see that a very broad range is covered, though there is a definite lack of laser active materials within the visible spectrum. This gap is filled by means of second harmonic generation of Ti^{3+}:sapphire lasers.

Frequency of a laser can be doubled by passing through an additional crystal featuring nonlinear light–matter interaction. This crystal can be placed either outside or inside the laser cavity. For example, a green laser pointer (Figure 6.6) routinely used in many applications has an active medium, a piece of Nd-doped crystal, featuring optical gain at 1.06 μm according to the four-level scheme; a crystal for frequency doubling to get visible radiation with 530 nm wavelength; a semi-transparent output mirror forming a cavity together with the rear mirror which is highly reflective at 1.06 μm but is transparent otherwise; and an infrared (IR) laser diode with wavelength 808 nm pumping the active medium through the rear mirror. Finally an output color filter is attached to absorb undesirable main frequency

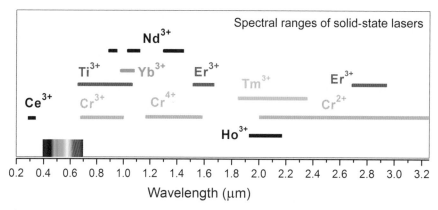

Figure 6.5 Spectral ranges of commercial solid-state lasers with optical pumping. All lasers can operate in continuous as well as Q-switching and mode-locking regimes to get CW, nanosecond, and picosecond radiation, respectively.

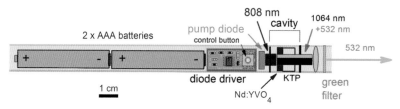

Figure 6.6 Design of a handheld green laser pointer. KTP is potassium titanyl phosphate ($KTiOPO_4$) crystal for frequency doubling.

radiation that has not been converted to green. Frequency conversion efficiency is typically much less than 1% for low-power laser radiation and can be as high as tens of percent for high-power lasers. Note that frequency doubling is a nonlinear process defined by radiation power but not energy. Therefore, low-energy but short-pulse lasers can feature very high power at pretty low energy, e.g., 1 MW per pulse is routinely feasible at 10 mJ pulse energy, which for a repetition rate about 100 Hz gives continuous power of about 1 W only.

6.2.2 Q-switching and mode-locking regimes

To get high-power short pulses, a saturable absorber can be inserted in a laser cavity. In this case, laser dynamics is defined not only by processes in the optically pumped active medium, but also by intensity-dependent transparency of the saturable absorber. This gives the possibility to pump the active medium hard until optical gain multiply exceeds losses without the saturable absorber taken into account, and then upon bleaching of the saturable absorber a short and powerful laser pulse will be generated. This mode of operation is termed *Q-switching* and is routinely used to generate nanosecond pulses. Q-switching enables squeezing of pump energy into a short burst of monochromatic radiation. Another regime with a saturable absorber is based upon keeping the laser very close to the threshold without saturable absorber losses taken into account, and using a saturable absorber that brings only small losses into the laser cavity. In this case, ultrashort picosecond or even subpicosecond pulses develop sequentially, leaving the cavity with time spacing defined by the radiation round trip in the cavity. This is termed the *mode-locking regime.* Laser dynamics both in Q-switching and in mode-locking are defined by a complex interplay of intensity- and time-dependent population dynamics both in the active medium and in the saturable absorber, and its analysis is beyond the scope of this chapter. It is important to note that glasses doped with semiconductor nanocrystals represent the definite type of nonlinear optical material suitable for application as saturable absorbers both for Q-switching and mode-locking. The first experiments on Q-switching using a commercial cutoff filter, i.e., CdS_xSe_{1-x}-semiconductor-doped glasses, were reported by G. Bret and F. Gires in 1964 for a ruby laser, though at that time quantum size effects on linear and nonlinear optical properties of glasses containing semiconductor nanocrystals had not been identified and even nanometer-size semiconductor crystallites had not been documented.

In 1990, Ursula Keller and co-workers proposed a fully semiconductor-based monocrystalline multilayer structure for both Q-switching and mode-locking in compact lasers. It is referred to as a semiconductor saturable absorber mirror (SESAM), or just a saturable absorber mirror (SAM). It represents a semiconductor monocrystalline Bragg mirror with high reflection covered by a thin film of a fast semiconductor saturable absorber, often a multiple quantum well structure. Figure 6.7 shows an example of a SESAM application in a visibly emitting praseodymium picosecond laser (Gaponenko et al. 2014). An active medium is pumped by a frequency-doubled semiconductor laser.

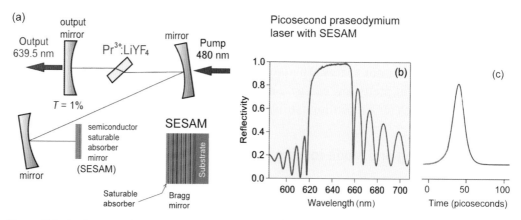

Figure 6.7 A picosecond praseodymium laser pumped with a frequency-doubled semiconductor laser and mode-locked with a semiconductor saturable absorber mirror consisting of a multilayer semiconductor nanostructure. (a) Laser setup; (b) reflectivity of the SESAM; (c) temporal shape of a laser pulse. Courtesy of M. Gaponenko.

6.3 SEMICONDUCTOR LASERS

Solid-state lasers need optical pumping by flash lamps, semiconductor LEDs, or another laser. However, for many practical applications electric pumping is needed. Since solid-state lasers are based on slightly doped dielectric crystals or glasses, they cannot conduct current and cannot convert electric energy into optical radiation. Such conversion is possible for *gas lasers* using gas discharge to pump atoms heavily. Carbon dioxide (CO_2, IR, approximately 10 μm), helium–neon (Ne ions, 632 nm), argon (Ar, 488 and 514 nm), nitrogen (N, 337 nm), and He–Cd (Cd ions, 325 and 440 nm) are used commercially in many applications. Probably the most powerful commercial material processing lasers are based on CO_2 as an active medium owing to efficient heating by IR radiation, and these form a big portion of the laser market. However, because the concentration of atoms in gases is low, the optical gain value is low as well, and a length of active medium of the order of 1 m is necessary to get reasonable power. Semiconductor lasers use condensed matter to develop optical gain and enable simultaneously high gain values and current injection pumping. There is no doubt that progress in semiconductor lasers revolutionized the laser industry, and nowadays the semiconductor laser market, by value, is approximately equal to the rest of the laser market, including gas lasers, solid-state lasers, and obsolete dye lasers. Semiconductor lasers made possible efficient optical communication (fiber to the home; FTTH), laser surgery devices, CD players, CD-ROM memory devices, DVD players and DVD rewritable data storage devices, their evolution to the Blu-ray standard, laser printers and related copy–fax devices, laser pointers, etc. Semiconductor lasers measure a few cubic millimeters or less, and can be used as coherent light sources for direct applications as well as for solid-state laser pumping.

Population dynamics: development of optical gain

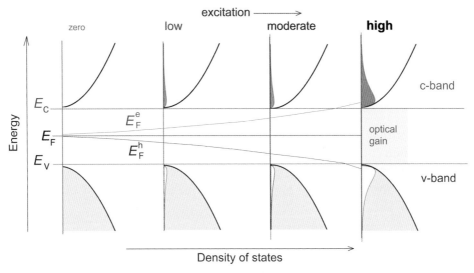

Figure 6.8 Change in population of electrons and holes and their quasi-Fermi levels under zero, low, moderate, and high excitation conditions. Gain develops for photon energies $E_g < \hbar\omega < E_F^e - E_F^h$ at high excitation levels provided that spacing between quasi-Fermi levels exceeds the band gap energy E_g. The extreme limit for quasi-Fermi level spacing equals the energy of excitation radiation quantum $\hbar\omega_{exc}$.

To understand the origin of optical gain in semiconductors, consider optical pumping of a semiconductor crystal by radiation with photon energy exceeding the band gap, $\hbar\omega > E_g$ (Figure 6.8). It is instructive to recall first Figure 3.2, in which electron distribution within the c-band is explained.

At zero or extremely low optical excitation levels, concentration of electrons in the c-band and holes in the v-band are zero or negligibly small. The Fermi level is in the very middle of the band gap. Then, upon increasing optical intensity, light absorption results in photogenerated electrons and holes in the c- and v-band, respectively. The semiconductor crystal emits light, with optical absorption being the same as for a non-excited crystal. At moderate incident light intensity, electron and hole concentrations increase and are no longer negligible compared to the available number of states. Then, optical absorption coefficient falls. Non-equilibrium electrons and holes obey the Fermi–Dirac distribution function within the c- and v-bands, respectively, but upon excitation electron gas and hole gas are characterized by separate Fermi energies, i.e., quasi-Fermi level for electrons, E_F^e, and quasi-Fermi level for holes, E_F^h. The separation between these energies, $\Delta E = E_F^e - E_F^h$ grows with carrier density and can serve as a measure of pump level. At moderate excitation levels, concentrations of electrons and holes are comparable with the available number of states within the band extremes (c-band bottom and v-band top) and optical absorption falls further. The optical absorption spectrum shifts to higher photon energies; this phenomenon for semiconductors is often referred to as the *dynamical*

Burstein–Moss shift to emphasize the analogy of heavily pumped semiconductors with heavily doped ones where absorption shifts to blue because of high equilibrium carrier concentration. When pump level is high, *quasi-Fermi level separation can exceed the band gap energy. This is the condition for optical gain in a semiconductor crystal.* Gain develops for photon energies

$$\text{Optical gain in semiconductor} \quad E_g < \hbar\omega < E_F^e - E_F^h \tag{6.8}$$

and equals zero at $\hbar\omega = E_F^e - E_F^h$. The extreme high-intensity value of zero gain (i.e., also zero absorption coefficient) at $\hbar\omega = E_F^e - E_F^h$ is the definite analog to an extremely pumped two-level system featuring equal populations of the excited and the ground states. The extreme limit for quasi-Fermi levels spacing equals the energy of excitation radiation quantum $\hbar\omega_{exc}$.

At high concentrations, electrons and holes no longer form an ideal gas, but essentially transform into *electron–hole plasma*. This gives rise to band gap shrinkage approximately by the amount of the average electron–hole Coulomb attractive interaction energy, i.e.,

$$E_g^* \approx E_g - \frac{1}{4\pi\varepsilon_0} \frac{e^2}{\varepsilon \bar{r}}, \quad \bar{r} = (n_e + n_h)^{-\frac{1}{3}}. \tag{6.9}$$

For correct calculation, one must account for change in dielectric permittivity with electron–hole concentration, i.e., for any given electron–hole pair, Coulomb interaction is partially screened by other electrons and holes. If $\bar{r} < a_B$, the interaction for every electron–hole pair becomes stronger than in a hydrogen-like exciton, and then excitons can no longer be identified. There is no difference between "free" and "coupled" electron–hole pair states. Upon growing excitation level (and electron–hole gas density), exciton absorption bands first broaden because of exciton–exciton, exciton–electron, and exciton–hole collisions, and then vanish to form a plain absorption edge without resolved peaks. It is for this reason that excitons are not presented in Figure 6.8. Additionally, for the sake of simplicity, band gap shrinkage is not shown in Figure 6.8.

Now we can discuss the design of the first laser diodes. Originally, a laser diode was designed as a p–n homojunction with degenerate p- and n-layers in the upper part of a bulk semiconductor (Figure 6.9). Optical gain developed in the narrow junction area (shadowed gray in the "voltage on" panel) upon direct bias. In this area, a condition of optical gain (Eq. (6.8)) can be met, and stimulated downward transitions will dominate absorption from stimulated upward transition. Laser radiation develops along the p–n junction plane with the positive feedback from the polished crystalline facets.

The first semiconductor homojunction lasers were made in 1962 by several USA groups (R. Hall et al., General Electric; M. Nathan et al., IBM; T. Quist et al., MIT) using GaAs crystals with a near-IR emission band. The first visible semiconductor laser was reported in the same year by N. Holonyak and S. Bevacqua at General Electric, based on a $Ga(As_{1-x}P_x)$ compound. In the USSR in 1963 the first laser was reported by V. S. Bagaev et al. (Lebedev Physical Institute).

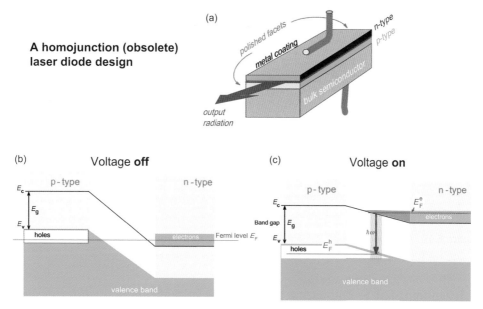

Figure 6.9 Design of the first semiconductor lasers based on a p–n homojunction. (a) The general layout; (b) energy diagram without voltage; and (c) energy diagram with voltage. When voltage is "on," the Fermi level breaks into expanding quasi-Fermi levels; when the condition in Eq. (6.8) is met, optical gain develops and light generation occurs owing to positive feedback from parallel crystal facets.

6.4 DOUBLE HETEROSTRUCTURES AND QUANTUM WELLS

Homojunction lasers were very inefficient, with the need for high current density because of poor localization conditions for injected electrons and holes in the p–n junction area, and also because of high absorptive losses in the thick p- and n-layers. Though these losses could be partially reduced or avoided owing to absorption saturation and band gap shrinkage at high electron and hole densities, the efficiency was well below 0.1% and current densities exceeded 10^3 A/cm^2. In 1963, Z. I. Alferov and R. F. Kazarinov at Ioffe Institute in Leningrad (USSR, now St-Petersburg, Russia) and H. Kremer at Varian Ass. (Palo Alto, USA) independently proposed the way to gradual improvement of semiconductor laser diodes. They suggested using the thin double heterostructure instead of the classical p–n junction (Figure 6.10; see also the box in Section 3.3). A thin layer of narrow-gap semiconductor material provides high localization of electrons and holes in space, and at the same time eliminates absorptive losses outside since surrounding wide-gap semiconductor materials were transparent at the operation frequency. Al$_x$Ga$_{1-x}$As/GaAs/Al$_x$Ga$_{1-x}$As with good lattice matching was used in the first double heterostructure lasers. This structure, with certain improvements, was widely used in the first CD players and CD-ROM drivers.

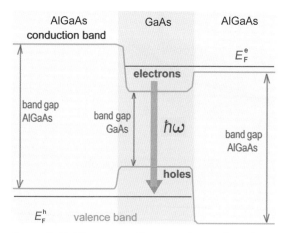

Figure 6.10 The energy diagram of a semiconductor heterojunction laser.

It is important to emphasize that progress in semiconductor laser design and the relevant semiconductor materials/structures toward higher efficiency and wider application range has become possible not only owing to ingenious optoelectronic ideas, but essentially owing to the progress in controllable growth of epitaxial multilayer heterostructures of complex semiconductor compounds meeting lattice matching, desirable operation wavelength, and potential heterojunction barrier requirements. It is the result of dramatic improvement in molecular beam epitaxy (MBE) and metal–organic chemical vapor deposition (MOCVD) technologies that semiconductor lasers have become an important part of our daily lives, covering optical communication, laser printing, optical data storage (CD and DVD players, memory devices), medical and analytical instruments, entertainment, etc. For optical communication needs, two ranges, one near 1.3 μm and the other near 1.5 μm, are the best because of the optical fiber transparency windows. A transition from ternary AlGaAs to quaternary system InGaAsP in semiconductor lasers was important (Lebedev Physical Institute, 1974). Now this is the most widespread injection laser for long-distance optical communication lines.

The double heterostructure design made it possible to reduce threshold currents by more than two orders of magnitude: from 10^5 A/cm^2 to less than 10^3 A/cm^2 during a few years from the mid-1960s to the beginning of the 1970s. A double heterostructure in a semiconductor laser diode not only enables strong localization of injected electrons and holes, but also gives rise to the optical confinement effect owing to higher refractive index of the narrow-band active middle layer. To further enhance optical confinement, a separate confinement five-layer heterostructure was first proposed and performed in 1973 at Bell Labs. This optical confinement promotes waveguiding of radiation in the desirable direction at the expense of inhibited radiation spreading otherwise (Figure 6.11).

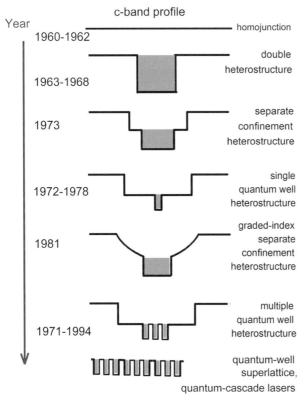

Figure 6.11 Time evolution of semiconductor lasers shown in terms of the conduction band profile. When two dates are indicated, the first corresponds to the theoretical prediction and the second to experimental realization.

Further ideas on electron and hole confinement have resulted in elaboration of a quantum well laser. It was suggested by Dingle et al. in 1974 and implemented for the first time in 1978 by N. Holonyak and co-authors (Dupius et al. 1979). The threshold current went down to 100 A/cm². The advantages of quantum well lasers are so considerable that this type of laser nowadays comprises essentially the entire commercial market for laser diodes. Notably, since electron and hole de Broglie wavelengths in semiconductors are much smaller than laser radiation wavelengths, quantum well-based structures can be spatially embedded into the graded-index design (Tsang 1981), allowing for further reduction in threshold current density. Z. I. Alferov and co-workers (1988) suggested and experimentally implemented a laser structure in which a single quantum well is surrounded by few-period superlattices from both sides. A few-period superlattice not only provides graded refractive index profile, but also forms a barrier for undesirable dislocation motions to the active quantum well area. Thus threshold currents below 50 A/cm² have become possible.

6.5 SURFACE-EMITTING SEMICONDUCTOR LASERS

The geometry of both homojunction and heterojunction semiconductor laser diodes presented in Figures 6.9 and 6.10 appear to be becoming obsolete. The main feature of these devices is light emission along the substrate. This solution has a number of drawbacks. For example, laser mirrors cannot be controlled since their reflection is fully defined by the semiconductor refractive index. Further, when lasers are fabricated in large quantities on a wafer, inspection and quality checking is not possible. Furthermore, light emitted by a thin active layer escapes outside it, to inevitably result in losses.

Nowadays, edge-emitting lasers are being replaced by surface-emitting lasers (Figure 6.12). In this design, a laser generates light across the active layer, normally to the substrate plane. To compensate for decreased optical gain resulting from a very thin gain layer, cavity mirrors need to be highly reflective and low-loss. The mirrors are fabricated as distributed Bragg reflectors (i.e., as a 1D periodic structure, considered in detail in Section 4.2). The advantage is the possibility to fabricate a high-quality, fully monocrystalline mirror by epitaxial growth of semiconductor thin layers. However, since the chemical compositions are of the alternating layers close to each other to ensure lattice matching and technological compatibility, the number of layers can be rather high (many tens of periods) because reflectivity of no less than 99% is typically required. Another fine aspect of the design, though not apparent at first glance, is the fact that multilayer Bragg reflectors are fully involved in current flow, as can be seen from Figure 6.12. Therefore, not only lattice matching, low absorption losses, and refractive index contrast requirements should be met, but also adequate doping to give either n- or p-type conductivity is mandatory.

The design shown in Figure 6.12 has acquired the notation VCSEL from Vertical Cavity Surface-Emitting Laser. It was proposed for the first time and realized by the Japanese group headed by K. Iga (Tokyo Institute of Technology) in 1979, first with metal mirrors.

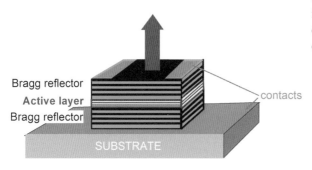

Figure 6.12 A sketch of VCSEL design. The real number of periods in multilayer Bragg reflectors can be considerably larger than sketched to ensure high reflectivity.

VCSEL is essentially a laser-on-chip device. Though its design might look more complicated compared to classical in-plane emitting diodes, in mass production VCSELs appear to be cheaper owing to options to automate dicing into single chips across a wafer of the desirable square (i.e., power) and to allow for efficient on-chip inspection and testing in the fabrication process after dicing but prior to cutting a wafer into individual lasers. VCSELs can be readily diced and cut into linear arrays or into two-dimensional arrays if necessary. Therefore, high-power arrays of VCSELs can be easily fabricated. A short cavity ensures large mode spacing; therefore, single-mode operation is easier than for the classical edge-emitting devices. VCSEL can be made cylindrical to shape the output beam in a controllable way.

6.6   QUANTUM DOT LASERS

Laser diodes warm up since their efficiency is always less than 1. Regretfully, warming up results in higher threshold current, J_{th}, that in turn promotes further heating of a diode. This relationship limits the maximal power a diode can generate for a given cooling regime. In 1982, Y. Arakawa and H. Sakaki examined the theoretically possible relationship of temperature-dependent threshold current and space dimensionality. They found that threshold current rises with temperature because of carriers' wider spreading on the energy axis in accordance with the product of the Fermi–Dirac distribution function and density of states (DOS) (see Figure 3.2). Since lower dimensionality offers a lower number of states, they arrived at the conclusion that lower dimensionality gives a smaller boost of J_{th} with temperature and makes it vanish entirely for zero-dimensional systems (i.e., quantum dots). For the dependence of normalized threshold current,

$$J_{\text{th}}^* = \frac{J_{\text{th}}(T)}{J_{\text{th}}(0)} = \exp\left(\frac{T}{T_0}\right),$$

(6.10)

they found progressively rising T_0 with falling dimensionality and arrived at temperature-independent threshold in the limit of zero-dimensional structures, i.e., an ideal quantum dot (Figure 6.13). They suggested that quantum dot-based lasers will therefore feature extreme temperature stability.

The first experimental evidence for lasing in semiconductor quantum dots was reported by Egorov et al. (1994) at 77 K at Ioffe Institute (Leningrad, former USSR). Nowadays, laser diodes based on quantum dots are commercially produced for the optical communication wavelength range around 1.3 μm. Quantum dot lasers are discussed in detail in Chapter 9.

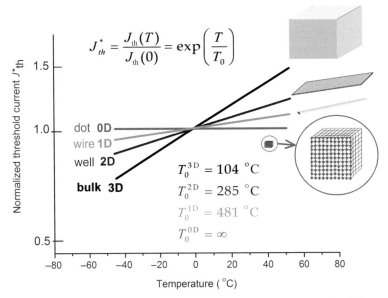

Normalized threshold current J^*_{th}

$$J^*_{th} = \frac{J_{th}(T)}{J_{th}(0)} = \exp\left(\frac{T}{T_0}\right)$$

dot **0D**
wire **1D**
well **2D**
bulk 3D

$T_0^{3D} = 104\ °C$
$T_0^{2D} = 285\ °C$
$T_0^{1D} = 481\ °C$
$T_0^{0D} = \infty$

Temperature ($°C$)

Figure 6.13 Temperature dependencies of threshold current for different dimensionalities calculated by Arakawa and Sakaki (1982).

6.7 QUANTUM CASCADE LASERS

Quantum cascade lasers are multiple semiconductor quantum well devices generating IR electromagnetic radiation owing to intraband downward transitions with multiple repetition of acceleration-emission processes by the same electrons. In a cascade laser, every injected electron can generate stimulated emission of several photons in the course of a multistage cascade process shown in Figure 6.14. A quantum cascade laser consists of a number of periodically arranged active regions and injector regions. Every active region contains discrete energy levels for electrons, and every injection region contains at least one miniband with a quasi-continuous spectrum. The energy alignment of every miniband with an adjacent active region enables an electron leaving low level 1 in a higher energy active region (left-hand part of Figure 6.14) after acceleration in the injector region to find itself at the high-energy level 2 in the low-energy active region (right-hand part in Figure 6.14). Electron transfer between an active region and an injector area occurs by means of resonance tunneling (see Section 2.2, Figure 2.8), and electron acceleration inside the injection region becomes possible owing to a continuous-like energy miniband formed by a quantum well superlattice.

The idea of intraband optical gain in a multiple quantum well system was first suggested by R. Kazarinov and R. Suris at Ioffe Institute (Leningrad, former USSR) in 1971, and was experimentally realized in 1994 at Bell Labs (Murray Hill, USA) by the group of

Quantum cascade laser

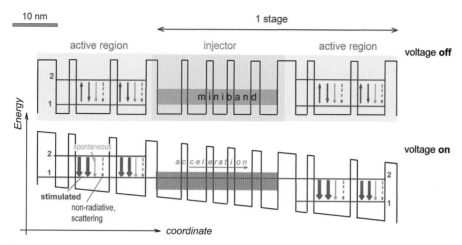

Figure 6.14 Conduction band energy diagram showing a portion of a quantum cascade laser. With external applied voltage, electrons after downward transition from level 2 to level 1 in the left-hand active region experience acceleration in the injector region with continuous increase in kinetic energy, and then again perform downward transitions from level 2 to level 1 in the right-hand active region.

Federico Capasso, with the term *quantum cascade lasers* being coined by the same group (Faist et al. 1994). Though the original idea appears quite clear and elegant, the experimental realization of a quantum cascade laser is not so straightforward. The main problem is fast excited electron nonradiative leakage by means of interaction with the lattice (electron–phonon scattering). Therefore, because of inevitable high internal losses, optical gain is quite low, contrary to interband optical transitions in which just a single quantum well can be an active medium of a laser. To overcome this problem, a long total active medium layer is necessary. Typical quantum cascade lasers contain 20–50 periods (stages) of an injector area plus an active region, comprising in total several hundred semiconductor layers. Another problem is the perfect optimized alignment of electron levels and minibands in a complex multiple quantum well structure. Here, e.g., an additional narrow well is added in the active region to promote higher tunneling probability from an injection area for which a slightly aperiodic superlattice is typically used.

Quantum cascade lasers have a number of important advantages. First, they generate powerful coherent radiation in mid- and even far-IR radiation. The operation wavelength is controlled by the material properties and by the quantum well width (layer thickness). The typical materials involved are AlGaAs/GaAs (operation wavelength from 2.6 to 14 μm), and GaInAs/AlInAs (operation wavelength from 70 to 250 μm); other materials under investigation are GaInAsP-based heterostructures and SiGeSn-based ones. Quantum cascade lasers present an efficient tool for spectral analysis of gas content, including first of all ecological and medical issues.

6.8 SEMICONDUCTOR LASERS ON THE MARKET

In 2015, the total value in the worldwide laser market was USD 10 billion, with laser diodes representing 43% of this value. Among semiconductor lasers, the VCSEL market can be estimated as approximately USD 1 billion, and is expected to grow to over USD 2 billion by 2020. The principal applications of laser diodes are optical data storage and optical communications. Others include solid-state laser pumping, medicine, barcode scanning, inspection and measurement, material processing, entertainment, and scientific research. Semiconductor lasers today cover continuously the wide range from near-UV (370 nm) to middle-IR (to 10 μm), with the only break in the range of 540–600 nm.

The total revenue of semiconductor lasers is foreseen to grow faster than for the laser industry as a whole, and is predicted to reach more than USD 9.5 billion by 2024 (Semiconductor Laser Market Analysis 2016). The fiber-optic semiconductor laser segment is expected to grow faster than other sectors owing to growing demand for optical communication services. North America and Europe will probably lose a portion of the market, whereas the Asia-Pacific region will gain owing to rapid industrial development in this region. Along with communications, industrial applications (mainly high-power diodes) and optical storage will drive the growth of the semiconductor laser market.

Conclusion

- Every two-level system under a high enough excitation rate exhibits absorption saturation, the absorbed power tending to the ultimate limit defined by the matter parameters, whereas optical transmission rises to 1.

- Systems with three or more levels, upon excitation by high-frequency radiation, can feature population inversion resulting in the optical gain for transition(s) corresponding to lower frequencies.

- A medium with optical gain combined with positive feedback by means of a resonator becomes an optical quantum generator, a laser provided that optical gain exceeds optical losses.

- Solid-state lasers cover a wide spectrum range from 286 nm to 2940 nm; however, gaps exist in the ranges 330–660 nm and 1670–1850 nm. No commercial solid-state laser exists that generates violet, blue, green, yellow, and orange light in the fundamental frequency, though frequency doubling of radiation of different solid-state lasers can give different colors in the visible at the expense of efficiency losses.

- Solid-state lasers can operate in continuous mode, Q-switched regime (nanosecond pulses), and mode-locked regime (picosecond pulses) using intracavity saturable absorbers based on semiconductor quantum dots, doped materials, or a thin semiconductor monocrystalline layer coupled with a multilayer Bragg mirror; the latter is referred to as SESAM (semiconductor saturable absorption mirror).

- Recently, solid-state microchip lasers have become routinely available with different operation modes owing to development of highly concentrated gain media and semiconductor Q-switching and mode-locking components.

- In semiconductors, the condition of optical gain is met if the difference between electron and hole quasi-Fermi levels exceeds band gap energy. This condition is an analog to population inversion in a system with discrete energy levels.

- Modern semiconductor lasers are grown epitaxially and emit light in the vertical direction (VCSEL, vertical cavity surface-emitting laser). These are based on an electrically pumped p–n junction and therefore are referred to as laser diodes or injection lasers, and contain (a) double heterostructure(s) in the middle, often forming a single or a few quantum wells. Modern laser diodes cover the range from 370 to 15,000 nm and can be used either as is or as pump sources for solid-state lasers. Pulse regimes can be obtained by current modulation, Q-switching, or mode-locking.

- To avoid undesirable temperature-dependent threshold currents that require higher currents and lead to higher losses upon semiconductor heating, quantum dot lasers have been suggested and commercially designed, currently for the wavelength range 1000–1300 nm.

- In quantum well superlattices, optical gain in the longwave range can be obtained resulting in quantum cascade lasers operating in the IR from 3 to 15 μm, and possibly for longer wavelengths toward microwaves.

- Nanostructures are used in lasers as Q-switches, mode-lockers, Bragg mirrors, and active media in laser diodes. Additionally, quantum well-based LED arrays can be used as efficient pump sources in solid-state lasers. In all cases except Bragg mirrors, electron and hole confinement resulting in quantum size effects are of principal importance. In Bragg mirrors (1D photonic crystals) the lightwave confinement matters.

Problems

6.1 Explain why matter bleaches at high radiation intensity and why this phenomenon is called absorption saturation.

6.2 Explain why optical gain is not possible in a two-level system but becomes feasible for higher numbers of levels.

6.3 Explain what the quasi-Fermi level is. Why is it intensity-dependent?

6.4 Explain why band gap shrinks upon optical excitation. Estimate band gap shrinking for carrier density 10^{17}, 10^{18}, and 10^{19} cm^{-3} for GaAs and GaN.

6.5 Explain what dynamic Burstein shift is.

6.6 Describe population inversion analog for semiconductor materials.

6.7 Consider how electron and lightwave confinement phenomena can be used in lasers.

6.8 Highlight applications of quantum dot structures in lasers.

6.9 Highlight applications of quantum wells and superlattices in lasers.

6.10 Explain the benefits of replacement of flash lamps by LED arrays in solid-state lasers.

Further reading

Chow, W. W., and Koch, S. W. (2013). *Semiconductor-Laser Fundamentals: Physics of the Gain Materials*. Springer Science & Business Media.

Coleman, J. J. (2012). The development of the semiconductor laser diode after the first demonstration in 1962. *Semicond Sci Technol*, **27**, 090207.

Gaponenko, S. V. (2005). *Optical Properties of Semiconductor Nanocrystals*. Cambridge University Press.

Gmachl, C., Capasso, F., Sivco, D. L., and Cho, A. Y. (2001). Recent progress in quantum cascade lasers and applications, *Rep Prog Phys*, **64**, 1533–1601.

Hall, R. N., Fenner, G. E., Kingsley, J. D., Soltys, T. J., and Carlson, R. O. (1962). Coherent light emission from GaAs junctions. *Phys Rev Lett*, **9**, 366–368.

Keller, U. (2010). Ultrafast solid-state laser oscillators: a success story for the last 20 years with no end in sight. *Appl Phys B*, **100**, 15–28.

Koechner, W. (2013). *Solid-State Laser Engineering*. Springer.

Ledentsov, N. N. (2011). Quantum dot laser. *Semicond Sci Technol*, **26**, 014001.

Sennaroglu, A. (ed.) (2006). *Solid-State Lasers and Applications*. CRC Press.

Svelto, O., and Hanna, D. C. (1998). *Principles of Lasers*, 4th edn. Plenum Press.

Ustinov, V. M., Zhukov, A. E., Egorov, A. Y., and Maleev, N. A. (2003). *Quantum Dot Lasers*. Oxford University Press.

References

Alferov, Z. I. (1998). The history and future of semiconductor heterostructures. *Semiconductors*, **32**, 1–14.

Arakawa, Y., and Sakaki, H. (1982). Multidimensional quantum well laser and temperature dependence of its threshold current. *Appl Phys Lett*, **40**, 939–941.

Bret, G., and Gires, F. (1964). Giant pulse laser and light amplifier using variable transmission coefficient glasses as light switches. *Appl Phys Lett*, **4**, 175–176.

Dingle, R., Wiegmann, W., and Henry, C. H. (1974). Quantum states of confined carriers in very thin $Al_xGa_{1-x}As$-GaAs-$Al_xGa_{1-x}As$ heterostructures. *Phys Rev Lett*, **33**, 827–830.

Dupuis, R. D., Dapkus, P. D., Chin, R., Holonyak, N., and Kirchoefer, S. W. (1979). Continuous 300 K laser operation of single quantum well $Al_x Ga_{1-x} AsGaAs$ heterostructure diodes grown by metalorganic chemical vapor deposition. *Appl Phys Lett*, **34**, 265–267.

Egorov, A. Y., Zhukov, A. E., Kop'ev, P. S., et al. (1994). Effect of deposition conditions on the formation of (In, Ga) As quantum clusters in a GaAs matrix. *Semiconductors*, **28**, 809–811.

Faist, J., Capasso, F., Sivco, D. L., et al. (1994). Quantum cascade laser. *Science*, **264**(5158), 553–556.

Gaponenko, M., Metz, P. W., Härkönen, A., et al. (2014). SESAM mode-locked red praseodymium laser. *Opt Lett*, **39**, 6939–6941.

Kazarinov, R. F., and Suris, R. A. (1971). Possibility of amplification of electromagnetic waves in a semiconductor with a superlattice, *Sov Phys Semicond*, **5**, 707–709.

Semiconductor Laser Market Analysis (2016). Semiconductor laser market analysis by laser type, by application, and segment forecasts to 2024. www.reportbuyer.com/product/4144263, accessed May 2018.

Tsang, W. T. (1981). A graded-index waveguide separate-confinement laser with very low threshold and a narrow Gaussian beam. *Appl Phys Lett*, **39** 134–137.

7 Energy transfer processes

In this chapter, we introduce the general phenomenon of excitation energy transfer, explain radiative and nonradiative types of energy transfer, and derive the basic processes of energy transfer. We look at the Förster resonance energy transfer (FRET) in particular. We also describe Dexter energy transfer, charge transfer, exciton diffusion, and exciton dissociation. Finally, we summarize the modifications of FRET when using nanostructures with mixed dimensionalities and in different assemblies.

7.1 INTRODUCTION

Excitation energy transfer is a directional process that takes place from the excited energy state of a donor (D) to the ground state of an acceptor (A) (see Figure 7.1). This is a common phenomenon that occurs in nature (e.g., in plant leaves). For this basic energy transfer to happen, the emission spectrum of the donor should overlap the absorption spectrum of the acceptor (at least partially). This transfer of the excitation energy from D to A can be simply represented as

$$(D^*, A) \rightarrow (D, A^*), \tag{7.1}$$

where * marks the excited states. Here D and A can in principle be various combinations of atoms (e.g., rare earth ion dopants), molecules (e.g., dyes, proteins), nanostructures (e.g., semiconductor nanocrystals, 2D materials), or pieces of solid (e.g., semiconductors). Some of the common D–A pairs are dye–dye, dye–protein, nanocrystal–nanocrystal, nanocrystal–2D material, and nanocrystal–semiconductor film.

In the most general sense, the excitation processes involved in the energy transfer can be *radiative*, *nonradiative*, or both. In the case of radiative transfer, a photon is emitted by the donor and then absorbed by the acceptor. On the other hand, in the case of nonradiative transfer, there is no "real" photon emitted in the process. Instead, during the energy transfer from the donor to the acceptor, the photon is still bound to the material, which is referred to as a "dressed" (or virtual) photon to be distinguished from real photons. This nonradiative character of the process allows for the possibility of high efficiency.

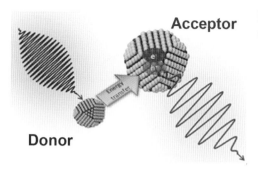

Acceptor

Donor

Figure 7.1 Illustration of energy transfer from a donor (D) to an acceptor (A).

If D and A are chemically different species, the energy transfer is known as a *hetero-transfer*. Otherwise, if they are the same, then this is a *homotransfer*. In the case of homotransfer, the process of energy transfer takes place back and forth and may extend over several species (i.e., resonates among several species).

7.2 RADIATIVE AND NONRADIATIVE ENERGY TRANSFERS

7.2.1 Radiative energy transfer

Radiative energy transfer is based on the emission of a photon by the donor followed by its optical absorption by the acceptor, and is effective when the average distance between the donor and the acceptor is larger than the wavelength. This process does not necessitate any direct interaction in the D–A pairs. Instead, this depends on the spectral overlap and the concentration. On the other hand, nonradiative energy transfer occurs at distances shorter than the wavelength and without the emission of photons; and it is the result of specific short- and long-range interactions between the D–A pair. For example, the nonradiative energy transfer by dipole–dipole interaction can reach distances up to (typically) nearly 20 nm. This provides a tool for determining distances of a few nanometers between D and A precisely.

Radiative energy transfer can be expressed as a two-step process in which first a photon is emitted by the donor and then is absorbed by the acceptor:

$$
\begin{aligned}
&1. \quad D^* \rightarrow D + h\mu \\
&2. \quad h\mu + A \rightarrow A^*
\end{aligned}
\tag{7.2}
$$

This process is usually considered a trivial transfer because of its simplicity; however, the quantitative description is actually complicated because it depends on the sample's size and its configuration with respect to excitation and observation. The fraction f of photons emitted by the donor and absorbed by the acceptor can be expressed as

$$
f = \frac{1}{Q_D} \int_0^\infty I_D(\lambda) \left[1 - 10^{-\varepsilon_A(\lambda)C_A l} \right] d\lambda,
\tag{7.3}
$$

where C_A is the molar concentration and $\varepsilon_A(\lambda)$ is molar absorption coefficient of the acceptor, l denotes the thickness of the sample, Q_D is the donor quantum efficiency (in the absence of acceptor), and $I_D(\lambda)$ is the donor emission intensity subject to the normalization condition that

$$Q_D = \int_0^\infty I_D(\lambda)\,d\lambda. \tag{7.4}$$

If the optical density is not too large, the fraction f can then be approximated by

$$f = \frac{2.3}{Q_D}C_A l \int_0^\infty I_D(\lambda)\varepsilon_A(\lambda)\,d\lambda, \tag{7.5}$$

where the integral gives the spectral overlap between the donor emission and the acceptor absorption, as is required for the radiative energy transfer to occur.

7.2.2 Nonradiative energy transfer

Nonradiative energy transfer requires a specific interaction between the donor and the acceptor species. For example, it can occur if the emission spectrum of the donor overlaps the absorption spectrum of the acceptor such that several vibronic transitions of the donor and the acceptor couple and are thus in resonance. This type of transfer is known as the *resonance energy transfer* (RET).

There are different types of interactions that can be involved in such nonradiative energy transfer. These interactions may, for example, include Coulombic coupling or intermolecular orbital overlap. The Coulombic interactions consist of the long-range dipole–dipole interactions (referred to as the *Förster mechanism*) and short-range multipolar interactions. The interactions due to the intermolecular orbital overlap, which includes electron exchange (referred to as the *Dexter mechanism*) and charge resonance interactions, are short-range. For the allowed transitions of the D–A pair, the Coulombic interaction dominates, even at short distances. For the forbidden transitions between D and A, the exchange mechanism prevails. The interaction distance range for the exchange mechanism is usually <1 nm (Dexter mechanism), whereas this is usually <20 nm (Förster mechanism) in the case of the Coulombic interactions.

7.3 BASIC PROCESSES OF ENERGY TRANSFER

To identify the basic processes of energy transfer, let us look at simple dipole–dipole interactions. For that we can consider two interacting electric dipoles and examine the energy transfer from the first dipole, which is thus the donor, to the second dipole, which is thus the acceptor. Here, for the basic description of the processes, we will specify the regions of interest in the energy transfer. We will start with the donor in the excited state. Therefore, the donor will be an oscillating dipole and radiate electric field. Meanwhile the acceptor dipole will be initially at rest, corresponding to the acceptor in the ground state, which will then absorb the field generated by the donor dipole (Born and Wolf 1999).

First, for the donor, consider the electric field of an oscillating dipole (in vacuum):

$$\mathbf{E}(r,t) = \frac{p(t')}{4\pi\varepsilon_0}\left\{\left[3(\mathbf{n}\cdot\mathbf{d})\mathbf{n}-\mathbf{d}\right]\left(\frac{1}{r^3}-\frac{ik}{r^2}\right)+\left[(\mathbf{n}\cdot\mathbf{d})\mathbf{n}-\mathbf{d}\right]\frac{k^2}{r}\right\}. \tag{7.6}$$

Here, $\mathbf{n}$ is the unit vector in the D–A direction, $\mathbf{d}$ is the unit vector along the donor dipole moment, $k=\omega/c$ is the wave number where c is the speed of light, r is the distance from the donor, and $p(t)=p_0\cos(\omega t)$ is the time-dependent electric dipole moment of the donor with an amplitude p_0 at an oscillation angular frequency ω and $t'=t-r/c$.

Second, for the acceptor, consider a passive absorber that is placed at a distance r from the donor dipole. The power absorbed by the acceptor is then

$$P' = \frac{1}{2}c\varepsilon_0 E_0^2 \sigma \tag{7.7}$$

where E_0^2 is the squared amplitude of the electric field generated by the donor at r distance away from it and σ is the absorption cross-section of the acceptor. Here it is useful to note that the acceptor has to have a nonzero absorption cross-section at the oscillating frequency of the donor to allow for the dipole–dipole interaction (whereby the spectral overlap of the acceptor absorption with the donor emission is required) and hence make the energy transfer possible.

Using Eq. (7.6), taking the square and averaging for all orientations, one can obtain E_0^2 of the donor dipole as follows:

$$E_0^2 = 2\left(\frac{p_0}{4\pi\varepsilon_0}\right)^2\left(\frac{k^4}{3r^2}+\frac{k^2}{3r^3}+\frac{1}{r^6}\right). \tag{7.8}$$

The power radiated by the donor dipole is given by

$$P^0 = \frac{p_0^2\omega^4}{12\pi\varepsilon_0 c^3}. \tag{7.9}$$

Combining Eqs. (7.7)–(7.9), we arrive at the general description for the power absorbed by the acceptor as a function of the distance between the acceptor and the donor, given the absorption cross-section of the acceptor and the power at the acceptor location radiated by the donor:

$$P' = \frac{\sigma}{4\pi r^2}\left[1+\left(\frac{\lambda}{r}\right)^2+3\left(\frac{\lambda}{r}\right)^4\right]P^0 \tag{7.10}$$

where $\lambda = \lambda/(2\pi)$.

For far enough distances, $r \gg \lambda$, one can see that Eq. (7.10) simply boils down to

$$P' = \frac{\sigma}{4\pi r^2}P^0. \tag{7.11}$$

Based on the simple geometric interpretation, this corresponds to the radiative transfer.

Therefore, we express the power emitted by the donor in the presence of an acceptor as the sum of

$$P = \left\{ 1 + \frac{\sigma}{4\pi r^2} \left[\left(\frac{\lambda}{r} \right)^2 + 3 \left(\frac{\lambda}{r} \right)^4 \right] \right\} P^0. \tag{7.12}$$

To examine the distance dependence, we can look closely at the electric field expression of the donor dipole in Eq. (7.6). Here, we find that indeed there are two main characteristic zones defined by the dipole's distance dependence.

1. Near-zone of $r \ll \lambda$: Here, the r^{-3} term is dominant, where the angular dependence is identical to a static dipole with longitudinal and transversal components.

2. Far-zone of $r \gg \lambda$ (also known as radiative or wave zone): Here, the r^{-1} term dominates, which corresponds to a spherical wave. The electric field is thus always perpendicular to the transversal field in this zone.

From Eq. (7.12), we observe that when the acceptor is located in the near-zone, P exceeds P^0. This means that the energy stored in the near-field is larger. Consequently, in the near-zone, there is stronger energy feeding to the acceptor and the donor decay rate then increases in the presence of the acceptor.

Using Eq. (7.11), the absorbed power of the acceptor in the near-zone can be rewritten as

$$P' = \frac{3\sigma}{64\pi^5} \left(\frac{\lambda^4}{r^6} \right) P^0. \tag{7.13}$$

We can express this equation in terms of the radiative rate k_r and the transfer rate k_T by dividing both sides by $h\upsilon$ and relating σ to the molar absorption coefficient ε_A:

$$k_T = k_r \left(\frac{3 \ln 10 \varepsilon_A \lambda^4}{64\pi^5 N_A n^4} \right) \frac{1}{r^6}. \tag{7.14}$$

Here, N_A denotes Avogadro's number, n is the medium refractive index, and $k_r = Q_D / \tau_0$, where τ_0 is the donor lifetime (in the absence of acceptor) while Q_D is the donor quantum efficiency.

Introducing the donor's emission spectrum characterized by $F_D(\lambda)$, we arrive at

$$k_T = \frac{1}{\tau_0} \left(\frac{Q_D 3 \ln 10}{64\pi^5 N_A n^4} \right) \left(\frac{1}{r^6} \right) \int_0^\infty F_D(\lambda) \varepsilon_A(\lambda) \lambda^4 \, d\lambda = \frac{1}{\tau_0} \left(\frac{R_0}{r} \right)^6. \tag{7.15}$$

Here, R_0 is referred to as the *Förster radius*, which is named after Theodor Förster, a German scientist. He first derived this expression using a quantum mechanical approach (Förster 1948) and the classical treatment (Förster 1951) of the dipole–dipole interaction of a D–A pair.

BOX 7.1 FÖRSTER RESONANCE ENERGY TRANSFER

Förster resonance energy transfer (FRET), is frequently employed to measure nanoscale distances and detect molecular interactions. Förster resonance energy transfer is named after the German physical chemist Theodor Förster. Among his greatest achievements are his contributions to the outstanding concept of FRET. Notably, in the late 1940s he developed the first theoretical model that explains the energy transfer between molecules over long distances beyond their orbital contacts. In this model, he used Coulombic coupling between electrons or the excited donor molecule and the acceptor molecule initially in the ground state, employing the coupling of their respective transition dipole moments. His model applies to highly fluorescent singlet-excited molecules. Today this model is commonly used in biolabeling and related proximity analyses in the understanding of biological pathways.

7.4 FÖRSTER RESONANCE ENERGY TRANSFER

One of the most important examples of nonradiative energy transfer is the *Förster resonance energy transfer* (FRET), which is also known as fluorescence resonance energy transfer. FRET is an electrodynamic phenomenon and is a direct result of the long-range dipole–dipole interactions between the donor and the acceptor discussed above. The rate of energy transfer k_T is dependent on the coupling of the donor and acceptor dipoles, which is simultaneously determined by a number of factors, including the extent of spectral overlap of the donor emission and the acceptor absorption, the donor quantum yield, the spatial distance between the donor and the acceptor, and the orientation of the donor and acceptor transition dipoles. The nonradiative process of this energy transfer can be expressed as a transition between the two states of

$$(D^*, A) \xrightarrow{\ k_T\ } (D, A^*) \tag{7.16}$$

where D is the donor in the ground state, D^* is the donor in the excited state, A is the acceptor in the ground state, and A^* is the acceptor in the excited state. Here, k_T denotes the rate of FRET between the D–A pair. In this process, the donor absorbs an external photon, generating an excited state. Subsequently, the donor transfers its excited energy via this nonradiative process to the acceptor, leading to its excited state.

Förster was the first scientist to theoretically explain this process correctly (Förster 1946, 1949). He developed the formulation of the FRET rate and efficiency, which is described in detail in various textbooks and reviews (Clegg 1996; Bredas and Silbey 2009; Clegg 2009; Lakowicz 2010).

From Förster's theory, Eq. (7.15) gives the rate of energy transfer from the donor to the acceptor

$$k_{\mathrm{T}}(r) = \frac{1}{\tau_D}\left(\frac{R_0}{r}\right)^6. \tag{7.17}$$

Once again, here τ_D is the donor photoluminescence decay lifetime (in the absence of acceptor), r is the distance between the donor and the acceptor, and R_0 is the *Förster radius*. We can clearly see from Eq. (7.17) that the rate of energy transfer depends strongly on the distance and scales down with r^{-6}. We also find that the rate of energy transfer is equal to the decay rate of the donor $1/\tau_D$ when $r = R_0$.

In a more detailed study of FRET, the rate of transfer for a single donor and a single acceptor separated by a distance r can be written as (Saricifcti et al. 1992; Clegg 1996),

$$k_{\mathrm{T}}(r) = \frac{Q_D \kappa^2}{\tau_D r^6}\left(\frac{9000(\ln 10)}{128\pi^5 N_A n^4}\right)\int_0^\infty F_D(\lambda)\varepsilon_A(\lambda)\lambda^4\, d\lambda. \tag{7.18}$$

Here, again, Q_D is the donor quantum yield (in the absence of acceptor), n is the refractive index, N_A is Avogadro's number, r is the distance between the donor and the acceptor, and τ_D is the lifetime of the donor (in the absence of an acceptor). The term κ^2 is the factor describing the relative orientation of the transition dipoles of the donor and the acceptor in space. κ^2 is taken as two-thirds for dynamic random averaging of the donor and the acceptor. $F_D(\lambda)$ is the normalized fluorescence intensity of the donor in the wavelength range from λ to $\lambda + \Delta\lambda$, with the total intensity (area under the curve) normalized to unity. $\varepsilon_A(\lambda)$ is the extinction coefficient of the acceptor at λ, which is typically in units of $\mathrm{M}^{-1}\mathrm{cm}^{-1}$.

The overlap integral $J(\lambda)$ expresses the degree of spectral overlap between the donor emission and the acceptor absorption:

$$J(\lambda) = \int_0^\infty F_D(\lambda)\varepsilon_A(\lambda)\lambda^4\, d\lambda, \tag{7.19}$$

$$J(\lambda) = \frac{\displaystyle\int_0^\infty F_D(\lambda)\varepsilon_A(\lambda)\lambda^4\, d\lambda}{\displaystyle\int_0^\infty F_D(\lambda)\, d\lambda}. \tag{7.20}$$

$F_D(\lambda)$ is taken as dimensionless. When calculating $J(\lambda)$, one should use the corrected emission spectrum with its area normalized to unity, or normalize the calculated value of $J(\lambda)$ by the area. The most common units of $J(\lambda)$ include $\mathrm{M}^{-1}\mathrm{cm}^3$ if $\varepsilon_A(\lambda)$ is expressed in units of $\mathrm{M}^{-1}\mathrm{cm}^{-1}$ and λ is in centimeters; and $\mathrm{M}^{-1}\mathrm{cm}^{-1}\mathrm{nm}^4$ if $\varepsilon_A(\lambda)$ is expressed in units of $\mathrm{M}^{-1}\mathrm{cm}^{-1}$ and λ is in nanometers $\left(\mathrm{M} = \dfrac{\mathrm{mol}}{\mathrm{L}}\right)$.

For practical reasons, it is easier to think in terms of the spatial distance rather than transfer rate. Thus, Eq. (7.18) can be expressed in terms of the Förster radius R_0. From Eqs. (7.17) and (7.18), we can obtain:

$$R_0^6 = \left(\frac{9000\,(\ln 10)\,Q_D \kappa^2}{128\pi^5 N_A n^4} \right) \int_0^\infty F_D(\lambda)\,\varepsilon_A(\lambda)\,\lambda^4\,d\lambda. \qquad (7.21)$$

This expression is useful for calculating the Förster radius from the spectral properties of the donor and the acceptor and the donor quantum yield. For typical values, the Förster radius most commonly ranges from 1 to 10 nm.

The efficiency of energy transfer (ξ) can be defined as the fraction of photons absorbed by the donor for which excitation energy is transferred to the acceptor. This fraction is given by

$$\boxed{\xi = \frac{k_T(r)}{\tau_D^{-1} + k_T(r)},} \qquad (7.22)$$

which is the ratio of the transfer rate to the total decay rate of the donor in the presence of the acceptor. From Eq. (7.22), we can see that, when the transfer rate is much faster than the decay rate, the resulting energy transfer is efficient. In contrast, when the transfer rate is slower than the decay rate, the energy transfer is inefficient because only a little transfer occurs during the excited state lifetime.

The efficiency of energy transfer can be rewritten as a function of the distance by substituting Eq. (7.17) into Eq. (7.22):

$$\xi = \frac{R_0^6}{R_0^6 + r^6}. \qquad (7.23)$$

This equation clearly shows that the transfer efficiency is strongly dependent on the distance when the D–A distance is near to R_0 (see Figure 7.2). When r is equal to R_0, the resulting transfer efficiency is 50%. From this observation, one can also define the Förster radius as the distance at which FRET is 50% efficient. At this distance $(r = R_0)$, the donor emission decreases to half its intensity in the absence of acceptors.

From Figure 7.2, we also observe that the efficiency quickly increases to unity as the D–A distance decreases below R_0. Conversely, the efficiency quickly diminishes if r is greater than R_0. Here, it is worth noting that the transfer efficiency is already a few percent (1.5%) at $r = 2R_0$, and it is almost unity (98.5%) at $r = 0.5R_0$.

FRET is a highly distance-sensitive process owing to the inverse sixth power (r^{-6}) dependence of the separation distance in the case of point-to-point dipole coupling. Therefore, FRET was proposed to be used as a nanoscale ruler (Stryer and Haugland 1967). FRET has been widely exploited in various applications in molecular biology, for sensing, labeling, nanoscale distance measurements, and understanding of molecular-level interactions. For these biological systems, typically in solution, point-to-point-like interaction is effective, and thus, r^{-6} dependence is commonly valid. Recently, FRET has been

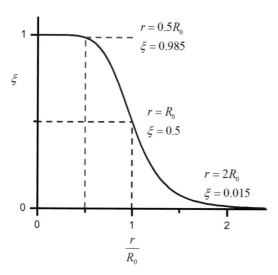

Figure 7.2 FRET efficiency (ξ) as a function of the D–A distance r normalized with respect to the Förster radius R_0.

shown to be useful for optoelectronic technologies toward the purpose of creating highly efficient lighting and solar energy-harvesting systems. For this aim, exciton energy transfer in the quantum dot, wire, and well-based nanostructure can be employed to improve and control the photonic properties for light-generation and -harvesting systems. In these systems, with dimensionality of nanoparticle systems and their assemblies, one has to be careful about the distance dependence. In the following sections, we will discuss the extended theory of FRET in the nanostructures of mixed dimensionality and in various assemblies.

Here we discussed FRET within the dipole–dipole approximation. Beyond this approximation, in the case of multipole Coulomb interactions, such as dipole–quadrupole and quadrupole–quadrupole interaction, it worth pointing out that the FRET rates become proportional to higher orders of $1/r$ (to r^{-8} in the case of dipole–quadrupole coupling and to r^{-10} in the case of quadrupole–quadrupole coupling; Baer and Rabani 2008). Clearly, the energy transfer rate then becomes increasingly strongly dependent on distance, and thus more spatially sensitive. In addition, the interaction range shrinks substantially. Therefore, the dominant term is the dipole–dipole interaction term, and higher other poles may be considered for larger species and/or when the donor and the acceptor are in very close proximity.

7.5 DEXTER ENERGY TRANSFER, CHARGE TRANSFER, EXCITON DIFFUSION, AND EXCITON DISSOCIATION

Among other nonradiative processes, we also have Dexter energy transfer, charge transfer, exciton diffusion, and exciton dissociation. *Dexter energy transfer* (Dexter 1953), which is also known as the charge (electron) exchange energy transfer, relies on the wave function overlap of the electronic states between different molecules in the near-field.

Dexter energy transfer is a short-range energy transfer, unlike FRET which is known to be a long-range energy transfer owing to the working distances that are on the order of 10 nm. Dexter energy transfer is only effective for the $D–A$ separations that are typically on the order of 1 nm or shorter. Dexter energy transfer can also occur between non-emissive electronic states of the materials, such as spin-forbidden triplet states, whereas it is currently believed that these excitons cannot be transferred via FRET because they have negligible oscillator strengths (Köhler and Bassler 2009). These exchange mechanisms are based on the *Wigner spin* conservation rule. Thus, the spin-allowed processes are: (1) singlet–singlet energy transfer, $^1D*+{}^1A \xrightarrow{k_{\text{Dexter}}} {}^1D+{}^1A*$; and (2) triplet–triplet energy transfer, $^3D*+{}^1A \xrightarrow{k_{\text{Dexter}}} {}^1D+{}^3A*$. Dexter energy transfer has exponential distance dependence as compared to the $k_T \propto r^{-3}-r^{-6}$ distance dependencies in the long-range FRET processes (Valeur 2002).

Another important excitonic process is the *exciton diffusion*. The exciton diffuses in a material in the broadened density of states of the same material; this process is called energy migration. Exciton diffusion has been widely studied for organic semiconductors in the search for materials with large diffusion lengths to increase the probability of charge separation at the $D–A$ hetero-interfaces in organic solar cells (Bredas and Silbey 2009). In addition to organic materials, exciton diffusion is crucial in bulk and quantum-confined semiconductor structures, i.e., quantum well, wire, and dot assemblies. Excitons can be transported in these quantum-confined materials; however, these systems should be well understood and controlled because defects can trap the diffusing excitons such that the emission efficiencies are significantly reduced due to the increase of nonradiative recombination channels of the excitons. This picture is also valid for organic semiconductors.

Exciton dissociation is the separation of the bound electron–hole pairs into free carriers. This dissociation is a central step for excitonic solar cells (Gregg 2003), including bulk-heterojunction (Saricifcti et al. 1992) and dye-sensitized (O'Regan and Grätzel 1991), because the generation of free charge carriers is required to realize the photovoltaic operation. In excitonic solar cells, dissociation of the excitons is facilitated by the interfaces that have type-II band alignments to physically break the excitons into free charges. The resistance against the dissociating of excitons in terms of energy stems from the exciton binding energy. Materials with larger exciton binding energy have more stable excitons because it is difficult to overcome this large Coulomb energy between the electron–hole pairs.

Lately, excitonic processes such as *multi-exciton generation* (MEG), *Auger recombination*, and *exciton–exciton annihilation* have been studied in quantum-confined semiconductors. MEG, also dubbed carrier multiplication, is the generation of multi-excitons upon the absorption of a high-energy photon $hv \geq 2 \times E_{\text{Gap}}$. It has been shown that semiconductor quantum dots can be efficient in terms of converting higher-energy photons into multi-excitons (Nozik 2008; Beard 2011). Related to the multi-exciton phenomena, Auger recombination becomes severe because excitons are spatially very close to each other. In Auger recombination, the energy of the recombining exciton is transferred to another already excited charge carrier in the material, such that this charge is excited into higher energy states (i.e., hot carrier). This hot carrier quickly thermalizes to the respective band edge by losing its energy to the phonon vibrations; therefore, Auger recombination can significantly decrease the multi-exciton operation in quantum-confined structures (Klimov et al. 2000).

7.6 NANOSTRUCTURES OF MIXED DIMENSIONALITIES

FRET distance dependency is altered for different acceptor geometries. For example, small molecules and 3D-confined quasi-zero-dimensional quantum dot or nanoparticle (NP) acceptors are considered to be infinitesimal transition dipoles, which lead to the classical r^{-6} dependence in the case of a single donor to a single acceptor. On the other hand, 2D and 1D confined nanowire (NW) and quantum well (QW) acceptors lead to distance dependences that vary with r^{-5} and r^{-4}, respectively (Agranovich et al. 2011; Hernández-Martínez et al. 2013). Basically, quantum confinement of the acceptor modifies the distance dependency of the FRET. Also, note that different assemblies of the acceptors can also alter the distance dependence, as in the case in which a 2D-like assembly of the semiconductor QDs (i.e., a monolayer of QDs on a QW donor) act as a 1D-confined structure, which consequently results in the distance dependence having the form of r^{-4}, similar to QWs.

Here, we look at the energy transfer rate for the cases of X → NP (nanoparticle), X → NW (nanowire), and X → QW (quantum well), where X is a single NP, NW, or QW. Table 7.1 illustrates the distance dependency for FRET: (1) when the acceptor is an NP, FRET is inversely proportional to d^{-6}; (2) when the acceptor is an NW, FRET is proportional to d^{-5}; and (3) when the acceptor is a QW, FRET is proportional to d^{-4}. This indicates that the donor dimensionality does not affect the functional distance dependency.

In all cases, the FRET's distance dependency is given by the acceptor geometry and it is independent of the donor's geometry. The effective dielectric constant, however, depends on the donor's geometry. Therefore, the FRET's distance dependency is dictated by the geometry of the acceptor nanostructure, whereas the donor's contribution to the FRET appears through the effective dielectric constant.

Figure 7.3 presents the energy transfer efficiency for the FRET as a function of d/d_0. These dependencies are important to understand FRET and are valid for the cases when

Table 7.1 **Distance dependency of the FRET rate for different acceptor's geometries. This list includes the effective dielectric constant effect, which is a function of the donor's geometry. Here X = NP (nanoparticle), NW (nanowire), or QW (quantum well)**

α-direction	Donor			Coefficients		Acceptor distance dependency
	NP	NW	QW			X→NP
x	$\varepsilon_{\text{eff}_D} = \dfrac{\varepsilon_{\text{NP}_D} + 2\varepsilon_0}{3}$	$\varepsilon_{\text{eff}_D} = \dfrac{\varepsilon_{\text{NW}} + \varepsilon_0}{2}$	$\varepsilon_{\text{eff}_D} = \varepsilon_0$		$b_x = \frac{1}{3}$	$\gamma_{\text{NP}} \propto \dfrac{1}{d^6}$
y	$\varepsilon_{\text{eff}_D} = \dfrac{\varepsilon_{\text{NP}_D} + 2\varepsilon_0}{3}$	$\varepsilon_{\text{eff}_D} = \varepsilon_0$	$\varepsilon_{\text{eff}_D} = \varepsilon_0$		$b_y = \frac{1}{3}$	
z	$\varepsilon_{\text{eff}_D} = \dfrac{\varepsilon_{\text{NP}_D} + 2\varepsilon_0}{3}$	$\varepsilon_{\text{eff}_D} = \dfrac{\varepsilon_{\text{NW}} + \varepsilon_0}{2}$	$\varepsilon_{\text{eff}_D} = \varepsilon_0$		$b_z = \frac{4}{3}$	
	NP	NW	QW			X→NW
x	$\varepsilon_{\text{eff}_D} = \dfrac{\varepsilon_{\text{NP}_D} + 2\varepsilon_0}{3}$	$\varepsilon_{\text{eff}_D} = \dfrac{\varepsilon_{\text{NW}} + \varepsilon_0}{2}$	$\varepsilon_{\text{eff}_D} = \varepsilon_0$	$a_x = 0$	$b_x = 1$	$\gamma_{\text{NW}} \propto \dfrac{1}{d^5}$
y	$\varepsilon_{\text{eff}_D} = \dfrac{\varepsilon_{\text{NP}_D} + 2\varepsilon_0}{3}$	$\varepsilon_{\text{eff}_D} = \varepsilon_0$	$\varepsilon_{\text{eff}_D} = \varepsilon_0$	$a_y = \frac{9}{16}$	$b_y = \frac{15}{16}$	
z	$\varepsilon_{\text{eff}_D} = \dfrac{\varepsilon_{\text{NP}_D} + 2\varepsilon_0}{3}$	$\varepsilon_{\text{eff}_D} = \dfrac{\varepsilon_{\text{NW}} + \varepsilon_0}{2}$	$\varepsilon_{\text{eff}_D} = \varepsilon_0$	$a_z = \frac{15}{16}$	$b_z = \frac{41}{16}$	
	NP	NW	QW			X→QW
x	$\varepsilon_{\text{eff}_D} = \dfrac{\varepsilon_{\text{NP}_D} + 2\varepsilon_0}{3}$	$\varepsilon_{\text{eff}_D} = \dfrac{\varepsilon_{\text{NW}} + \varepsilon_0}{2}$	$\varepsilon_{\text{eff}_D} = \varepsilon_0$		$b_x = \frac{3}{16}$	$\gamma_{\text{QW}} \propto \dfrac{1}{d^4}$
y	$\varepsilon_{\text{eff}_D} = \dfrac{\varepsilon_{\text{NP}_D} + 2\varepsilon_0}{3}$	$\varepsilon_{\text{eff}_D} = \varepsilon_0$	$\varepsilon_{\text{eff}_D} = \varepsilon_0$		$b_y = \frac{3}{16}$	
z	$\varepsilon_{\text{eff}_D} = \dfrac{\varepsilon_{\text{NP}_D} + 2\varepsilon_0}{3}$	$\varepsilon_{\text{eff}_D} = \dfrac{\varepsilon_{\text{NW}} + \varepsilon_0}{2}$	$\varepsilon_{\text{eff}_D} = \varepsilon_0$		$b_z = \frac{3}{8}$	

the donor–donor and acceptor–acceptor separation distance is larger compared to the donor–acceptor separation distance. However, this condition is difficult to achieve experimentally and most of the experiments (in solid phase) are set using an assembly of nanostructures. Therefore, it is crucial to understand FRET for the cases when the nanocrystals (NP and NW) are assembled into arrays (e.g., chains and films). This aspect is explained in the following section.

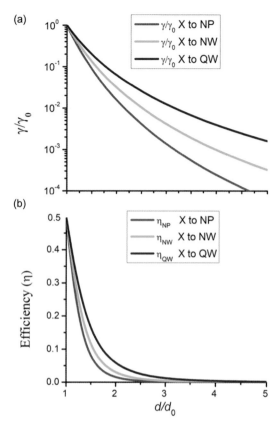

Figure 7.3 (a) Generic FRET rate distance dependency. Energy transfer rates are depicted as a function of d/d_0, where d_0 is the characteristic distance, which satisfies the asymptotic condition. (b) Energy transfer efficiency for the FRET. The red line shows the energy transfer efficiency for the D–A pair, when the acceptor is NP. The green line depicts the energy transfer efficiency for the D–A pair when the acceptor is NW. The blue line gives the energy transfer efficiency for the D–A pair when the acceptor is QW. Here, X = NP, NW, or QW.

7.7 ASSEMBLY OF NANOSTRUCTURES

Here, we describe the framework of generalized Förster-type nonradiative energy transfer between one-dimensional (1D), two-dimensional (2D), and three-dimensional (3D) assemblies of nanostructures consisting of mixed dimensions in confinement, namely, nanoparticles (NPs) and nanowires (NWs). Also, the modification of the FRET mechanism with respect to the nanostructure serving as the donor versus the acceptor is listed, focusing on the rate's distance dependency. Table 7.2 summarizes the transfer rates in the dipole approximation for all combinations with mixed dimensionality (NP, NW, QW) in all possible arrayed architectures (1D NP, 2D NP, 3D NP, 1D NW, 2D NW). This table illustrates the functional distance dependency for the FRET rates: (1) when the acceptor is a 1D NP assembly, the FRET rate is proportional to d^{-5}; (2) when the acceptor is a 2D

Table 7.2 Generic distance dependence for the FRET rates, with equivalent cases of arrayed nanostructures in terms of *d* dependence (Hérnandez-Martínez et al. 2013)

Generic distance dependence	FRET donor (*D*) → acceptor (*A*)
$\gamma \propto \dfrac{1}{d^6}$	X → ●
$\gamma \propto \dfrac{1}{d^5}$	X → ▬ ≡ X → ◉
$\gamma \propto \dfrac{1}{d^4}$	X → ▭ ≡ X → ▤ ≡ X → ▤
$\gamma \propto \dfrac{1}{d^3}$	X → ▭ ≡ X → ▦ ≡ X → ▦

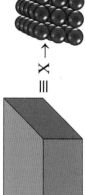

X = ● , ▬ : equivalent.

d: separation distance between *D* and *A*, ≡ : equivalent.

NP assembly, the FRET rate is proportional to d^{-4}; and (3) when the acceptor is a 1D NW assembly, the FRET rate is proportional to d^{-4}.

We see that the donor dimensionality (NP, NW, QW) does not affect the functional dependency on the distance. In all cases, the FRET rate distance dependence is given by the acceptor geometry and acceptor array architecture, and it is independent of the donor's geometry. Table 7.2 summarizes the FRET rate generic distance dependence with equivalent cases in terms of the distance dependence. It is worth pointing out that the effective dielectric constant depends only on the donor's geometry. Therefore, we find that the FRET's distance dependency is dictated by the confinement degree of the acceptor nanostructure and its stacked array dimensions, whereas the donor's confinement affects the modification of effective constant.

Conclusion

- Energy transfer is a directional process that occurs from the excited energy state of a donor, D, to the ground energy state of an acceptor, A: $(D^*, A) \rightarrow (D, A^*)$.

- The energy transfer can be radiative or nonradiative. The radiative transfer involves photon emission. In the case of nonradiative transfer, there is no real photon emitted in the process.

- Nonradiative energy transfer requires the emission spectrum of the donor to overlap with the absorption spectrum of the acceptor such that optical transitions of the donor and the acceptor couple and are in resonance. This type of transfer is thus also referred to as the resonance energy transfer (RET).

- The Förster resonance energy transfer (FRET), also known as fluorescence resonance energy transfer, is a direct result of the long-range dipole–dipole interactions between a donor and acceptor pair, where the excitation energy (in the form of an exciton) is transferred from the donor to the acceptor. It is different than the Dexter process, which allows for charge transfer.

- The energy transfer efficiency depends on the coupling of the donor and acceptor dipoles. This in turn depends on the extent of spectral overlap of the donor emission and the acceptor absorption, the donor quantum yield, the spatial distance between the donor and the acceptor, and the orientation of the donor and acceptor transition dipoles.

- Förster radius is the separating distance between the donor and acceptor dipoles at which the energy transfer efficiency is a half.

- In the case of coupling with a point-like dipole acceptor (e.g., quasi-zero dimensional (0D) quantum emitter acceptor), FRET scales down with d^{-6}, where d is the separating distance between the donor and the acceptor. When the acceptor is a one-dimensional (1D) quantum emitter (e.g., a nanowire), this scaling is modified to d^{-5}, and in the case of using a two-dimensional (2D) quantum emitter as the acceptor (e.g., quantum well), this becomes d^{-4}.

Problems

7.1 Assume the Förster radius for the following FRET pairs is exactly the same: (1) a quantum dot donor and a quantum wire acceptor, and (2) a quantum wire donor and a quantum dot acceptor. Which pair has the highest energy transfer efficiency when the separation between them is half of the Förster radius?

7.2 Plot the generic distance dependencies of the energy transfer rate as a function of the distance between a donor–acceptor pair for the cases of the acceptor being (1) quasi-zero dimensional (0D), (2) one-dimensional (1D), and (3) two-dimensional (2D) quantum emitters. Discuss why there is a specific trend.

7.3 State the generic distance dependencies of the energy transfer rate between a donor–acceptor pair for the cases of the acceptor being (1) a two-dimensional array of quasi-zero dimensional (0D) quantum emitters, (2) a one-dimensional array of one-dimensional (1D) quantum emitters, and (3) a single two-dimensional (2D) quantum emitter. Discuss why.

7.4 Consider two dipoles. In which orientation is the energy transfer the maximum and in which orientation is it the minimum?

Further reading

Agranovich, V. M., Gartstein, Y. N., and Litinskaya, M. (2011). Hybrid resonant organic–inorganic nanostructures for optoelectronic applications. *Chem Reviews*, **111**, 5179–5214.

Clegg, R. M. (2009). Förster resonance energy transfer – FRET what is it, why do it, and how it's done. In: T. W. J. Gadella (ed.), *Laboratory Techniques in Biochemistry and Molecular Biology*, vol. **33**. Academic Press.

Govorov, A., Hernández-Martínez, P. L., and Demir, H. V. (2016). *Understanding and Modeling of Förster-type Resonance Energy Transfer (FRET)*, vols. I–III. Springer.

Valeur, B., and Berberan-Santos, M. N. (2012). *Molecular Fluorescence: Principles and Applications*, 2nd edn. Wiley-VCH.

References

Agranovich, V. M., Gartstein, Y. N., and Litinskaya, M. (2011). Hybrid resonant organic–inorganic nanostructures for optoelectronic applications. *Chem Reviews*, **111**, 5179–5214.

Baer, R., and Rabani, E. (2008). Theory of resonance energy transfer involving nanocrystals: the role of high multipoles. *J Chem Phys*, **128**, 184710.

Beard, M. C. (2011). Multiple exciton generation in semiconductor quantum dots. *J Phys Chem Lett*, **2**, 1282–1288.

Born, M., and Wolf, E. (1999). *Principles of Optics*. 7th edn. Cambridge University Press.

Bredas, J.-L., and Silbey, R. (2009). Excitons surf along conjugated polymer chains. *Science*, **323**, 348–349.

Clegg, R. M. (1996). Fluorescence resonance energy transfer. In: X.F. Wang and B. Herman (eds.), *Fluorescence Imaging Spectroscopy and Microscopy*. John Wiley & Sons, 179–252.

Clegg, R. M. (2009). Förster resonance energy transfer – FRET what is it, why do it, and how it's done. In: T. W. J. Gadella (ed.), *Laboratory Techniques in Biochemistry and Molecular Biology*, vol. **33**. Academic Press.

Dexter, D. L. (1953). A theory of sensitized luminescence in solids. *J Chem Phys*, **21**, 836–850.

Förster, Th. (1946). Energieanwendrung und fluoreszenz. *Naturwissenschaften*, **6**, 166–175.

Förster, Th. (1948). Zwischenmolekulare energiewanderung und fluoreszens. *Annalen der Physik*, **437**, 55–75.

Förster, Th. (1949). Expermentelle und theoretische untersuchtung des zwischengmolekularen über-gangs von elektronenanregungsenergie. *Z Elektrochem*, **53**, 93–100.

Förster, Th. (1951). *Fluoreszenz Organischer Verbindungen*. Vandenhoeck & Ruprecht.

Hernández-Martínez, P. L., Govorov, A. O., and Demir, H. V. (2013). Generalized theory of Förster-type nonradiative energy transfer in nanostructures with mixed dimensionality. *J Phys Chem C*, **117**, 10203–10212.

Klimov, V. I., Mikhailovsky, A. A., Xu, S., et al. (2000). Optical gain and stimulated emission in nano-crystal quantum dots. *Science*, **290**, 314–317.

Köhler, A., and Bassler, H. (2009). Triplet states in organic semiconductors. *Mater Sci Eng*, **R66**, 71–109.

Lakowicz, J. R. (2010). *Principles of Fluorescence Spectroscopy*. 3rd edn. Springer.

Nozik, A. J. (2008). Multiple exciton generation in semiconductor quantum dots. *Chem Phys Lett*, **457**, 3–11.

O'Regan, B., and Grätzel, M. (1991). A low-cost, high efficiency solar cell based on dye-sensitized colloidal TiO$_2$ films. *Nature*, **353**, 737–740.

Saricifcti, N. S., Smilowitz, L., Heeger, A. J., and Wudl, F. (1992). Photoinduced electron transfer from a conducting polymer to buckminsterfullerene. *Science*, **258**, 1474–1476.

Stryer, L., and Haugland, R. P. (1967). Energy transfer: a spectroscopic ruler. *PNAS*, **58**, 719–726.

Valeur, B. (2002). *Molecular Fluorescence: Principles and Applications*. Wiley-VCH.

Part II

Advances and challenges

8 Lighting with nanostructures

This chapter considers application of various nanostructures in lighting devices and components. It includes an introduction to photometry and color perception by humans, and discussion of colloidal quantum dots as spectral converters in display devices and lighting sources; advances in development of colloidal quantum dot light-emitting diodes (LEDs); epitaxial quantum well structures as the core technology in solid-state lighting; the potential for metal nanostructures to improve efficiency of light sources; and the outlook for challenging issues in this field. The new trends in eye-friendly customized lighting and the lighting adapted to human biorhythms are covered, and possible application of residential lighting systems for wireless optical communication (Li-Fi) are discussed. Before reading this chapter, it is advisable to recall the content of Chapter 3, where basic principles of electron confinement phenomena and modern LEDs were introduced, and Chapter 5 (especially Sections 5.1 and 5.8), where spontaneous light emission and its enhancement with metal nanostructures were considered.

8.1 HUMAN VISION AND PHOTOMETRY

Lighting applications need materials and devices with visible light emission and which should meet a number of requirements to fit human vision characteristics. Remarkably, our perceptual ability to distinguish colors and enjoy a rich color palette with myriad tints comes from just the three different cones types in the human retina (Figure 8.1(a)). The daylight-adapted human sensitivity spectrum (referred to as the *photopic* sensitivity curve) is defined by cone properties and extends from 400 nm to 700 nm, peaking at 555 nm (green–yellow). Daylight-adapted sensitivity definitely fits the range in which sunlight contains the highest portion of radiation energy (compare Figure 8.1(b) and the sunlight spectrum given in the left-hand panel of Figure 5.3). The apparent correlation of human eye sensitivity with the sunlight spectrum is indicative of evolutional adaptation of humans to environmental conditions.

In addition to *radiometric* measures expressing energy, power, or intensity emitted by a radiator, a set of *photometric* measures is essential in lighting to quantify human-dependent visual effects rather than observer-independent physical parameters. Spectral features of human vision give rise to a number of measures that are based on physical values introduced to account for our perceptual ability.

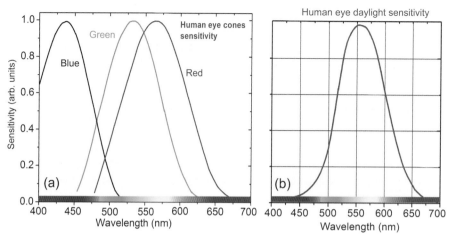

Figure 8.1 Human vision spectral characteristics. (a) Normalized spectral sensitivity of the three types of cones (red, green, and blue) enabling daylight vision. (b) Human eye daylight-adapted (photopic) sensitivity spectrum, the *standard luminosity function.* Source: based on Commission Internationale de l'Éclairage (1931).

BOX 8.1 GLOSSARY: PHOTOMETRY

Candella, cd, is the SI base unit, defined as luminous intensity of 1/683 W/sr at 555 nm.

Lumen, lm, is the SI-derived unit of luminous flux, 1 lm = 1 cd·sr.

Luminance efficacy of radiation, LER, characterizes luminosity of a light source per watt of radiated power, in lm/$W_{optical}$.

Color rendering index, CRI, is a measure of a light source's ability to reveal the correct colors of an illuminated object with respect to a reference (ideal) light source whose CRI is set at 100.

Gamut is a portion of color space reproducible in certain process, e.g., TV, display, photofilm, or printing. Photofilm features a larger gamut than TV or PC monitor screen and color printing, but still less than many art paintings.

Correlated color temperature, CCT, is the temperature of an ideal black body emitter that radiates light of comparable hue to that of the light source under consideration.

2000 K	candle flame, sunset, sunrise
2500 K	incandescent lamp
3000 K	warm light LED lamp and CFL
5000 K	horizon daylight, cool white LED lamps and CFL
6000 K	vertical daylight

The primary photometric value is an SI base unit, the *candela*, a unit of *luminous intensity*. It is defined as the luminous intensity of a light source that emits 1/683 watt/steradian at wavelength (in vacuum or in air) of 555 nm (corresponds to maximal daytime eye sensitivity). It approximately equals the luminosity of a true handheld *candle*, hence the term.

Another important value is *lumen* (lm). This is the SI-derived unit of luminous flux. It relates to candela as 1 lm = 1 cd·sr. The luminous flux is a measure of the total quantity of visible light emitted by a source. A familiar but obsolete incandescent lamp of approximately 75 W electric power gives roughly about 1000 lm, whereas compact fluorescent lamps (CFLs) and LED lamps consume only 15–20 W of electric power to give the same visible output.

An important parameter is *luminance efficacy of optical radiation* (LER). It shows how much luminous intensity is produced per watt of emitted power, in units of lm/W_{opt}. The highest LER is inherent, by definition, to a monochromatic light source emitting at 555 nm; its LER will be 683 lm/W_{opt}. However, monochromatic green light to which our eyes are most sensitive cannot provide comfortable lighting conditions since it fits neither to sun radiation nor to candle or fireplace emission. An incandescent lamp (temperature about 3000 K) can give up to 15 lm/W_{opt}, and an ideal black body emitter (close to real sunlight in the daytime) measures up to 95 lm/W_{opt}. Recalling the properties of a hot black body type of emitter (Figure 5.2) one can see that rising temperature from low values to approximately 6000 K increases the portion of radiation in the visible because the maximum of its spectrum shifts from infrared (IR) to the visible. However, for temperatures higher than approximately 6000 K, further rises in temperature lowers LER value because the further shift to higher frequencies (smaller wavelengths) moves the emission spectrum into the ultraviolet (UV) range.

The human sense of comfortable lighting implies that color recognition and color image perception will be close to those when objects are exposed to the sun in the daytime, or 1–2 hours after sunrise and 1–2 hours prior to sunset. The two latter cases, by the way, give spectral features close to a fireplace or a candle-light spectrum. In all the above examples, visual images appear as a result of illumination by *a continuous spectrum*. Light that is similar to that emitted by a candle or a fireplace is often termed "warm," probably to emphasize that its perception suggests heat coming from the above sources. It corresponds to an incandescent lamp or a black body radiation whose temperature is about 3000 K. Contrary to this case, radiation of mercury-based luminescent lamps (CFL and office-type tubular bulbs) is said to be "cool" or "cold," though its spectral content, while being not continuous, corresponds to higher temperatures when compared to black body ones because of the large portion of shortwave blue–green light dominating the orange–yellow light.

To characterize spectral energy distribution within the visible for lighting devices, the term *correlated color temperature* (CCT) is used. Description of its calculation is beyond the scope of this book and should be located in more specialist literature. The CCT value indicates the closest black body-like emitter (e.g., sunlight for higher temperatures and tungsten lamps for lower ones) on the Planck spectrum (see Figure 5.2) that has a similar hue to the source in question. It provides a reasonable estimate for light sources on the conventional "warm–cool" scale as is shown in Box 8.1.

A further important notion is the *color space*. It is used to assign numerical values to colors and to provide useful links between colors that can be used in color mixing.

The CIE (Commission Internationale de l'Éclairage, i.e., International Commission for Illumination) elaborates the following approach. First, the three color-matching functions, shown in Figure 8.2(a) are introduced. The first function, $\bar{x}(\lambda)$, characterizes eye sensitivity in the blue, the second, $\bar{y}(\lambda)$, reproduces strictly daylight eye sensitivity, and the third one, $\bar{z}(\lambda)$, appears with the two maxima in the blue and in the red. With these basic functions for a given spectral distribution of intensity, $I(\lambda)$, the three functions are then calculated as follows:

$$X = \int_{380 \text{ nm}}^{780 \text{ nm}} I(\lambda)\bar{x}(\lambda)d\lambda,$$

$$Y = \int_{380 \text{ nm}}^{780 \text{ nm}} I(\lambda)\bar{y}(\lambda)d\lambda, \tag{8.1}$$

$$Z = \int_{380 \text{ nm}}^{780 \text{ nm}} I(\lambda)\bar{z}(\lambda)d\lambda,$$

which in turn are used to calculate the *chromaticity coordinates, x, y*, and *z*,

$$x = \frac{X}{X+Y+Z}, \quad y = \frac{Y}{X+Y+Z}, \quad z = \frac{Z}{X+Y+Z} \equiv 1-x-y. \tag{8.2}$$

This approach allows for the two-dimensional (xy-plane) color space to be used to give the CIE (1931) chromaticity diagram shown in Figure 8.2(b). In the same system, the Planck locus is plotted, i.e., the curve showing the apparent color for emission spectrum following the Planck formula, which describes radiation spectral density of an ideal hot body (black body) with temperature T (see Chapter 5, Figure 5.2, and Eqs. (5.3) and (5.4)) and the set of straight lines used to evaluate the CCT value for light sources.

The chromaticity diagram reflects the regularities of color perception by humans and contains the full gamut of human vision. A useful summary of its properties is given in Figure 8.3. It contains all monochromatic colors represented by wavelength in nanometers (black border line), all multitudes of colors inside, and purple colors (mixture of red and violet–blue) that are not monochromatic, in the bottom. The white has coordinates centered around $x = y = z = 0.333$.

Every time all three cone types are excited in certain proportions, we feel the light is white. Interestingly, since the cones' sensitivity curves overlap, three cones can be stimulated with only two narrow-band or monochromatic sources (see Figure 8.1(a)). Thus, white light can be generated with only two monochromatic or narrow-band sources. In Figure 8.3, three light gray lines connect sample pairs of wavelengths providing white light

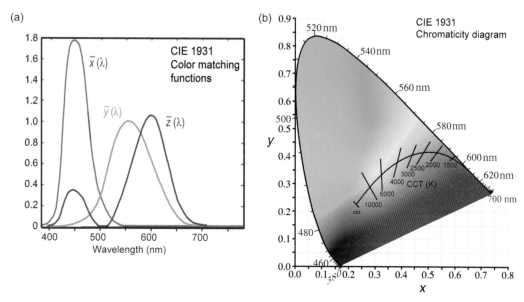

Figure 8.2 CIE (1931) color space. (a) Color matching functions. (b) The chromaticity diagram with Planck radiation locus (describes an ideal black body or an incandescent lamp) for the temperature range from $T = 1500$ K to infinity, and the straight lines that are used to evaluate the correlated color temperature of the light sources. Colors along the Planck locus have been discussed in Box 8.1. Adapted from CIE (1932).

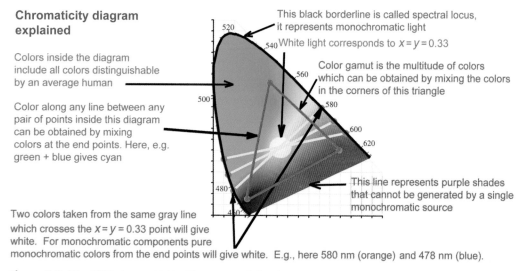

Figure 8.3 The CIE chromaticity diagram explained.

perception. Every line crossing the center white light point in the CIE chromaticity diagram ($x = y = 0.33$) and the spectral locus (the black borderline) defines two wavelengths (see explanation in Figure 8.3) which, presented together to an observer, give the impression of white light. However, for quality lighting, the whole wealth of colors is necessary and the light source should provide controllable stimuli to each of the three cones in human eyes. The same is crucial for high color rendering index (CRI) values. Therefore, promising luminophores must provide controllable emission with relatively narrow bands fitting the perception features of human cone cells.

8.2 QUANTUM DOT EMITTERS AND LIGHT SOURCES

8.2.1 Excitation, relaxation, and emission

In this section we consider applications of semiconductor nanocrystals (quantum dots) as active components of *luminophores*, i.e., photoluminescent materials that emit light of the desirable spectral range upon excitation by optical radiation. In photoluminescence, the emission spectrum is typically shifted to longer wavelengths with respect to excitation wavelength since various energy relaxation processes occur in the matter and the emitted photon energy is typically lower than the absorbed one. In most semiconductors relaxation occurs through a continuous set of states within c- and v-bands, as was shown in Figure 2.14, whereas in molecules, atoms, or quantum dots relaxation occurs through a number of discrete states (Figure 8.4). Longwave (low-frequency) shift of photoluminescence emission spectrum versus excitation wavelength is referred to as *Stokes shift*, named after George Stokes (1819–1903), who is known for pioneering studies of light emission by optically excited matter. Photoluminescent materials are widely used as light converters in luminescent lamps to convert ultraviolet and blue mercury vapor radiation into the visible spectrum and in white LEDs to complement blue LED emission to generate white light. Because of frequency downshifting, transformation of the light spectrum by luminescent materials is often referred to as *down-conversion*. There can also be a case of *anti-Stokes luminescence*, e.g., either by

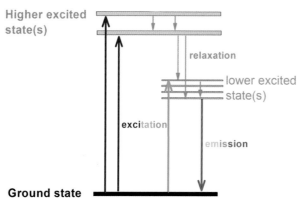

Figure 8.4 Illustration of excitation, relaxation, and emission processes in a hypothetical multilevel quantum system (atom, molecule, or quantum dot).

means of two-photon excitation or owing to thermal activation processes – then emitted radiation frequency exceeds frequency of absorbed light and the process is referred to as *up-conversion*.

Luminescent light sources have completely substituted obsolete energy-wasting incandescent lamps, and are widely used in indoor illumination, LCD display backlighting, landscape lighting, traffic lights, and car lights. It seems that in a decade traditional mercury-based luminescent bulb-like and tubular lamps (including CFLs) will be completely replaced by solid-state lighting sources based on semiconductor LEDs and luminophores. The latter, with respect to white LEDs, is often termed *phosphor*. Taking into account that mercury-based bulbs are being phased out, we shall consider semiconductor quantum dots as potential luminophores integrable with LED-based lighting devices.

8.2.2 Colloidal quantum dots as emerging luminophores for lighting

Quantum dots (semiconductor nanocrystals with size of the order of 3–10 nm) can be developed based on the following approaches:

1. commercial glass-based technologies known for decades;

2. colloidal synthesis in solutions and polymers developed since the 1980s;

3. epitaxial growth using surface self-organization processes in submonolayer strained heterostructures, discovered in the 1990s.

The first approach, glass technology, though very old and commercially available for decades, can only provide reliable cutoff filters for general applications and saturable absorbers for Q-switching (to get nanosecond pulses) and mode-locking (to get picosecond pulses) in lasers. The main reason for semiconductor-doped glass uselessness for light-emitting applications is the impossibility of controlling the interface between a dot and a matrix during growth and absence of processes/treatments to modify interface properties afterwards. Because of indefinite and uncontrollable surface structure, two undesirable processes occur. First, there is a typical efficient recombination channel formed by surface states and often seen as a wide emission band with very large Stokes shift (see Figure 3.21). Second, there is photoinduced development of fast recombination channel(s), which reduce quantum yield drastically upon prolonged illumination. A typical illumination dose to quench luminescence of a semiconductor-doped glass sample is of the order of 1 J/mm^2. This corresponds to a few hours of illumination by bright sunlight. Under this "treatment," luminescence quenches and recombination time goes down by two orders of magnitude, from approximately 10^{-8} s to approximately 10^{-10} s.

Epitaxial dots developed in strained heterostructures are used as active components of near-IR laser devices for optical communication purpose in the range around 1.3 μm. These structures will be considered in detail in Chapter 9, but not here as this type of quantum dot is not promising for lighting at present.

Therefore, in what follows we concentrate on colloidal semiconductor nanocrystals, sometime called nanocrystal quantum dots (to distinguish from epitaxial quantum dots), as promising active components for lighting. Advances in colloidal quantum dot synthesis can offer a number of approaches toward novel luminophores for lighting. This becomes possible owing to (1) size-dependent narrow emission spectrum; (2) high stability; and (3) broad-band excitation spectrum.

Based on the quantum mechanical description of confined electrons and holes (see Section 3.5, Figures 3.17 and 3.19), one can readily foresee that the whole visible spectrum can be covered entirely by nanocrystals of the compounds whose band gap correspond to the near-IR. There are at least two II–VI compounds (CdSe and CdTe), and also some III–V compounds (InP, GaAs) with the bulk band gap in the near-IR. Among these, CdSe and InP have become the first colloidal quantum dot materials for commercial display applications.

Figure 8.5 shows an example of three-color narrow-band emission from CdSe colloidal quantum dots. Depending on size (growing from the left to the right) blue, green, or red emission can be obtained. Remarkably, all three emitters can be excited by the same deep-blue, violet, or near-UV light source. The reader is requested to estimate the size of nanocrystals based on the formulas and/or experimental data provided in Section 3.5 (see Problem 8.5). One can see that within the whole visible spectrum, CdSe colloidal quantum dots feature emission bandwidth 30–50 nm (full-width at half-maximum), which is a very important property for luminophores, enabling versatile design of an emission spectrum by combination of a number of bands. Another important feature of

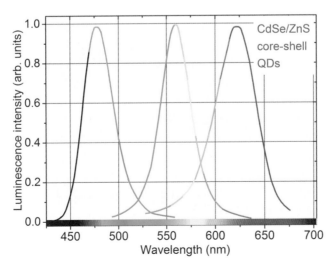

Figure 8.5 Normalized photoluminescence emission spectra of colloidal CdSe nanocrystals in a solution of different sizes (size grows from left to right). All samples were excited by the same lamp with near-UV radiation.

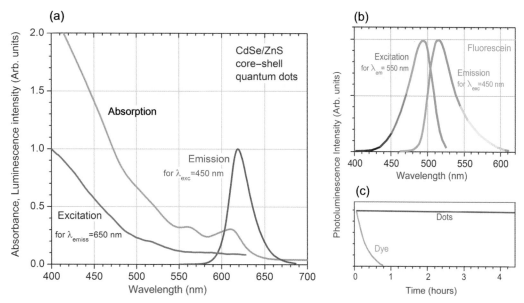

Figure 8.6 Photoluminescence properties of semiconductor colloidal quantum dots versus organic dyes. (a) Optical absorption, emission, and excitation spectra of a sample colloidal solution of CdSe core–shell quantum dots. (b) Emission and excitation spectra of fluorescein solution. (c) Degradation of photoluminescence upon prolonged illumination for dots and a dye.

quantum dot emitters is their wide *excitation spectrum*. Excitation spectrum is photoluminescence intensity dependence at a given wavelength versus excitation wavelength. In the case of quantum dots, the excitation spectrum starts from the emission wavelength and extents toward shorter wavelengths in direct correlation with the absorption spectrum (Figure 8.6).

When compared to organic dyes (Figure 8.6(b)), one can see that dyes exhibit similar emission spectra, which can be shifted through the visible by means of dye change (e.g., coumarines in the blue, fluorescein in the green, rhodamines in yellow–orange–red), but in all cases the dye excitation spectrum is defined by the absorption one and has the width close to the emission spectrum. Excitation and emission spectra of dye molecules show nearly symmetric shapes. That means, e.g., that using blue excitation (like a commercial GaInN LED), only blue or green emission can be obtained, whereas for orange or red emission green–yellow excitation is needed. This is the critical drawback of organic dyes as luminophores when compared to semiconductor quantum dots. For semiconductor quantum dots the full visible spectrum can be covered by a single excitation source. Furthermore, organic dyes can be used neither in display devices nor in lighting because of pronounced *irreversible photobleaching* that causes prompt decay of luminescence intensity upon continuous illumination (Figure 8.6(c)).

In colloidal quantum dot luminophore development, invention of *core–shell* colloidal quantum dot growth in the mid-1990s was a marked milestone (Hines and Guyot-Sionnest 1996). Since then, many groups over the world have adopted this approach. As shown in

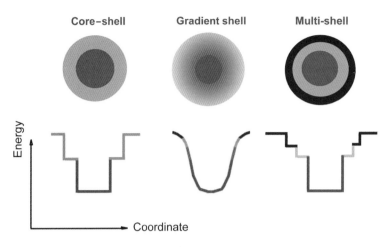

Figure 8.7 Core–shell, gradient shell, and multi-shell (core–shell–shell) quantum dot structures.

Figure 8.7, a nanometer-size crystallite of CdSe is capped by a nanometer-thick shell of wider-gap semiconductor, here ZnS, to ensure surface passivation (inhibits surface trap state formation) and improved electron–hole confinement (better electron–hole wave functions overlap increases the probability of their direct recombination). An additional positive effect is isolation of the core crystallite from undesirable chemical impact of an ambient medium, e.g., oxygen is known to promote photodegradation of II–VI nanocrystals. Many groups reported on core–shell CdSe/ZnS colloidal dots with quantum yield up to 90% or even close to 100%.

Typical II–VI core–shell colloidal quantum dots feature much higher stability than do organic dyes. However, there is the undesirable basic phenomenon of photoluminescence degradation upon exposition time/dose that has prevented commercial application of quantum dot emitters in display and lighting devices, although their application for bioimaging and fluorescent labeling has been recognized. The undesirable process responsible for photodegradation upon prolonged illumination (i.e., large number of absorption–emission events) comes from *Auger recombination* (Figure 8.8).

8.2.3 Multi-shell structures against Auger recombination

Auger recombination (after Pierre V. Auger, a French physicist) is known for electron–hole plasma in semiconductors and also in atomic plasmas. This is a three-particle process (Figure 8.8) in which an electron and a hole recombine but, instead of photon emission the energy released is picked up by another electron. For this process to occur, energy and quasi-momentum conservation laws should be rigidly met as is shown in Figure 8.8(a). One can see these constraints are pretty tough. An additional prerequisite is strong spatial overlap of wave functions for all three particles whose original and final states meet the energy and quasi-momentum conservation rules. This is shown in Figure 8.8(b) in the energy–coordinate diagram. For these reasons, in bulk semiconductors Auger processes become noticeable at higher electron–hole pair density ($>10^{18}$ cm^{-3}) that needs hard

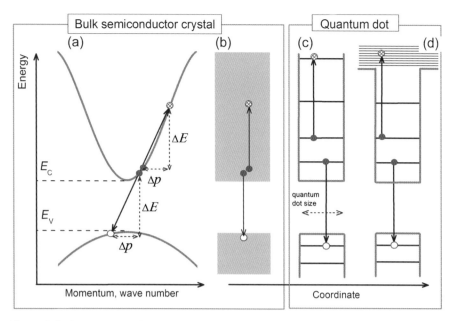

Figure 8.8 Auger recombination (a,b) in a bulk semiconductor crystal, and (c,d) in a quantum dot.

excitation by pico- or femtosecond lasers ($> 10^8$ W/cm^2 per pulse) because longer pulses will destroy the sample surface before these concentration levels can be reached.

A quantum dot situation is rather different. First, there is no quasi-momentum conservation for dots. As was shown in Chapter 2, quasi-momentum conservation is the consequence of the translational space symmetry inherent in a bulk crystal. Lack of translation symmetry in a dot removes the quasi-momentum conservation requirement. Therefore, only energy-matching conditions are to be met. Second, strong spatial confinement by definition ensures a priori necessary overlap of wave functions of quasiparticles excited in a dot. It should be noted that the number of electron–hole pairs in every dot is always discrete (0, 1, 2, …), and even at low or moderate excitation levels when the average number of electron–hole pairs per dot is well below one, a certain portion of dots will have more than one pair. Then, the Auger process is promoted. If energy acquired by an electron is large enough to rise over a potential barrier, a dot will ionize and then will not luminesce until a backward neutralization event happens. Ionization (charging) of dots further promotes nonradiative recombination of electron–hole pairs, since extra carriers to pick-up energy are already present in a dot. Absorption and emission properties of charged dots also change because a photon absorbed in a charged dot, instead of creating an electron–hole pair (exciton), now will create a *trion*, a three-particle state consisting either of two electrons and one hole or of two holes and one electron, depending on the dot charge sign.

The crucial role of the Auger process in degradation of luminescent quantum dots was recognized in the 1990s (Chepic et al. 1990). V. Klimov and co-workers revealed the universal $1/(a^3)$ growth of the Auger recombination rate for nanocrystals of various semiconductors (Robel et al. 2009), which is evidence of the role of spatial overlap of confined electrons and holes on Auger probability in a dot. The Auger-related photoluminescence degradation mechanism was the main obstacle to commercial application of quantum dots in display and lighting devices. To inhibit the Auger process, in addition to the simple core–shell structure, Dmitry Talapin and co-workers (2004) suggested and implemented core–shell–shell CdSe/CdS/ZnS and CdSe/ZnSe/ZnS quantum dots (Figure 8.7). Other groups suggested thick shell structures (Chen et al. 2008; Mahler et al. 2008) and gradient shell structures with a smooth potential barrier (Bae et al. 2009; Lim et al. 2011). In all cases, pronounced enhancement of photostability has been reported and attributed to inhibition of Auger recombination.

Cragg and Efros (2010) have elegantly explained that a parabolic spherical potential barrier can inhibit the Auger process in quantum dots. The reasoning is as follows. To inhibit the Auger process, overlap of the wave function describing an electron in the lower state in the c-band with that for an electron in the higher state in the c-band should be minimized since this overlap promotes energy pick-up by an extra electron from the recombining electron–hole pair. The higher energy state corresponds to a high wave number value and therefore is an oscillating function with many periods within a dot. The lower state, vice versa, is a function with a single maximum within a dot. Therefore, overlapping should be kept to a minimum by means of lower-state wave function shaping. Further, since parabolic potential gives rise to exponential wave function, it will get smaller overlap as compared to a rectangular barrier featuring a sine profile of the wave function (see Chapter 2). Probably, any barrier smoother than rectangular will have the same property, including multi-shell ones as an approximation (as shown in Figure 8.7).

Recently, indium phosphide colloidal core–shell quantum dots were synthesized; these possess high quantum yield and tunability from green to red (Figure 8.9). Together with a blue LED, these nanocrystals can be used to obtain white light emission. However, the emission spectrum is wider than that of CdSe dots and quantum yield is still lower (about 80% for InP dots versus 98% for CdSe dots). InP quantum dots are considered as a Cd-free alternative to CdSe dots in luminescent applications and it is believed they will reach the

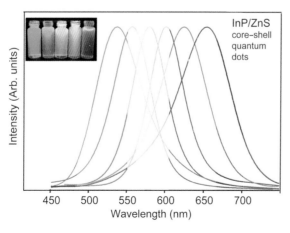

Figure 8.9 Photoluminescence of InP core–shell quantum dots of different sizes. Adapted from Zan and Ren (2012) with permission of the Royal Society of Chemistry.

same efficiency and spectral width parameters as CdSe dots in a few years. InP and CdSe colloidal core–shell quantum dots have become the first quantum dot materials integrated in commercial *liquid crystal display* (LCD) devices, including computer monitors and televisions, to get better color reproduction by means of improved white *backlight* spectrum.

Liquid crystal displays. LCD screens dominate in modern TV sets and PC monitors over, e.g., plasma display devices. Laptops, tablets, and cell phones are based exclusively on LCDs. An LCD screen is a matrix of voltage-controllable cells that filter the light emitted by a backlight source. Liquid crystals have the ability to change their refractive index in a certain spectral range for linearly polarized light. This property is used to make display pixels in the form of a voltage-tunable filter. An LCD panel should be backlit by linearly polarized white radiation. Formerly, fluorescent bulbs were used for backlight; now, white LEDs dominate in the display industry. The colorful shades of an LCD screen are therefore preconditioned by the emission spectrum of the backlight. The wrong backlight spectrum will result in poor color reproduction. White LEDs are also supposed to revolutionize residential and possibly street and traffic lighting. At present the most popular and cost-efficient design of white LEDs is based upon yellow rare earth phosphor to complement a blue InGaN electroluminescent chip.

8.2.4 Properties of rare earth luminophores

Rare earth elements got their name because of their rare appearance as mineral resources. Their industrial preparation and processing is complicated and expensive, especially careful separation from complex mixtures and compounds. These are 15 elements of group III and period 6 of the Periodic Table, called *lanthanides*, starting from lanthanum (number 57) to lutetium (number 71). The principal property of lanthanides is incomplete internal $4f$-shell. This gives rise to a multitude of optical transitions used to get luminescence and optical gain and lasing. Since the inner f-shell is covered and screened by the outer s- and p-shells, optical properties of lanthanides are very stable with respect to ambient dielectric matrix and impurities; however, transition energies do depend on crystal matrix properties and this is used to slightly tune absorption and emission spectra. All lanthanides form 3^+-ions, and a few of them exist also as 2^+- or 4^+-ions. Different ion charge can also change absorption and emission spectra. Properties of lanthanides are sketched in Box 8.2. Phosphors containing cerium, europium, and terbium, for example, are widely used as wavelength-converting coatings on blue LED chips to generate broad-band photoluminescence in the visible spectrum, peaking at yellow, red, or green wavelengths, depending on the rare earth ions used in the phosphors. A number of lanthanides are used as active media of IR lasers (see also Figure 6.5).

In modern solid-state lighting, the basic design of a typical, most cost-efficient white LED comprises an electrically pumped blue LED covered with a luminophore to complement blue electroluminescence, with the rest of the spectrum covered by means of photoluminescent frequency down-conversion. Widely used Ce-doped YAG-crystallites

BOX 8.2 LANTHANIDES (RARE EARTH ELEMENTS)

Lanthanides are elements in the Periodic Table with the atomic numbers from 57 (lanthanum) to 71 (lutetium). They belong to group III and period 6 of the table. For these elements, electron transitions occur within the incomplete f-shell, providing a multitude of transitions and extreme stability of optical properties with respect to ambient material. Crystals and glasses doped with lanthanides are used as active media in solid-state lasers (Pr is used for red lasers, Nd, Ho, Er, Tm, and Yb for IR lasers). Ce-, Eu-, and Tb-doped dielectrics are important yellow, red, and green luminophores, respectively.

58Ce	59Pr	60Nd	63Eu	65Tb	67Ho	68Er	69Tm	70Yb
Cerium	Praseodymium	Neodymium	Europium	Terbium	Holmium	Erbium	Thulium	Ytterbium
$4f^2\,6s^2$	$4f^3\,6s^2$	$4f^4\,6s^2$	$4f^7\,6s^2$	$4f^9\,6s^2$	$4f^{11}\,6s^2$	$4f^{12}\,6s^2$	$4f^{13}\,6s^2$	$4f^{14}\,6s^2$

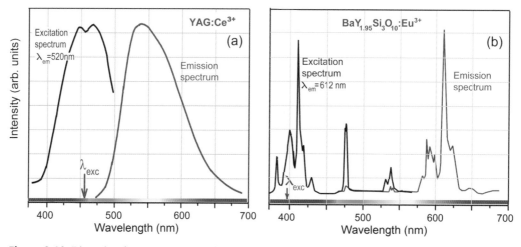

Figure 8.10 Photoluminescence properties of the two representative rare earth luminophores. (a) Yttrium aluminum garnet (YAG) doped with Ce^{3+} ions in microcrystalline form is currently the main material in commercial white LEDs. (b) $BaY_{1.95}Si_3O_{10}$ microcrystals doped with Eu^{3+} ions. Eu^{3+} ions feature pronounced red emission and are considered as potential additions to commercial Ce-based luminophores to improve coloristic parameters of commercial white LEDs. Adapted with permission from Shi et al. (2014) copyright OSA, and Zhou and Xia (2015), copyright RSC.

produce white light, with the spectrum featuring a lack of red component to obtain a high CRI. The typical spectrum of a commercial white LED was shown in Figure 5.3. Figure 8.10(a) presents YAG:Ce^{3+} photoluminescence data, including the emission and excitation spectra. One can see that the emission spectrum is broader than a typical quantum dot spectrum, whereas the excitation spectrum is narrower. However, a combination of excitation–emission spectra is very favorable for white light generation with blue LEDs emitting at 450–460 nm.

If only Ce^{3+} ions are used for frequency conversion, CRI appears to be low for high-quality color reproduction because of the lack of a red component. Changing matrix composition allows shifting the emission spectrum to the red, but then the overall efficiency drops because of large average photon energy losses in conversion. Addition of a small amount of narrow-band red emitter to standard Ce^{3+}–phosphor is considered as a reasonable approach to raise white LED CRI, with Eu^{3+} ions being the best candidate for this function (Figure 8.10(b)). However, Eu^{3+} ions have no excitation band fitting the commercial blue LED (440–460 nm) used in white LEDs. There is a narrow excitation band near 480 nm that is also capable of exciting Ce^{3+} luminescence, but then a combination of 480 nm cyan light with longwave emission will feature a lack of deep-blue and violet and in this case CRI will be low. A possible compromise solution is sought to find a mediator like co-doping with addition of a third rare earth ion to enable energy transfer for more efficient Eu^{3+} ion excitation.

The above discussion briefly highlights the problems researchers face when designing high-quality solid-state lighting sources. It explains why quantum dots can appear competitive and advantageous with respect to rare earths, owing to cheap production and high flexibility of tunable emission spectra, the excitation one always being wide enough to be compatible with blue LEDs.

8.2.5 First applications of quantum dots in display devices

The Samsung Advanced Institute of Technology pioneered quantum dot application in a backlight unit of an LCD TV screen in 2010. Jang et al. (2010) used a spectral converter on top of a standard blue InGaN LED chip. The converter consisted of a mixture of two different sizes of CdSe core quantum dot with multiple shells, the emission maxima being at 540 nm (green) and 630 nm (red). The quantum yield was close to 100%. White LED radiation enabled full coverage of the NTSC color gamut requirements. A total of 960 white LEDs were used to make the backlight for a 46-inch TV screen. There is a possibility to enhance the color gamut toward the full human vision range well beyond the HDTV standard using quantum dot LEDs (Figure 8.11).

Several companies have reported on quantum dot manufacturing for TV backlight units (Nanoco, Nanosys, Nexxus Lighting, and QD Vision). Though CdSe dots are considered as the major material (e.g., Nanosys, QD Vision), InP is also used (Nanoco). Nanosys proposed a film containing green and red LEDs (QDEF™, Quantum Dot Enhancement Film) that can be directly embedded in an LCD screen using the existing design (Figure 8.12). Instead of discrete white LEDs, quantum dot film is placed on top of a light guiding plate coupled to a side blue LED array. One more advantage is essential for quantum dots in the context of

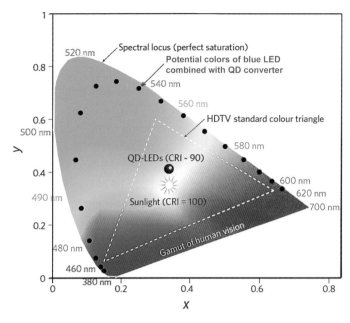

Figure 8.11 Optical advantages of colloidal quantum dots for display applications. In the CIE chromaticity diagram the spectral purity of quantum dots enables a color gamut (dotted line) larger than the high-definition television (HDTV) standard (dashed line). Reprinted with permission from Macmillan Publishers Ltd.; Shirasaki et al. (2013).

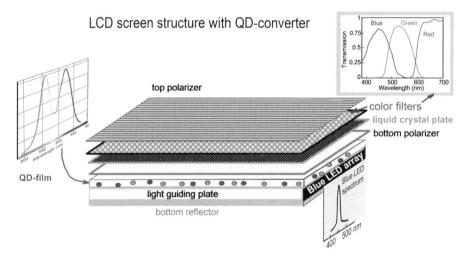

Figure 8.12 LCD screen with a color-converting quantum dot film.

LCD applications. Their narrow emission bands allow perfect fit to currently used RGB (red–green–blue) filters in every LCD pixel. Thus, undesirable color cross-talk can be eliminated.

Quantum dots applications for better color reproduction have been advertised by global TV and display device manufacturers (Samsung Quantum Dot TV, Sony Triluminous™, Philips QD Vision Color IQ™), as well as mobile phones and tablet PCs. Many analysts consider that quantum dot-based LED backlight units may push aside OLED technology owing to higher efficiency and better color reproduction.

8.2.6 New materials for backlight quantum dot units

There are certain possible side issues in Cd utilization from the customer display devices and even potential carcinogenic properties of InP. Therefore, extension of the potential material list for quantum dot backlight units is reasonable. During recent years a new class of visible quantum dot compounds has been proposed, organometal halide *perovskite* colloidal nano-particles (Zhang et al. 2015; Docampo and Bein 2016; Gonzales-Carrero et al. 2016; Xing et al. 2016;). Perovskite nanoparticles enable efficient narrow-band photoluminescence in the visible tunable range by means of chemical composition. They have the chemical formula $APbX_3$, where A is an organic cation (methylammonium or formamidinium) or cesium cation, and X is Cl, Br, or I. However, the presence of lead atoms in the compounds may create even more difficult recycling issues compared to Cd-containing dots.

Copper indium sulfide (CIS) quantum dots have also been recently suggested as a potential trend in quantum dot lighting devices. CIS–ZnS core–shell quantum dots have been reported to exhibit efficient green and red emission and were shown to offer high-quality white light when integrated with blue LEDs (Song and Yang 2012; Anc et al. 2013). The first reports are to be followed by further studies to optimize quantum yield and spectral width for good fitting to LCD spectral filters.

Last but not least, doped carbon quantum dots are also considered as a potential alternative to the Cd-containing approach (Reckmeier et al. 2016). High chemical stability, bright luminescence, and customizable surface functionalization of carbon dots make them an important novel nanomaterial for lighting applications.

8.3 QUANTUM WELL-BASED BLUE LEDS

The key component of solid-state lighting is blue-emitting diodes that are made of GaN/InGaN quantum wells (Figure 8.13). Typically, in these LED structures the multiple quantum wells are sandwiched between the n-doped region (commonly doped with Si) and the p-doped region (commonly doped with Mg). The epitaxy of this LED structure usually follows an unintentionally doped GaN region, preceded by low-temperature buffer and interlayer growth, commonly on patterned sapphire substrate (PSS), as shown in Figure 8.13(a). The resulting as-grown epitaxial wafer can be optically characterized as a whole, for example, for peak emission wavelength across the full epitaxial wafer. Figure 8.13(b) shows an exemplary map of the emission peak wavelength, here with a nonuniformity level of <1%. Similar optical characterizations are possible for emission full-width half-maximum and intensity profiles, which altogether provide useful information about the quality of the epi-growth. In fact, it is also possible to track in real time the growth using an *in-situ* reflectance spectrum, as presented in Figure 8.13(c) showing the real-time recording during the epitaxial growth of each LED layer. By closely following every layer, one can carefully follow the proper formation of epitaxial layers in the LED design. In

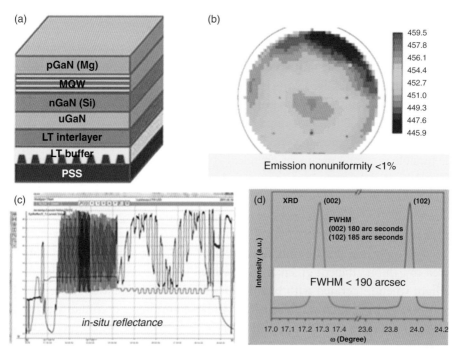

Figure 8.13 (a) Typical LED structure consisting of InGaN/GaN multiple quantum wells sandwiched between the *n*-doped region (*n*-GaN, doped with Si) on the bottom and *p*-doped region (*p*-GaN, doped with Mg) on the top, which follows the unintentionally doped GaN region preceded by low temperature buffer and interlayer on PSS. (b) Peak emission wavelength profile across a full epitaxial wafer exhibiting emission nonuniformity <1%. The emission map is color-coded and the color scale represents the emission wavelength on the right. (c) *In-situ* reflectance spectrum showing real-time recording during the epitaxial growth of LED layers. (d) X-ray diffraction spectroscopy showing a full-width half-maximum of <190 arcsec.

addition, structural characterization, which is also very commonly performed, is useful as well, one of which is, for example, x-ray diffraction (XRD) spectroscopy, depicted in Figure 8.13(d), showing high crystal quality with an XRD peak full-width half-maximum of <190 arcsec. Using alloys of $In_xGa_{1-x}N$ in the quantum wells it is possible to tune the peak emission from near-UV to green. The most commonly used emission is in the range 450–460 nm (blue–cyan) for solid-state lighting.

Basically, the steps of the LED-making process span from materials growth to device fabrication and packaging. In the first step, starting with a substrate (for example, sapphire that can be pre-patterned, as shown in Figure 8.14, or a native substrate of GaN), typically a metal–organic chemical vapor deposition (MOCVD) system is used to grow the epitaxial structure in a very controlled way almost layer by layer. The MOCVD system can have multiple substrate carriers; several tens of substrates are common in mass production. During the growth, three critical parameters – the composition, doping, and thickness of each layer – are precisely controlled, resulting in wafers that carry the full LED epitaxy on the starting substrate – called together the epitaxial wafer, or epi-wafer, pictured in Figure 8.14. One can see the blue light

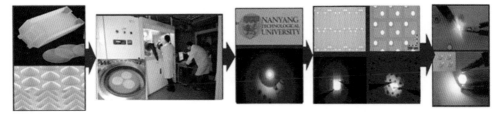

Figure 8.14 Steps of the LED-making process: starting with sapphire substrates that are pre-patterned as shown in the microscopic image; continuing with the epitaxial growth using a MOCVD system, with multiple carriers, resulting in wafers that have LED epitaxy on the starting substrate, known as the epi-wafer; subsequently fabrication of LED devices on the epi-wafer; and finally packaging of the individual LED chips.

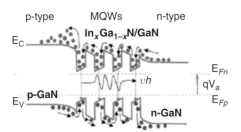

Figure 8.15 Energy band diagram of a GaN/InGaN LED with multiple quantum wells of $In_xGa_{1-x}N/GaN$ sandwiched between p-GaN and n-GaN. The band diagram illustrates electron injection from the n-region and hole injection from the p-region. Injected electrons and holes captured in the quantum wells radiatively recombine to emit light.

emission coming out of the as-grown epi-wafer locally wherever there is current flow through its layers internally from the p-region to the n-region, driven by the external driving circuitry. This corresponds to the spot where the epi-wafer holes are fed into the p-region and electrons into the n-region, as illustrated in Figure 8.15. For the electron–hole injection, it is not necessary to have device mesa definition, as shown in Figure 8.16. Finally, following the epi-growth comes the fabrication of LED devices on the epi-wafer and subsequently the packaging of individual LED chips, as demonstrated in the last two steps of Figure 8.14.

In operation of an LED, electrons are injected from the n-region and holes from the p-region when the current is driven in the forward bias, as sketched in the energy band diagram in Figure 8.15. These injected electrons and holes are captured in the $In_xGa_{1-x}N/$ GaN quantum wells and recombine radiatively to emit light. This is the electroluminescence of the LED. In this operation, one of the main device performance advantages is efficiency, which can be defined in various ways. Table 8.1 provides a list of commonly used efficiencies and their relations. Among them, one of the most common is the external quantum efficiency (EQE), which is the product of internal quantum efficiency (IQE) and light extraction efficiency (LEE). EQE simply corresponds to the ratio of the number of photons emitted by the LED to the number of electrons injected into it. Here, IQE depends in turn on the injection efficiency (IJE), which is the ratio of the electrons injected into the active region to those injected into the LED, and the radiative efficiency (RDE), which is the ratio of the radiative recombination to the total. On the other hand, LEE gives the ratio of the photons emitted by the LED to those emitted from the active region.

Figure 8.16 Photograph of a full epi-wafer after completing the LED epitaxy. Although the epi-wafer is transparent in the visible, because of the patterning of the substrate, it is highly scattering and looks semi-transparent. In operation, however, even simply by driving current through the epi-wafer (using some indium bumps to make a temporary contact), the epi-wafer provides nice blue electroluminescence.

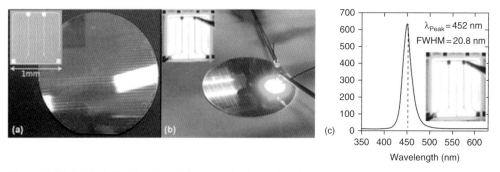

Figure 8.17 (a) A full epi-wafer with LED devices after fabrication, showing a single LED device in the inset; (b) the same epi-wafer with a single device electrically driven, along with an inset of the single device in operation; and (c) electroluminescence spectrum of this device (with an emission peak of 452 nm and a full-width half-maximum of 20.8 nm).

In addition to EQE, the overall wall-plug efficiency (WPE) is also used commonly, which corresponds to the ratio of the optical output power coming out of the LED to the electrical input power going into the LED. Defining voltage efficiency (VTE) as the photon energy of the emission per driving electrical energy of a single injected carrier, one can express the WPE as the product of IJE, RDE, LEE, and VTE. A typical problem with EQE (and thus WPE) is that it drops with increasing current density. This is referred to as the efficiency droop. The reasons include thermal heating (thermal droop), as well as an increase in Auger recombination and electron leakage at the increased current densities.

Figure 8.17(a) shows a fully processed epi-wafer; the micro-patterning of the devices on the epi-wafer can be easily seen. The inset shows an optical microscopy image of a single LED device on the epi-wafer (with 1 mm side length in this case). Figure 8.17(b) shows the

Table 8.1 **A summary list of various metrics used to quantify device efficiency including EQE, IQE, IJE, RDE, LEE, WPE, and VTE. The list also provides relations between these efficiency metrics. Along with the list is shown device schematics to illustrate P_{out} and P_{active}.**

External quantum efficiency (EQE)

$$\eta_{EQE} = \frac{\text{number of photons emitted out of LED per second}}{\text{number of electrons injected into LED per second}} = \frac{P_{out}/hv}{I_{in}/e} = \eta_{IQE}\eta_{LEE}$$

Internal quantum efficiency (IQE)

$$\eta_{IQE} = \frac{\text{number of photons emitted from active region per second}}{\text{number of electrons injected into LED per second}} = \frac{P_{active}/hv}{I_{in}/e} = \eta_{INJ}\eta_{RAD}$$

Injection efficiency (IJE)

$$\eta_{INJ} = \frac{\text{number of electrons injected into active region per second}}{\text{number of electrons injected into LED per second}} = \frac{I_{active}/e}{I_{in}/e}$$

Radiative efficiency (RDE)

$$\eta_{RAD} = \frac{\text{radiative total recombination rate}}{\text{total recombination rate}} = \frac{R_{rad}}{R_{rad} + R_{nonrad}}$$

Light extraction efficiency (LEE)

$$\eta_{LEE} = \frac{\text{number of photons emitted out of LED per second}}{\text{number of photons emitted from active region per second}} = \frac{P_{out}/hv}{P_{active}/hv}$$

Wall-plug efficiency (WPE)

$$\eta_{wall-plug} = \frac{\text{light power output from LED}}{\text{electrical power fed to LED}} = \frac{P_{out}}{I_{in}V} = \eta_{INJ}\eta_{RAD}\eta_{LEE}\eta_{VTG}$$

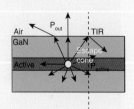

Voltage efficiency (VTE)

$$\eta_{VTG} = \frac{\text{energy of photons emitted from LED}}{\text{energy of electrons fed to LED}} = \frac{hv}{eV}$$

same epi-wafer (and the device in the inset) in operation when electrically driven, with its electroluminescence spectrum shown in Figure 8.17(c), with a peak emission wavelength of 452 nm and a full-width half-maximum of 20.8 nm here. Using an optical microscope, another LED design is shown on the epi-wafer in operation together with the other devices around it in Figure 8.18 and a zoomed microscopic image of another individual LED in Figure 8.19. Figure 8.20 presents four LED chips packaged using diced LEDs with phosphor dispensed on their top, along with one of them driven by DC current in operation.

Different generations of LED devices have been developed thus far. Examples of such devices are shown in Figure 8.21, which include the so-called lateral chip, flip-chip, vertical and reverse vertical architectures. These devices differ in terms of their substrates, that can be kept or removed, and their contacts that can be introduced to device mesas from the same side or different sides. Going from the lateral chip to the reverse vertical, the device complexity increases and larger numbers of lithography steps are needed, but at the cost of decreased device fabrication yield. With more sophisticated device architecture, the output power range increases accordingly as the most important benefit, and as a result, these devices target different market sectors that require varying levels of optical output power.

Figure 8.18 An optical microscopy image of a single micro-fabricated device when driven by a pair of probes on the wafer.

Figure 8.19 An optical microscopy image of a single LED chip.

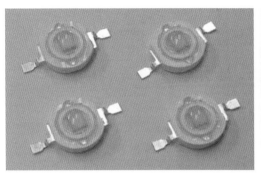

Figure 8.20 Four LED chips packaged using phosphor dispensed on the top and one of them in operation, driven by DC current.

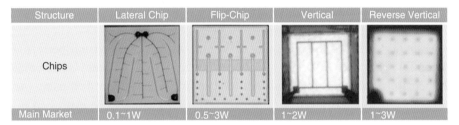

Figure 8.21 Four generations of LEDs with the device architectures of lateral chip, flip-chip, vertical and reverse vertical structure, with the targeted main markets specified by the output powers. Going from the lateral chip to the reverse vertical, the output power range increases.

8.4 QUALITY WHITE LIGHT WITH NANOCRYSTALS

Lighting consumes 15–20% of total electrical energy generation. Therefore, achieving energy efficiency in lighting is critical. However, it is not only about the energy saving; at the same time, increasing visual acuity and color perception is also essential. Solid-state lighting based on visible LEDs offers efficient lighting solutions, but for general lighting at a large scale, reaching the required color quality simultaneously is key. The spectral responses of the color-sensitive cells in the human eye (the so-called red, green, and blue cones) collectively determine color perception (see Figure 8.1). Clearly, from Figure 8.1, one can see that these sensitivity curves do not define spectrally a complete eigenfunction set; indeed, they have major spectral overlaps. This means that different combinations of spectral content may lead to exactly the same color perception, without the brain knowing the difference at all. The three color matching $x(\lambda)$, $y(\lambda)$, and $z(\lambda)$ curves (Figure 8.2(a)) also overlap. However, despite spectrally overlapping, the three sensitivity curves generate different levels of signal as a function of the optical wavelengths in the visible range (from 400 nm to 700 nm). It is indeed these signal differences that make color differentiation possible when processed in the human brain.

In Figures 8.2 and 8.3, the widely accepted CIE 1931 color gamut obtained as a result of these color sensitivity curves spans color space that can be generated in an infinite number of spectral combinations. The color gamut is represented as a function of the normalized chromaticity coordinates (in fact consisting of three coordinates, two of which are independent and thus suffice for color representation). On the chromaticity coordinates, the upper borders of the color gamut correspond to the monochromatic (pure) colors, starting from red–orange–yellow on the right, going through green–cyan on the top, and ending in blue–violet on the left. As can be seen on the color gamut of Figure 8.2(b), the pure spectral points on the borders are actually not equally spaced. Indeed, the color space is completely nonuniform across the entire gamut in terms of the spectral components, again as a direct consequence of the overlapping color matching curves $x(\lambda)$, $y(\lambda)$, and $z(\lambda)$ and their unequal signal levels. Also, on the gamut, note that the purer the spectral content is, the closer its corresponding color point is to the borders. In contrast, the white region is in the middle, covering shades of white.

Color quality includes three aspects of vision: visual acuity, visual performance, and visual comfort (Figure 8.22). For visual acuity, the device efficiency surely matters. But human perception also has to be taken into account. This is quantified using the LER, in units of lm/W_{opt}, which is given per optical power. This can also be expressed in terms of electrical input power, which would then be the LE in units of lm/W_{elect}. It is worth noting that, in addition to the cones that provide color differentiation, another group of light-sensitive cells in the human eye, known as *rods*, also contribute to vision (though they have no color sensitivity). Rods can function at low intensity levels, while cones require higher levels. Therefore, due to rods' additional visual contribution, the overall eye sensitivity curve shifts from the *photopic range* (photon-adapted vision) when there are a sufficient number of photons, taking the human eye into the *scotopic range* (dark-adapted vision) when there are very few photons available (see visual acuity in Figure 8.22). Going from the photopic to the scotopic, color perception is lost while the peak sensitivity moves from the typical peak at ~550 nm to ~500 nm by a substantial spectral shift of about 50 nm. Between the two regimes is the transition range known as *mesopic vision*, in which rods can contribute substantially to perceived brightness (Figure 8.23). This gives rise to spectrally enhanced lighting, which can be quantified by a metric known as the *scotopic-to-photopic ratio* (s/p ratio). The bigger the s/p ratio, the more enhanced the perceived brightness would be. The typical s/p ratio for white light sources falls below 2.5.

Visual performance depends on the shade of white light generated. This is given by the CCT in the unit K. The CCT of an operating color point is defined by the color temperature of a black body radiator closest to this point on the color gamut (Figure 8.2(b)). The color point in the white region moves toward the blue with increasing CCT and gains a bluish tint, referred to as cool white light. Conversely, it moves toward the yellow with decreasing CCT and gains a yellowish tint, referred to as warm white light. It is worth noting that the description of "warm" versus "cool" in white light is completely opposite to its corresponding actual black body temperature. The visual performance is vitally

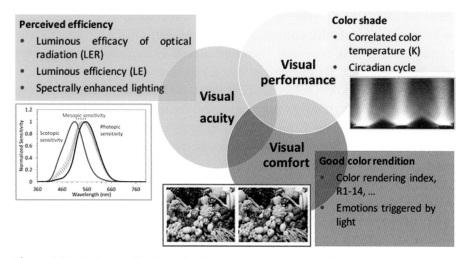

Figure 8.22 Color quality from the three aspects of vision: visual acuity, visual performance, and visual comfort.

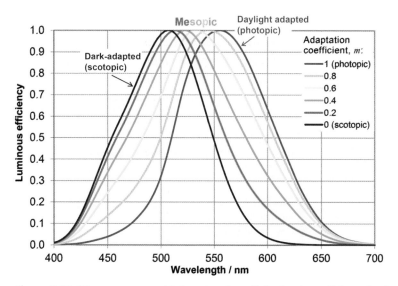

Figure 8.23 Human eye sensitivity at various light-level conditions. In daylight conditions, sensitivity follows the red curve (photopic vision), which is provided by the three types of cones enabling color perception. In nighttime conditions, dark-adapted (scotopic) vision is possible with higher sensitivity but without color distinction. Now vision is mediated by rods. At an intermediate illumination level the human eye sensitivity is characterized as mesopic and can be described in terms of the scotopic/photopic ratio as shown by the green–orange curves. Adapted with permission from CIE (2016).

important for avoiding undesired disruption on the biological rhythm of a human body. This is because too much blue content in white light disrupts the circadian cycle and can cause insomnia with long exposure. Therefore, especially for indoor lighting, warm white (i.e., with low enough CCT, preferably <3000 K) is essential.

Finally, as the third aspect of color quality, visual comfort relies on good color rendition. This is related to the capability of a light source to reflect the true colors of an illuminated object. As an indicator of the color rendering, CRI is commonly used. CRI takes values between –100 and +100, and is unitless. Black body radiators possess a perfect CRI of 100. As the CRI cannot capture the right rendering level for high-CCT light sources various metrics such as color quality scale (CQS) have been defined to provide corrected rendering. Alternatively, color rendering can be quantified comparing to Munsell reference samples (R1–14). This aspect of color science is opening a new field that investigates emotions triggered by color.

In solid-state lighting, there are two mainstream approaches to white light generation using LEDs (Figure 8.24). Those are the multichip and color conversion strategies. In the case of the multichip approach, red, green, and blue LEDs are used together, which are to be driven at the same time in different weighting intensities to generate a targeted white point. However, green LEDs suffer from low efficiency due to the green gap problem. While this approach can achieve high-quality color, it requires complex circuitry and suffers cost issues. As a result, today the multichip technology is not the most common one. Instead, color conversion LEDs presently dominate in general lighting applications. In color conversion, a short-wavelength (e.g., blue, cyan, or near-UV) LED optically pumps some color-converting materials, which are used to generate the longer-wavelength spectrum, for instance, in yellow or green and red. This is a low-cost approach, and thus is highly favored. However, this one suffers an efficiency penalty due to the down-conversion process (using high-energy photons from the LED electroluminescence to generate

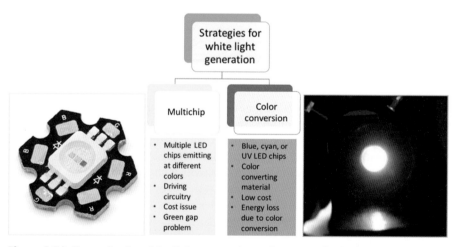

Figure 8.24 Strategies for white light generation using LEDs including the multichip and color conversion approaches.

low-energy photons from the photoluminescence of color conversion materials). Most of the commonly used color conversion materials include phosphors that contain emission centers of rare earth ion dopants. In the case of using only yellow phosphors, the resulting color rendering is typically low (in the range of 70). To reach a high CRI above 90, both green and red phosphors need to be employed in addition to the blue color component. However, red phosphors typically possess a wide emission spectrum, as a result of which their long-wavelength emission tail spills over the human eye sensitivity curve. This adversely leads to a substantially reduced LER (typically below 300 lm/W$_{opt}$). Therefore, one of the major issues with color-converting phosphors (especially in the red region) is the fundamental difficulty of controlling and tuning the details of their emission spectra. As a consequence, LER, CRI, and CCT cannot be optimized simultaneously using broad-emitter phosphors. Additionally, there are supply concerns for rare earth elements.

As an alternative to rare earth dopant-based phosphors, semiconductor nanocrystals, also known as colloidal quantum dots (CQDs), offer a promising solution for color conversion (Figure 8.25). Such CQDs exhibit favorable advantages including tunable

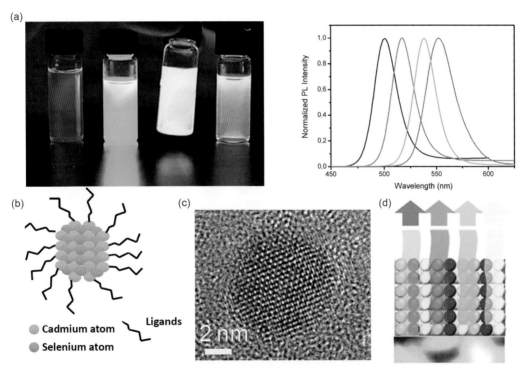

Figure 8.25 (a) Semiconductor nanocrystal quantum dots, with a photo of dispersion samples and their corresponding emission spectra. (b) Sketch of a single nanocrystal showing an inorganic core of Cd and Se atoms coated with organic ligand stabilizers and (c) transmission electron microscopy of such CdSe nanocrystals (with high crystal quality). (d) Illustration of multiple color-converting semiconductor nanocrystals integrated on blue LEDs.

absorption/emission, along with broad absorption and narrow emission spectra. It is possible to control the peak emission wavelength of CQDs with 1–2 nm precision using the quantum confinement. In addition to the size effect, their shape can be changed and composition tailored to control the resulting excitonic and optical properties. Also, being of semiconductor type, their absorption is broad-band, unlike phosphors; as a result, they can be effectively excited by pump LEDs below their band edge. Because of their narrow-band emission, CQDs provide color purity, which is key to achieving a highly optimized set of LER, CRI, and CCT at the same time. Therefore, semiconductor nanocrystals enable high-quality white light generation. This is a relatively low-cost solution.

With semiconductor nanocrystals, using various well-established chemical routes, it is routinely possible to reach high quantum efficiencies of >90%. The colloidal synthesis of semiconductor nanocrystals allows for the growth of high crystal quality, especially for II–VI groups. The best examples include Cd-containing CQDs, commonly reaching near-unity efficiency. However, due to recycling issues, the holy grail is to make Cd-free nanocrystals. As Cd-free options, InP-based nanocrystals and their heterostructure (e.g., InP/ZnS) hold great promise (Figures 8.9 and 8.26). Starting with an InP core, coating with ZnS significantly increases the stability. A recent study showed that the material system of InP/ZnS can cover the entire visible spectrum using the quantum size effect.

The scientific question is how it is possible to make high-quality white light using colloidal semiconductor quantum dots as color conversion materials in conjunction with blue LEDs. Clearly, it is advantageous to use such discrete combinations of narrow emitters compared to continuous and broad emitters of phosphors. To address this question, a

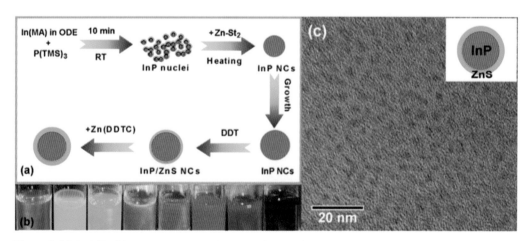

Figure 8.26 InP/ZnS hetero-nanocrystals as Cd-free options. (a) Chemical steps of nucleus formation, subsequent core growth and final shell coating; (b) emission color tuning across the visible; and (c) imaging of the synthesized nanocrystals.

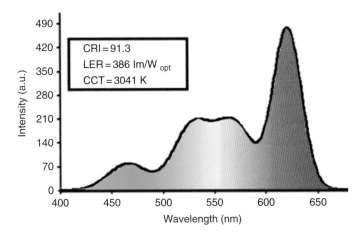

Figure 8.27 High-quality lighting spectrum using blue LEDs with red-, green-, and yellow-emitting CQDs, leading to CRI = 91.3, LER = 386 lm/W$_{opt}$, and CCT = 3041 K.

recent study systematically investigated nanocrystal-based color conversion LEDs over 230 million designs. This study found that only a few percent (about 2%) can reach the joint performance criteria of CRI ≥ 80, LER ≥ 300 lm/W$_{opt}$, CCT < 4000 K, whereas only a fraction (0.001%) of them can meet the ambitious performance of CRI ≥ 90, LER ≥ 380 lm/W$_{opt}$, CCT < 4000 K. Figure 8.27 depicts one such optimized emission spectrum using blue LEDs with red-, green-, and yellow-emitting CQDs, which is intended for photopic vision. This spectrum results in a high CRI of 91.3 and a high LER of 386 lm/W$_{opt}$ at a warm CCT of 3041 K, which surely exceeds the typical performance levels of conventional phosphors. Here, the key findings are that the red CQD component is the most critical one. Given the human eye sensitivity curve in the photopic regime, there is an optimal red wavelength that gives high enough CRI at warm enough CCT while keeping LER high enough at the same time, and this is 620 nm. This red color component needs to be highly precise – the penalty of spectrally mis-positioning the red CQDs is the highest in terms of reduction in LER and CRI.

This shows that using the right combination of narrow CQD emissions at the strategic wavelengths it is feasible to outperform conventional phosphors. Obtaining highly efficient white light sources requires spectra that are tuned very carefully. The CRI–LER tradeoffs include: (1) high CRI comes at the cost of reduced LER at all CCTs; and (2) CRI decreases more slowly with respect to LER at high CCT values (see Figure 8.28). Figure 8.29 and Figure 8.30 show the experimental proof-of-concept demonstration of nanocrystal-based color conversion LEDs for photon-adapted vision (intended for indoor lighting) and dark-adapted vision (intended for outdoor lighting), respectively. Similar to the tradeoff between CRI and LER, we see that s/p ratio drops with increasing CRI. While a typical phosphor-based color conversion LED leads to s/p = 2.03 with

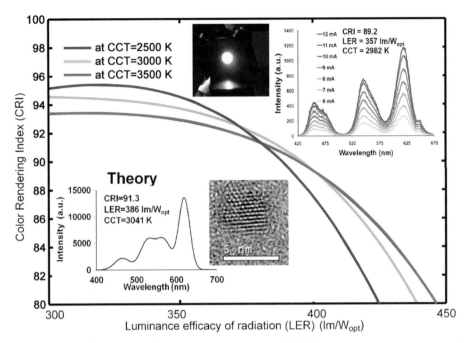

Figure 8.28 CRI versus LER for various values of CCT. The insets show a theoretically predicted spectrum (theory) and experimentally performed spectra for various current values along with the real image of a device (top) and electron microscope image of a colloidal quantum dot.

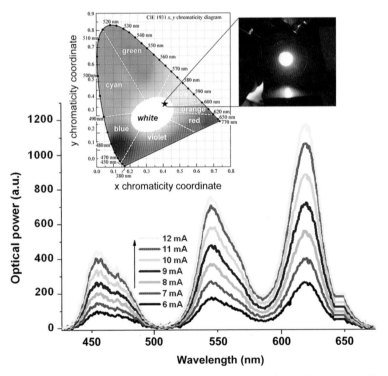

Figure 8.29 Photopic proof-of-concept demonstration of nanocrystal-based color conversion LED for photon-adapted vision (intended for indoor lighting).

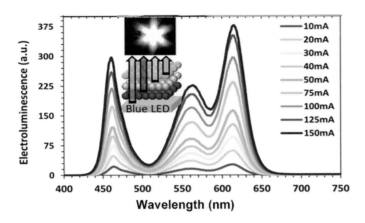

Figure 8.30 Scotopic proof-of-concept demonstration of nanocrystal-based color conversion LED for dark-adapted vision (intended for outdoor lighting).

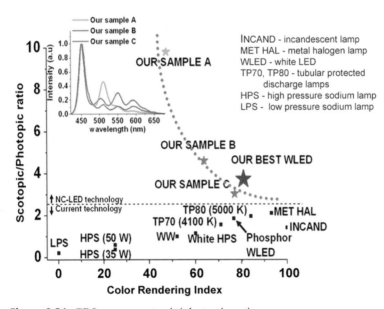

Figure 8.31 CRI versus scotopic/photopic ratio.

CRI ~70, the nanocrystal-based one can achieve s/p = 3.74 with CRI ~81. This gives 35.7% brighter light and 84.1% better nighttime vision at the same output power level (see Figure 8.31). In addition to integrating with LED chips, it is possible to use these nanocrystals in standalone films for remote color conversion and enrichment media, e.g., in displays (Figure 8.32).

Figure 8.32 Nanocrystal standalone color conversion films for remote color conversion and enrichment.

8.5 COLLOIDAL LEDS

Colloidal LEDs are very similar to organic LEDs, except for their organic light-emitting materials being replaced with the colloidal semiconductor nanocrystals (see Figure 8.33). As semiconductor emitters, nanocrystals offer narrow-band emission, which is conveniently tunable by design (using the quantum size effect). Furthermore, compared to their organic counterparts, these nanocrystals provide higher levels of photo- and thermal-stability, mainly as a result of their inorganic cores, in conjunction with potentially using various passivation strategies such as shell coating subsequent to core growth. These unique properties, together with fine-tuning and customization capabilities during synthesis, make nanocrystals highly attractive as the active emitters to be electrically driven in LEDs. Similar to polymer-based LEDs, the ability to make such colloidal LEDs using low-cost solution processing is a major technical incentive. In addition to being low cost, solution processing of colloidal LEDs potentially offers the possibility to scale up the device fabrication to roll-to-roll processing. While being highly promising, however, nanocrystals suffer from organic surfactants, commonly referred to as ligands, used as stabilizers on their outermost surfaces. These ligands make it difficult to inject

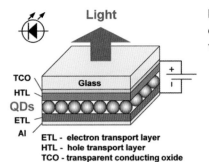

Figure 8.33 Typical colloidal LED structure consisting of a quantum dot layer, ETL, HTL, and contacts, at least one of which is a TCO.

ETL - electron transport layer
HTL - hole transport layer
TCO - transparent conducting oxide

carriers into the nanocrystals unless the surrounding layers in their LED structure are carefully chosen.

In a typical DC-driven colloidal LED, the CQD film is sandwiched between a hole injection layer (HTL) from the positive terminal side (anode) and an electron injection layer (ETL) from the negative terminal side (cathode) (see Figure 8.33). In operation, holes are injected from the positive terminal through the HTL into the quantum dot layer. Similarly, electrons are injected from the negative terminal through the ETL into the quantum dot layer. For this, HTL and ETL need to be carefully chosen to match the LUMO (lowest unoccupied molecular orbital) level with the ETL and HOMO (highest occupied molecular orbital) level with the HTL. Also, for effective charge injection, custom ligand exchange following colloidal synthesis is useful, as the ligands may otherwise form a barrier for the charges. The typical device architecture also includes at least one transparent contact, e.g., transparent conducting oxide (TCO), through which the emitted light is collected.

The device efficiency strongly depends on the ability to inject the carriers into the quantum dot layer. Electrons driven by the cathode go through the ETL and then, while some of them are effectively injected into the quantum dots through the charge injection, those that are not make it to the other side (see Figure 8.34). Similarly, holes driven by the anode go through the HTL and then while some of them are effectively injected into the quantum dots, those that are not make it to the other side. Those electrons and holes making it into the quantum dots come together and create excitons to lead to radiative emission; this layer where the radiative recombination takes place is known as the exciton recombination zone. On the other hand, the charges that are overdriven will make it through the quantum dot layer and thus completely miss the quantum dots. Thus, these charges, having not being injected into the quantum dots, would normally be wasted. However, some of these overdriven charges may take part in exciton formation in the nearby region outside the quantum dot layer. If this happens, then these newly formed excitons close to the quantum dot layer can be transferred into the active quantum dot layer via FRET. This will in turn boost the overall device efficiency. Therefore, some of these overdriven charges may also contribute to the electroluminescence of the active quantum dot layer if they are zipped back via FRET even when they leak out of the quantum dot layer.

QD-LED operation

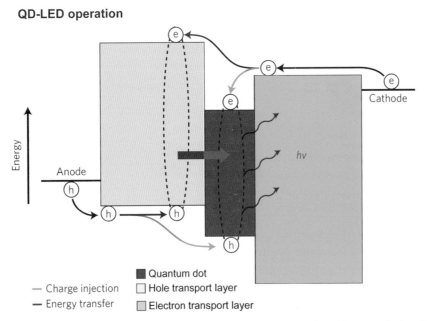

Figure 8.34 Illustration of colloidal LED operation. Reprinted by permission from Macmillan Publishers Ltd.; Shirasaki et al. (2013).

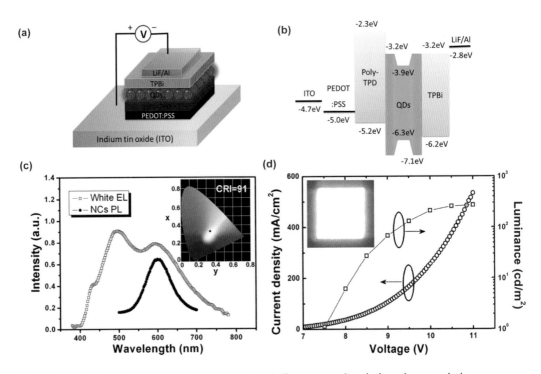

Figure 8.35 White colloidal LED structure, band diagram, and emission characteristics.

White colloidal LEDs using Cd-free quantum dots have been successfully fabricated. One proof-of-concept demonstration includes an InP/ZnS quantum dot layer sandwiched between a TPBi layer as the ETL and a poly-TPD and PEDOT:PSS as the HTL. The cathode is made of Al with LiF interface to TPBi. The anode is transparent ITO (indium tin oxide). The device structure is illustrated in Figure 8.35(a), along with the corresponding energy band diagram sketched in Figure 8.35(b). It is worth noting that this band diagram does not correctly capture the true energy levels of interfaces and only gives a rough idea of how the energy levels align, assuming their bulk values. However, this diagram is useful to see how the electron and hole injections would occur. From this picture, one can easily see that the electron injection is relatively easy, whereas the hole injection is difficult due to larger potential barriers for holes to overcome. With the right combination of quantum dots, one can generate the white spectrum from the electroluminescence of a fabricated InP/ZnS quantum dot LED, as shown in Figure 8.35(c), along with its corresponding white color point on the color gamut in the inset. Luminance–current density–voltage characteristics

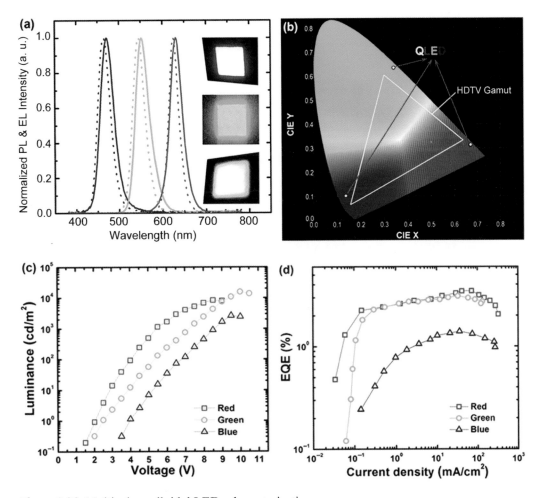

Figure 8.36 Multicolor colloidal LEDs characterization.

are presented in Figure 8.35(d), together with the picture of the resulting LED emitting white when electrically driven in the inset. Instead of white emission, it is also possible to make monochromatic colloidal LEDs, as depicted in the photoluminescence and electroluminescence spectra of Figure 8.36(a). Such multicolor colloidal LEDs correspond to the red, green, and blue regions in the color gamut and can be used in combination in principle for full-color display applications, spanning a large color space defined by these three corner color points (see Figure 8.36(b)). Their characteristic luminance–voltage and EQE–current density behavior are presented in Figure 8.36(c,d) for each color component. Such colloidal LEDs can also be constructed on flexible platforms, shown as a conceptual demonstration using Kapton tapes in Figure 8.37(a–d). It is useful to note that these LEDs can operate even in the flexed configuration, as shown with the T-bending test in Figure 8.37(e,f). This is surely a direct consequence of the solution-processing capability. Yet another advantage of the solution processing is the ability to introduce light extraction features conveniently directly on the colloidal LED platform, as illustrated on a typical colloidal LED structure (Figure 8.38(a)) processed step by step in Figure 8.38(b). The resulting fabricated light extraction features on the LED platform are shown with varied sizes in Figure 8.39(a–d).

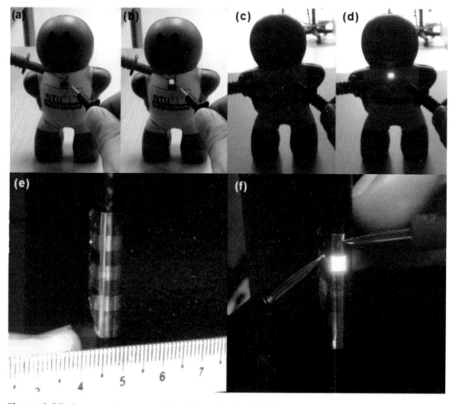

Figure 8.37 Demonstrators of flexible colloidal LEDs.

Surface patterning to enhance light extraction

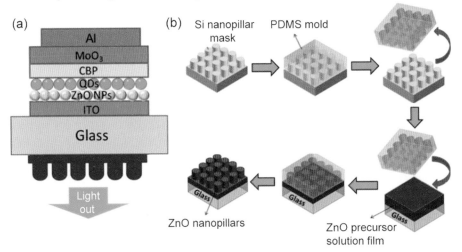

Figure 8.38 Structure and processing of light extraction features on colloidal LEDs: (a) device structure; (b) solution-processing steps of the light extraction features.

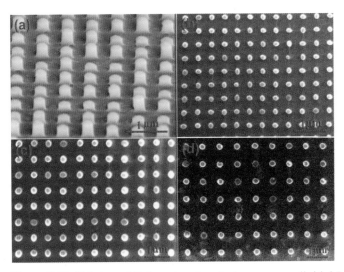

Figure 8.39 Fabricated light extraction features on a colloidal LED.

8.6 POSSIBLE PLASMONIC ENHANCEMENT OF LED PERFORMANCE

8.6.1 Photoluminescence intensity enhancement

Plasmonic phenomena in photonics are related to a number of effects coming from plasmon excitation in metal nanobodies or nanotextured metal surfaces. In Chapter 4 (Section 4.7), we saw that metal nanoparticles can enhance incident intensity of electromagnetic radiation roughly by a factor of ten for simple spherical particles and by a factor of 10^2 or even more for single spheroids or for coupled nanospheres. Enhancement occurs for radiation wavelength where extinction (absorption plus scattering) of a composite metal–dielectric medium is high. Therefore it depends on metal type, size, and shape of metal nanoparticles and dielectric permittivity of the ambient material. In Chapter 5 (Section 5.9), we saw that proximity of a metal nanobody gives rise to pronounced decrease in lifetime of excited states of matter bodies, promoting enhancement of both radiative (spontaneous decay) and nonradiative (undesirable losses bypassing the light emission process and resulting in metal heating) decay rates. It was emphasized that metal enhancement of radiative decay follows the extinction spectrum and, therefore, similar to local incident field enhancement, it depends not only on metal type but also on shape, size of metal nanoparticles, and on dielectric permittivity of an ambient environment. Unlike incident field enhancement and radiative rate enhancement, nonradiative losses depend mainly on metal type (metal dielectric permittivity) rather than on the extinction spectrum of metal–dielectric composite material. Moreover, metal-enhanced nonradiative losses feature much sharper distance dependence, vanishing at a distance typically more than 15–20 nm, whereas incident field and radiative rate enhancement are still noticeable in this range.

The results of plasmonic enhancement of photoluminescence (Eq. (5.37) and Figures 5.20 and 5.21) are fully applicable to enhancement of luminescence intensity of phosphors used in commercial white LEDs based on epitaxial InGaN quantum well heterostructures. Metal nanoparticles can be easily embedded in spectral light converters made as silicone compounds containing CQDs. Figure 8.40 presents the results of modeling for excitation wavelength typical for a blue LED and emission spectrum in the rest of the visible spectrum. The simplest case of spherical silver particles has been chosen, the emitter dipole moment orientation and incident light polarization being the best for luminescence intensity enhancement.

The real situation with random orientation of emitters will not give the ultimate enhancement factors up to 40–50 times for luminescence intensity as predicted in Figure 8.40(c,d), but still approximately ten-fold intensity enhancement is feasible. It can be commercially reasonable if metal nanoparticles are cheaper than core–shell quantum dots in mass production.

8.6.2 Decay rate enhancement

It is important that photoluminescence intensity enhancement always occurs simultaneously with the considerable enhancement of the decay rate. For example, in the cases shown in Figure 8.40(b), ten-fold or even greater increase in the decay rate is possible for green–yellow–red emitting quantum dots in white LEDs. Enhancement of decay rate diminishes the role of the Auger process since the probability of the Auger process should now be compared with the probability of the modified (i.e., accelerated) recombination of an electron–hole pair in a quantum dot. Therefore, the Auger process should become less pronounced for quantum dots placed in the vicinity of metal nanoparticles. Thus plasmonics appears to give rise to another positive effect on photoluminescence: enhanced photostability owing to diminished Auger recombination. However, this apparent effect is not

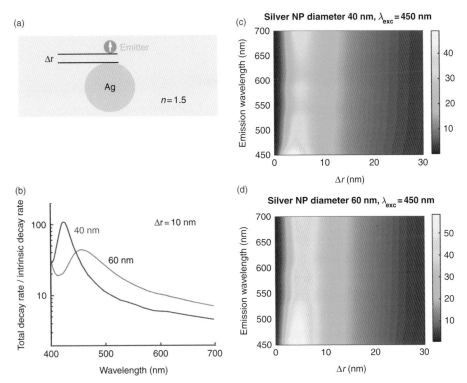

Figure 8.40 Calculated photoluminescence enhancement for 450 nm excitation wavelength for emitters near Ag nanoparticles embedded in a medium with refraction index $n = 1.5$. (a) An emitter versus metal nanoparticle displacement and orientation; (b) total decay rate (radiative + nonradiative) normalized to the intrinsic radiative decay rate enhancement versus emission wavelength for silver nanoparticle diameter 40 and 60 nm; (c,d) emission intensity enhancement versus emission wavelength and emitter–metal spacing Δr for silver nanoparticle diameter 40 and 60 nm. The emitter intrinsic quantum yield is equal to 1. Reprinted from Guzatov et al. (2018b).

easy to implement in practice. Note, photoluminescence intensity enhancement typically occurs simultaneously with, and to a large extent owing to, local enhancement of incident electromagnetic radiation. Therefore, an emitter (a quantum dot in our example) will experience higher excitation rate, i.e., more excitation–emission cycles per second. Thus the rate of the Auger process (it is the product of electron–hole pair generation rate and Auger probability) will rise accordingly. Therefore, for the Auger process to be diminished by the metal enhanced decay rate, the decay rate enhancement factor should dominate over the excitation rate enhancement factor, i.e., the combination of an emitter position, orientation, excitation, and emission wavelength should be chosen to ensure that the condition

$$\text{Photostabilty precondition}: \frac{(\gamma_{\text{rad}} + \gamma_{\text{nr}})}{\gamma_0} > \frac{|\mathbf{E}|^2}{|\mathbf{E}_0|^2} \qquad (8.3)$$

is safely met (see Section 5.9 for notations). This issue has not been especially addressed in either theory or experiments and therefore remains a subject for additional study.

8.6.3 Electroluminescence intensity enhancement

Electroluminescence enhancement is possible exclusively through quantum yield enhancement (see Section 5.9 and Eq. (5.38)) and therefore can happen only if the intrinsic quantum yield of an electrically pumped system is less than 1. For quantum yield enhancement, radiative decay rate should experience higher plasmonic enhancement than nonradiative ones and we have seen in Section 5.9 that there are plenty of experimental situations in which this condition can be met. In Figure 8.41, calculated enhancement of electroluminescence intensity is presented for various parameters of the experiments.

When comparing Figure 8.41 with Figure 5.19, one can see that for the target emission wavelengths 440–460 nm (commercial blue LEDs used in white LED production), increase in refraction index of an ambient medium from 1 to 1.5 essentially raises the plasmonic enhancement factors. This occurs owing to the extinction spectrum maximum shift toward the target wavelength for 50–60 nm particles in the case of silver. In this context, silver represents a unique plasmonic material for commercial blue LEDs. Even with randomly oriented emitters for the pretty wide spacing range from 3 nm to 15 nm, two- and four-fold intensity enhancement is possible for the intrinsic quantum yield of emitters $Q_0 = 0.25$ and 0.1, respectively. For aligned emitters enhancement rises by a factor of 1.5–2.

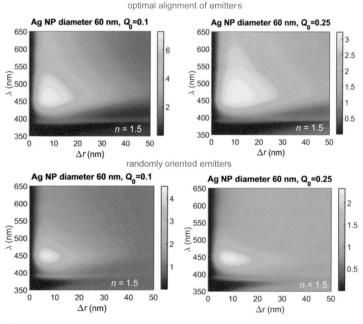

Figure 8.41 Calculated electroluminescence intensity enhancement with 60 nm Ag nanoparticles for aligned and random emitters with intrinsic quantum yield $Q_0 = 0.1$ and 0.25 versus emission wavelength λ and emitter–metal spacing Δr. Courtesy of D. V. Guzatov.

There are a few early experiments that confirm the general property of electroluminescence enhancement with metal nanoparticles: the lower the IQE, the higher is the
enhancement factor. For example, for an InGaN quantum well blue LED, Khurgin et al.
(2008) observed about five-fold enhancement for IQE = 0.1 and 20-fold for IQE = 0.01.
Plasmonic enhancement of electroluminescence has also been reported for CQDs. For
IQE = 0.25, Yang et al. (2015) observed 1.46 enhancement in reasonable agreement with
the theory (Figure 8.42), except for metal–emitter spacing which has been found in experiments to give the best results at 25 nm.

8.6.4 Decay rate enhancement in LEDs

Modern InGaN LEDs suffer from efficiency droop when current density exceeds a certain
critical value, dependent on LED design. This efficiency droop is the main obstacle preventing higher luminous output from the same chip size. The tentative reason for efficiency
droop is Auger recombination, which becomes more probable at higher electron–hole pair
densities in the LED active region. Thus, both quantum dot and quantum well-based
LEDs suffer from undesirable Auger processes bypassing radiative electron–hole recombination under high current (or optical excitation) conditions. As has been highlighted
above, *plasmonics always enhances excited state decay rate*. Therefore, in all cases when
metal particles are embedded in close vicinity of light emitters, they can prevent or diminish Auger processes. This can happen even if metal particles do not increase emission

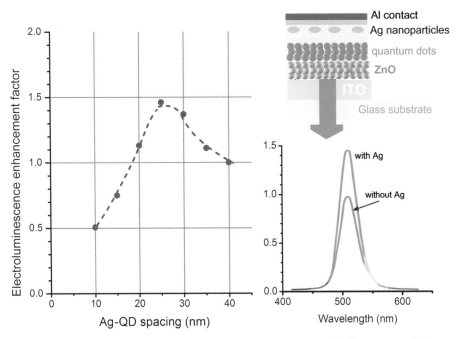

Figure 8.42 Plasmonic enhancement of a CQD green LED with Ag nanoparticles.

intensity. Thus, a favorable effect of metal nanoparticles on LED performance does not reduce the intensity enhancement but can also extend operation margins toward higher current densities without efficiency droop. This possible positive impact of metal nanoparticles on LED efficiency at high current densities still requires experimental testing.

8.7 CHALLENGES AND OUTLOOK

8.7.1 Lighting revolution

The advancement of durable and compact LEDs has made a major impact on lighting technologies. This is now triggering a lighting revolution, by the end of which LEDs are expected to be deployed on a large scale, essentially in any luminary or any system using illumination. However, historically, lighting technologies have been using either black body radiation, which by its nature provides a spectral continuum of emission (e.g., incandescent lamps) or a highly dense combination of emission line widths (e.g., rare earth elements), altogether resulting in broad emission. Therefore, the conventional approach to lighting has been most typically to base the light source designs on continuous and broad emission spectra to generate white light. However, it turns out that reaching high photometric quality with broad emission in which the details of spectral features are not controlled is technically challenging.

This challenge generates an opportunity for semiconductor nanocrystals that feature sharp emission spectra to be used along with LEDs. Using the quantum size effect, it is possible to precisely fine-tune their emission peaks as desired. Therefore, using combinations of nanocrystals as discrete emitters at strategic wavelengths, an important outlook is to generate quality lighting including high luminous efficiency and large color rendering at the target color temperatures (e.g., warm white shade). This type of emitter could include CQDs and quantum wells, among other possible choices.

These new trends will allow for customized eye-friendly lighting and also lighting adapted to the human biorhythm. The idea of using engineered quantum emitters will enable tailored lighting to address specific needs of human beings, such as spectrally enhanced road lighting (in the mesopic to scotopic ranges). This is now opening up new fields in lighting such as color science and perception of discrete emitters and emotional lighting.

The revolution happening in lighting also influences trends in electronic displays. Another interesting direction is color enrichment in displays, as it has been a major technical challenge to reach color richness in electronic displays that exist in nature. Therefore, for electronic displays, having primary colors as pure as possible is essential to span as large a color area as possible on the color gamut. For this purpose, once again, semiconductor nanocrystals are highly promising owing to their sharp emissions. This trend has already been taking place for the LCD industry, which has been most recently adapting nanocrystal-based color conversion either directly on their LED backlighting units or as additional standalone films in their backlight units.

The outlook on the use of this type of sharp quantum emitter is indeed more wide-reaching than typically acknowledged or recognized. The introduction of nano-emitters has in fact extended the lifetime of LCD technology and made it for the first time equally competitive and compatible with displays made of organic LEDs (OLEDs) in terms of color quality. With next-generation nanocrystal color enrichment, which will also feature nanopatterned pixelated films, LCDs will also be made substantially more efficient, in addition to their color superiority. Especially for large-format displays, as a result of these improvements, combined with cost, lifetime, and yield issues, LCDs may prevail over OLEDs.

As another trend in lighting and displays, there is a tendency to move toward using lasers. Future lighting systems that specifically require high output power and great efficiency at such high output power levels in their applications (e.g., headlamps) may employ lasing. This is again an opportunity for nanoemitters to be utilized for color conversion and enrichment in gain (e.g., using compact all-colloidal lasers) along with laser diode pumps (e.g., in blue, near-UV) in lighting and displays.

8.7.2   Li-Fi: lighting as a wireless communication platform

The very idea of wireless optical communication dates back to the famous A. G. Bell experiments in 1880 with sunlight, as mentioned in Chapter 1. Invention of durable blue LEDs and white solid-state nanoemitters used in conjunction with them will definitely result in replacement of residential and probably outdoor fluorescent lamps by LEDs. Development of residential solid-state lighting offers an opportunity to develop optical wireless indoor communication systems to be implemented provided that light sources can be properly modulated and all data processing equipment units are supplied with an optical transmitter/receiver module. This idea was suggested in 2000 by Y. Tanaka and co-workers, and the first demonstration of this type of in-room communication was made in 2012 by H. Haas and co-workers. They have also coined the term Li-Fi (light fidelity) by analogy with Wi-Fi networking in the radiofrequency domain. The Li-Fi paradigm easily overcomes radiofrequency bandwidth limitations that are emerging with a high density of users, enables security against tapping from outside the confined space covered by a specific Li-Fi hub, inhibits undesirable noise and cross-talk issues, and therefore not only can be readily integrated in residential communication services but also in special interference-sensitive areas like hospitals or aircraft. Current radiofrequency Wi-Fi networking is approaching the physical limit because of the growing datastreams related to bulky audio and video data from the global internet and the increasing density of users in the same space.

The Li-Fi concept (Figure 8.43) implies that in residential areas LED light intensity in ceiling luminaries and probably even in desktop lamps will be modulated to transfer useful data to different devices located within the illuminated area equipped with individual optical receiver/transmitter units. It is important to note that Li-Fi replacing Wi-Fi networks not only promises higher data transfer rates but will also eliminate electromagnetic radiation effects from active networking.

An optical communication channel allows data transfer rates in the gigabytes-per-second range data versus 54 megabytes per second of the Wi-Fi standard. Thus, Li-Fi can be considered as a reasonable solution for extension of the Internet of Things (IoT) to every home as well as unlimited access to online video and audio data available on the global internet. Free space optical communication, by its nature, may allow for massively parallel optical links, without increasing the overall cost of the system. Plasmonics can enhance modulation rate for both a core blue InGaN LED and a phosphor thereon, the latter being typically much slower than the former.

8.7.3 New nanostructures for enhanced light emission and flow management

Existing challenges are not only due to the quality- or efficiency-related issues of the devices and systems. They are also related to better management of the optical light generated. It is possible to manage different attributes of light used in the lighting system. These, for example, include the emission patterns and emission polarization. For instance, using photonic confinement in photonic crystals directly introduced on GaN LEDs, emission enhancement and tailored emission patterns have been shown (e.g., see Erchak et

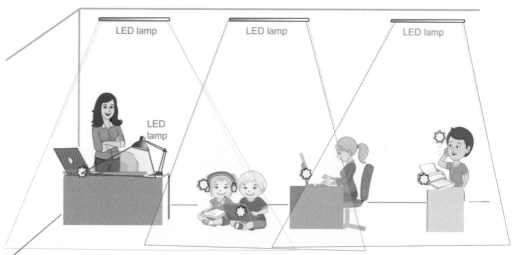

Figure 8.43 The Li-Fi concept. LED-generated light can be modulated to enable bulky data transfer to individual data processing or communication units (desktop and laptop computers, tablet PCs, audio and video systems, printers, fax and copy machines, cell phones, etc. Devices can communicate between each other directly through an optical data hub located on the ceiling above them. The latter can deliver data from distant sources received by Wi-Fi or fiber links and converted into light signals.

al. 2001; Zhmakin 2011). Similarly, there is a new trend of using nanostructures directly defined on LEDs. These nanostructures can be designed to possess a plasmonic resonance coincident with the emission wavelength of LEDs, for example, made of silver and gold. With careful design of such plasmonic structures, it is possible to increase the electroluminescence intensity, which is useful for general lighting purposes, and accelerate the emission rate, which is useful for modulation purposes such as in Li-Fi. In addition to plasmonic structures, another direction is to introduce metasurfaces defined on top of LEDs. Such metasurfaces can be dielectric materials. These could be defined even directly in the GaN surface of LEDs. Such nanostructured surfaces will surely enable superior photon management for controlling and choosing the specific emission polarization and directivity.

8.7.4 New nanomaterials for light generation

There are also new nanomaterials emerging recently, which will find use in lighting. Among these are semiconductor colloidal nanoplatelets and nanorods. With their structural high-aspect ratio, these nanomaterials enable intrinsically polarized emission. In addition to CQDs, there will be active devices, such as LEDs, made of colloidal quantum wells and wires, resulting in polarized electroluminescence. They also form densely packed solids. For example, colloidal quantum wells may stack into long chains. Such superstructures allow for additional routes to control excitonic properties. In the future, colloidal LEDs, with their additional solution-processing advantage, may replace LCD devices, if they can be fabricated with high yield over large surfaces and the lifetime and stability issues are addressed. In such a colloidal display architecture, the colloidal LEDs as active components in the display will define pixels consisting of three (RGB) components (see Bae et al. 2014). There are also other new material systems, including perovskites and carbon dots as well as new prospects for silicon–germanium (see Priolo 2014).

Conclusion

- The recent progress in quantum well-based InGaN blue LEDs paves the way for a lighting revolution in a decade, when all lighting sources will be based on LEDs. However, serious technological problems with high-efficiency green LEDs make the combination of a blue or violet LED with luminophores the most cost-effective solution. LEDs are already becoming the dominant light sources in TV screens and computer monitors based on liquid crystal matrices.
- Semiconductor nanocrystals (CQDs) feature size-controllable emission spectra. Advances in synthesis of efficient and photostable semiconductor quantum dots using the core–shell approach have made it possible to enter the commercial applications field as color-converting phosphors in LCD devices.

- Semiconductor nanocrystals offer the possibility to shape the light source spectrum to perfectly meet human vision, specifically including color rendering in the daytime and enhanced sensitivity in the nighttime. Adaptation of emission spectra to the so-called mesopic conditions (intermediate between the photon- and dark-adapted vision) play an important role in modern lighting trends. Modification of the light spectrum depending on the time of morning/evening will enable human biorhythm cycles to be taken into account, resulting in both cost-effective and comfortable lighting.

Problems

8.1 What is Stokes shift?

8.2 Explain why the energy efficiency of every luminophore is always less than 100%.

8.3 Explain why we need special photometric parameters in lighting in addition to standard characterization of light by spectral distribution of intensity.

8.4 Explain the origin of the RGB (red–green–blue) approach in color formation in light-emitting and display devices.

8.5 Estimate the size of CdSe nanocrystals (Figure 8.5) and InP nanocrystals (Figure 8.9) based on the formulas and/or experimental data provided in Section 3.5.

8.6 Explain what an excitation spectrum is and why it is important for luminophore applications. Compare and analyze excitation spectra for luminescing dyes, quantum dots, and rare earth ions.

8.7 Explain why Auger recombination occurs more readily in quantum dots compared to bulky parent crystals.

8.8 Recall every case in which nanostructures can be used in lighting systems. Trace what is currently more important: electron or lightwave confinement?

8.9 Explain why light sources should meet different criteria for backlight LCD devices and for indoor lighting.

8.10 Explain why LER value features non-monotonic dependence on black body temperature and why the maximum occurs at a temperature of about 6000 K.

8.11 Explain why metal nanoparticles cannot enhance electroluminescence intensity as easily as happens with respect to photoluminescence. What is the necessary prerequisite to expect plasmonic enhancement of electroluminescence?

8.12 Explain how metal nanoparticles can help to prevent Auger processes and why it is important for photo- and electroluminescent devices.

8.13 Explain the Li-Fi communication principles and analyze advantages and disadvantages versus Wi-Fi communication.

Further reading

Dimitrov, S., and Haas, H. (2015). *Principles of LED Light Communications: Towards Networked Li-Fi*. Cambridge University Press.

Docampo, P., and Bein, T. (2016). A long-term view on perovskite optoelectronics. *Acc Chem Res*, **49**, 339–346.

Erdem, T., and Demir, H. V. (2013). Color science of nanocrystal quantum dots for lighting and displays. *Nanophotonics*, **2**, 57–81.

Gaponenko, S. V. (1998). *Optical Properties of Semiconductor Nanocrystals*. Cambridge University Press.

Gaponenko, S. V. (2010). *Introduction to Nanophotonics*. Cambridge University Press.

Khan, T. Q., and Bodrogi, P. (eds.) (2015). *LED Lighting: Technology and Perception*. John Wiley & Sons.

Klimov, V. I. (ed.) (2010). *Nanocrystal Quantum Dots*. CRC Press.

Pietryga, J. M., Park, Y. S., Lim, J., et al. (2016). Spectroscopic and device aspects of nanocrystal quantum dots. *Chem Rev*, **116**, 10513–10622.

Schubert, E. F. (2006). *Light-Emitting Diodes*. Cambridge University Press.

Su, L., Zhang, X., Zhang, Y., and Rogach, A. L., (2016). Recent progress in quantum dot based white light-emitting devices. *Top Curr Chem*, **374**, 1–25.

Wood, V., and Bulović, V. (2010). Colloidal quantum dot light-emitting devices. *Nano Rev*, **1**, 5202–5210.

Wood, V., and Bulović, V. (2013). Colloidal quantum dot light-emitting devices. In G. Konstantatos and E. H. Sargent (eds.), *Colloidal Quantum Dot Optoelectronics and Photovoltaics*. Cambridge University Press.

Yang, X., Zhao, D., Leck, K. S., et al. (2012). Full visible range covering InP/ZnS nanocrystals with high photometric performance and their application to white quantum dot light-emitting diodes. *Adv Mater*, **24**, 4180–4185.

References

Anc, M. J., Pickett, N. L., Gresty, N. C., Harris, J. A., and Mishra, K. C. (2013). Progress in non-Cd quantum dot development for lighting applications. *ECS J Solid State Sci Technol*, **2**, R3071–R3082.

Bae, W. K., Kwak, J., Park, J. W., et al. (2009). Highly efficient green-light-emitting diodes based on CdSe@ZnS quantum dots with a chemical-composition gradient. *Adv Mater*, **21**, 1690–1694.

Bae, W. K., Lim, J., Lee, D., et al. (2014). R/G/B/natural white light thin colloidal quantum dot-based light-emitting devices. *Adv Mater*, **26**, 6387–6393.

Chen, Y., Vela, J., Htoon, H., et al. (2008). "Giant" multishell CdSe nanocrystal quantum dots with suppressed blinking. *J Amer Chem Soc*, **130**, 5026–5027.

Chepic, D. I., Efros, Al. L., Ekimov, A. I., et al. (1990). Auger ionization of semiconductor quantum drops in a glass matrix. *J Lumin*, **47**, 113–127.

CIE (1931). *Commission Internationale de l'Eclairage Proceedings, 1931*. Cambridge University Press.

CIE (2016). *The Use of Terms and Units in Photometry: Implementation of the CIE System for Mesopic Photometry*. CIE. Available at http://files.cie.co.at/841_CIE_TN_004-2016.pdf (accessed December 20, 2016).

Cragg, G. E., and Efros, A. L. (2010). Suppression of Auger processes in confined structures. *Nano Letters*, **10**, 313–317.

Docampo, P., and Bein, T. (2016). A long-term view on perovskite optoelectronics. *Acc Chem Res*, **49**, 339–346.

Erchak, A. A., Ripin, D. J., Fan, S., et al. (2001). Enhanced coupling to vertical radiation using a two-dimensional photonic crystal in a semiconductor light-emitting diode. *Appl Phys Lett*, **78**, 563–565.

Gonzalez-Carrero, S., Galian, R. E., and Pérez-Prieto, J. (2016). Organic–inorganic and all-inorganic lead halide nanoparticles [Invited]. *Opt Expr*, **24**, A285–A301.

Guzatov, D. V., Gaponenko, S. V., and Demir, H. V. (2018). Plasmonic enhancement of electroluminescence. *AIP Advances*, **8**, 015324.

Hines, M. A., and Guyot-Sionnest, P. (1996). Synthesis and characterization of strongly luminescing ZnS-capped CdSe nanocrystals. *J Phys Chem*, **100,** 468–471.

Jang, E., Jun, S., Jang, H., et al. (2010). White-light-emitting diodes with quantum dot color converters for display backlights. *Adv Mater*, **22**, 3076–3080.

Khurgin, J. B., Sun, G., and Soref, R. A. (2008). Electroluminescence enhancement using metal nanoparticles. *Appl Phys Lett*, **93**, 021120.

Lim, J., Bae, W. K., Lee, D., et al. (2011). InP–ZnSeS core–composition gradient shell quantum dots with enhanced stability. *Chem Mater*, **23**, 4459–4463.

Mahler, B., Spinicelli, P., Buil, S., et al. (2008). Towards non-blinking colloidal quantum dots. *Nature Mater*, **7**, 659–664.

Priolo, F., Gregorkiewicz, T., Galli, M., and Krauss, T. F. (2014). Silicon nanostructures for photonics and photovoltaics. *Nature Nanotechn*, **9**, 19–26.

Reckmeier, C. J., Schneider, J., Susha, A. S., and Rogach, A. L. (2016). Luminescent colloidal carbon dots: optical properties and effects of doping [Invited]. *Opt Expr*, **24**, A312–A340.

Robel, I., Gresback, R., Kortshagen, U., Schaller, R. D., and Klimov, V. I. (2009). Universal size-dependent trend in Auger recombination in direct-gap and indirect-gap semiconductor nanocrystals. *Phys Rev Lett*, **102**, 177404.

Shi, H., Zhu, C., Huang, J., et al. (2014). Luminescence properties of YAG:Ce, Gd phosphors synthesized under vacuum condition and their white LED performances. *Opt Mater Expr*, **4**, 649–655.

Shirasaki, Y., Supran, G. J., Bawendi, M. G., and Bulović, V. (2013). Emergence of colloidal quantum-dot light-emitting technologies. *Nature Photonics*, **7**, 13–23.

Song, W. S., and Yang, H. (2012). Fabrication of white light-emitting diodes based on solvothermally synthesized copper indium sulfide quantum dots as color converters. *Appl Phys Lett*, **100**, 183104.

Talapin, D. V., Mekis, I., Götzinger, S., et al. (2004). CdSe/CdS/ZnS and CdSe/ZnSe/ZnS core–shell–shell nanocrystals. *J Phys Chem B*, **108**, 18826–18831.

Xing, J., Yan, F., Zhao, Y., et al. (2016). High-efficiency light-emitting diodes of organometal halide perovskite amorphous nanoparticles. *ACS Nano*, **10**, 6623–6630.

Yang, X., Hernandez-Martinez, P. L., Dang, C., et al. (2015). Electroluminescence efficiency enhancement in quantum dot light-emitting diodes by embedding a silver nanoisland layer. *Adv Opt Mater*, **3**, 1439–1445.

Zan, F., and Ren, J. (2012). Gas–liquid phase synthesis of highly luminescent InP/ZnS core/shell quantum dots using zinc phosphide as a new phosphorus source. *J Mater Chem*, **22**, 1794–1799.

Zhang, F., Zhong, H., Chen, C., et al. (2015). Brightly luminescent and color-tunable colloidal $CH_3NH_3PbX_3$ (X = Br, I, Cl) quantum dots: potential alternatives for display technology. *ACS Nano*, **9**, 4533–4542.

Zhmakin, A. I. (2011). Enhancement of light extraction from light emitting diodes. *Phys Rep*, **498**, 189–241.

Zhou, J., and Xia, Zh. (2015). Luminescence color tuning of Ce^{3+}, Tb^{3+} and Eu^{3+} codoped and tridoped $BaY_2Si_3O_{10}$ phosphors *via* energy transfer. *J Mater Chem*, **C3**, 7552–7560.

9 Lasers

This chapter focuses on application of nanostructures in lasers. Nanostructures form active media in semiconductor lasers (quantum wells and quantum dots, epitaxial and colloidal ones); they appear also in multilayer mirrors (distributed Bragg reflectors [DBRs]) in solid-state lasers. Quantum wells and quantum dots coupled to a DBR and also quantum dots themselves dispersed in a transparent matrix can serve as laser Q-switching and mode-locking components for nano-, pico-, and femtosecond pulse generation. Using lightwave confinement in nanostructures known as photonic crystals gives rise to the smallest lasers, measuring no more than a few micrometers in all dimensions, which paves the way for integrated photonic circuits. Lasers containing nanostructures range from femtosecond to continuous wave (CW) regimes and from milliwatt microchips for optical communication to powerful multi-watt CW devices for multiple applications, from metrology to medicine. Lasers with nanostructures dominate in the market and remain to be an active field of research and developments promising new exciting records in the near future. An inexperienced reader is advised to recall the content of Chapters 3 and 6 prior to proceeding with this chapter.

9.1 EPITAXIAL QUANTUM WELL LASERS

Epitaxial *quantum well* lasers are the main type of commercial semiconductor laser. These are used for optical data storage, as pumps and transmission sources in optical fiber communications, in thermal and xerographic printing, in various research and analytical equipment, and in common laser pointers. A semiconductor diode can be extremely efficient. For example, a 940 nm wavelength laser diode exists with up to 70% wall-plug efficiency (WPE). The spectral range of quantum well lasers with a p–n junction spreads from 0.4 μm to 2.8 μm and can be further extended toward 15 μm using the *quantum cascade* design. Epitaxial *quantum dot* lasers are not common on the market and have entered the commercial realm only lately, filling the 1.3 μm niche of optical communication short-distance networks. Most of the modern commercial semiconductor lasers are edge-emitting devices (see Figure 6.10), which in turn break down into Fabry–Pérot and distributed feedback (DFB) types. In this section these types of quantum well lasers, as well as the quantum cascade ones, will be briefly discussed since the mature industrial scale of these lasers has already made them the subject of many textbooks and lecture notes. The main focus of this section will be *vertical cavity surface-emitting lasers* (VCSELs), which are the subject of extensive research. VCSELs occupy about 25% of the overall laser diode market and are forecast to gain 30% growth per annum for the period till 2024.

Epitaxial quantum well lasers are based on group III–V compounds and break down into three types depending on the basic substrate material, namely GaAs-, InP-, and GaN-based lasers. *GaAs-based* quantum well lasers are formed from alloys of group III (Ga, Al, In) and group V (As, P) elements grown in compositions that are lattice-matched to a GaAs substrate. They can emit at any wavelength from about 630 to about 1100 nm, the most common commercial ones being as follows: 635, 650, 680, and 780 nm for optical storage devices and displays; 785, 808, 830, 920, and 940 nm for various pumping and printing applications; 980 nm for pumping of fiber amplifiers in telecommunications. *InP-based quantum* well lasers are formed from alloys of the same group III (Ga, Al, In) and group V (As, P) components but in compositions that are lattice-matched to InP. They range from about 1100 to 2800 nm, but by far the most common are emitters at 1300, 1480, and 1550 nm which are used in fiber-optic communications. Extension to longer wavelengths needs the GaAs/AlGaAs n-type quantum cascade lasers based on *quantum well superlattices. GaN-based* quantum well lasers emit in the range 370–530 nm and use (Al, Ga, In) nitrides in the form of ternary compounds grown on GaN substrates.

9.1.1 Single-spatial-mode Fabry–Pérot edge-emitting lasers

Edge-emitting lasers with parallel polished facets forming a cavity are referred to as *Fabry–Pérot* lasers. The simplest and most common type is the single-spatial-mode Fabry–Pérot laser (Figure 9.1). *Single spatial mode* implies that the beam can be focused to a diffraction-limited spot. A relatively narrow waveguide is bounded by the cleaved facets. The narrow waveguide supports only a single optical mode. This limits its width to 2–5 μm. The light beam emitted by a single-mode laser at the output facet has typically 1 μm height and 3–4 μm width. Emitted radiation diffracts with a divergence of 20–30° in the vertical direction and 5–10° in the horizontal direction.

For current I below the threshold current I_{th}, a semiconductor laser features the properties of a light-emitting diode (LED), i.e., it emits luminescent radiation owing to dominating spontaneous transitions (Figure 9.2). At threshold current, cavity gain per round trip equals the total cavity losses and for $I > I_{th}$ the slope changes and the dependence becomes much steeper. Now, stimulated downward transitions dominate over the spontaneous one, and recombination rate (and the possible modulation rate) many times exceeds that for the spontaneous (luminescence) regime. Consider the simplified common relations between material and cavity parameters in a laser diode.

When current density, J, increases, carrier concentration, N, increases accordingly, and the relation is valid:

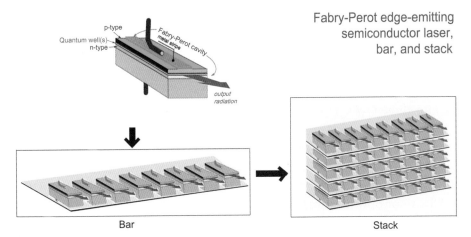

Fabry-Perot edge-emitting
semiconductor laser,
bar, and stack

Bar · Stack

Figure 9.1 A Fabry–Pérot edge-emitting individual laser, a laser bar, and a stack.

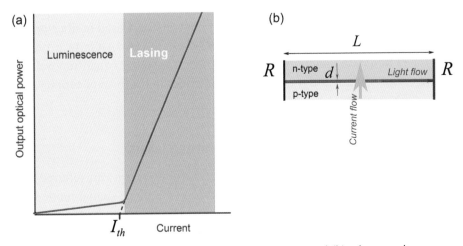

Figure 9.2 (a) Laser optical output power versus current and (b) a laser cavity.

$$\frac{N}{\tau_s} = \eta_i J \frac{1}{ed},$$
(9.1)

where τ_s is recombination lifetime, η_i is *injection efficiency*, often it is referred to as *internal efficiency*, d is gain medium thickness, and e is electron charge. With increasing current density absorption saturates (see Section 6.3) and at certain current density value J_0 absorption coefficient drops to zero. Then at further current growth optical gain develops. From Eq. (9.1), one can write

$$J_0 = N_0 \frac{ed}{\eta_i \tau_s},$$
(9.2)

where N_0 is referred to as *transparency carrier density* and J_0 is referred to as *transparency current*. N_0 is the property of a given *semiconductor structure*; it has an order of magnitude about 10^{18} cm^{-3}. For lasing to occur, gain should exceed the total cavity losses. Gain coefficient (in cm^{-1}) for a given cavity mode averaged over the cavity length L is referred to as *modal gain coefficient g*, and the dimensionless product gL is referred to as *modal gain*. Similar to Eq. (9.2), the threshold current density is related to the threshold carrier density N_{th},

$$J_{th} = N_{th} \frac{ed}{\eta_i \tau_s},$$

(9.3)

where N_{th} should be larger than N_0 by the amount enabling compensation for all cavity losses, and therefore N_{th} and J_{th} are defined by both the semiconductor properties and the cavity design. For this reason, it is convenient to write N_{th} in the form (Iga 2008),

$$N_{th} = N_0 + \frac{\alpha_a + \alpha_d + \alpha_m}{g_0 \xi},$$

(9.4)

where αs (in cm^{-1}) stand for different types of losses expressed as per unit length, namely: α_a is absorptive loss; α_d accounts for diffractive and scattering losses; and α_m is the mirror loss. The latter expresses a portion of intensity not reflected by cavity mirrors in terms of equivalent absorption losses per unit length in the course of the radiation round trip over the cavity, i.e.,

$$\exp(-2\alpha_m L) = R_f R_r, \quad \alpha_m = \frac{1}{2L} \ln \frac{1}{R_f R_r},$$

(9.5)

where R_f (R_r) are intensity reflection coefficient of the front (rear) mirror. The differential optical gain, g_0 in Eq. (9.4), is defined as dg/dN. This is the value defining the slope of the $g(N)$ curve, and in the first approximation, the relation is valid:

$$g(N) = g_0 (N - N_0).$$

(9.6)

Finally, ξ is the optical confinement factor defined by the laser design. It gives the fraction of radiation power in the cavity passing inside the gain medium.

To summarize, from Eqs. (9.3)–(9.5) one has the following expression for the threshold current density, J_{th}:

Threshold current density $\qquad J_{th} = \frac{ed}{\eta_i \tau_s} \left(N_0 + \frac{\alpha_a + \alpha_d + \dfrac{1}{2L} \ln \dfrac{1}{R_f R_r}}{g_0 \xi} \right).$

(9.7)

This relation apparently shows the ways to lower threshold current value. Active layer thickness d should be minimal (i.e., a double heterostructure is useful!); injection efficiency η_i should be as high as possible; diffraction, scattering, and absorption losses should be minimized; the active medium length should be longer; the mirror reflectance should be high; the $g(N)$ curve should be as steep as possible, and optical confinement should be perfect (i.e., close to 1).

When the condition for lasing is met, the dependence of output power P (watts) versus current for $I \gg I_{th}$ can be approximated as

$$P(I) = \eta_i \eta_e \frac{hv}{e}(I - I_{th}), \quad v = c/\lambda, \quad (9.8)$$

where η_e is *extraction efficiency*, which is equal to the portion of radiation power inside a cavity that is extracted outside. In the simplest case η_e is defined as

$$\eta_e = \frac{\text{front mirror loss}}{\text{total loss}} = \frac{\dfrac{1}{L}\ln\dfrac{1}{R_f}}{\alpha_a + \alpha_d + \dfrac{1}{2L}\ln\dfrac{1}{R_f R_r}}. \quad (9.9)$$

The product of injection efficiency and extraction efficiency is referred to as the *external differential efficiency*, η_{ext}:

$$\text{External differential efficiency } \eta_{ext} = \eta_i \eta_e. \quad (9.10)$$

Consider the external power efficiency $\eta_P = P/P_0 \equiv P/(IU_0)$, where U_0 is the applied external voltage. Taking into account Eqs. (9.8) and (9.9), and keeping in mind that I_{th} can be neglected for $I \gg I_{th}$, $hv \approx E_g$, the band gap energy, and hv/e numerically is equal to energy expressed in electronvolts, we arrive at the reasonable estimate for WPE (external power efficiency) in the form

$$\text{External power conversion efficiency } \eta_P = \eta_{ext}\frac{E_g(\text{eV})}{U_0(\text{V})} \approx \eta_{ext}, \quad (9.11)$$

where E_g is expressed in electronvolts. One can see that the overall power conversion efficiency from the wall-plug to the single-mode monochromatic radiation is defined by the external differential efficiency, that is by carrier injection efficiency (pumping) and by extraction efficiency (low losses, high optical confinement). In the best laser diodes it can be as high as 50%, or even higher! Neither a laser of other type nor other light source can offer such high efficiency. When comparing to LEDs, one can say that a laser diode features higher extraction efficiency by converting radiation power into a single mode.

The history of semiconductor diode lasers is actually the history of inventions toward lower threshold current density. In the early homojunction lasers it was of the order of 5000 A/cm², then, owing to double heterostructure invention, it dropped to below 1000 A/cm²; further progress owing to various optical confinement solutions (see diagrams in Figure 6.11) and quantum wells allowed reaching a value of 40 A/cm² at the end of the 1980s, which is probably close to the basic limit for two-dimensional gain media defined by the energy-independent two-dimensional electron–hole density of states (see Figure 3.1). The recent emergence of quantum dot gain media in laser design promises further downshifting of threshold current density. Thresholds below 10 A/cm² have been reported for research-grade quantum dot laser structures.

Modulation bandwidth is critically important for laser applications in optical communication. For laser output intensity modulated by injection current modulation, the modulation bandwidth (at the –3 dB level) equals $1.55 f_r$, with f_r being the laser relaxation frequency, which reads (Iga 2008)

$$f_r = \frac{1}{2\pi}\sqrt{\frac{1}{\tau_s \tau_p}\frac{I - I_{th}}{I_{th}}} \xrightarrow{I \gg I_{th}} \frac{1}{2\pi}\sqrt{\frac{1}{\tau_s \tau_p}\frac{I}{I_{th}}} = \frac{1}{2\pi}\sqrt{\frac{1}{\tau_s \tau_p}\frac{J}{J_{th}}}, \tag{9.12}$$

where τ_p is defined by intensity decay time in a cavity upon multiple round trips. It is equal to the round trip time $2Ln_{eff}/c$ (n_{eff} being the effective refractive index of the intracavity medium) divided by the factor describing intensity loss per round trip, i.e., absorptive, scattering, diffraction, and mirror losses. It is referred to as *photon lifetime in a cavity*. If mirror losses are the only ones, it reads

Photon lifetime in a cavity $\qquad \tau_p = -\dfrac{2Ln_{eff}/c}{\ln(R_f R_r)}. \tag{9.13}$

Using the total loss components (Eqs. (9.4), (9.5)) expressed per unit length, photon lifetimes becomes L-independent and reads

$$\tau_p = \frac{n_{eff}/c}{\alpha_a + \alpha_d + \alpha_m}. \tag{9.14}$$

Recalling Eq. (9.3) for threshold current density allows excluding τ_s in Eq. (9.12) and to arrive at the relation

Modulation frequency $\qquad f_r \approx \dfrac{1}{2\pi}\sqrt{\dfrac{1}{\tau_p}\dfrac{J\eta_i}{eN_{th}d}}$ for $J \gg J_0$. $\tag{9.15}$

This relation clearly shows the ways to higher bandwidth:

1. increase current density (may result in overheating);

2. reduce photon lifetime (cavity property; however, cavity should meet the lasing condition, i.e., mirrors should have reflectance dependent on medium gain);

3. reduce gain medium thickness.

For typical edge-emitting diodes, the gain medium thickness is $d = 3$ μm, cavity length $L = 300$ μm, mirror reflectance about 0.3, and photon lifetime is $\tau_p = 1$ ps to result in relaxation frequency about 5 GHz. Regretfully, the condition $J \gg J_0$ can hardly be met in most cases due to heat management problems.

9.1.2 Multimode lasers, bars, and stacks

Single-spatial-mode lasers are used when a diffraction-limited radiation beam is required, such as optical data storage, laser printers, and pump sources in single-mode optical fiber communication. Their power can be up to 1 W. The power can be raised using more sophisticated laser structures. In a Fabry–Pérot edge-emitting laser, higher power can be obtained simply by making the laser structure wider, raising the width from 3–4 μm to 50–100 μm. The wider lasers can deliver up to 5 W at the expense of losing spatial coherence. These lasers are called *broad-area* or *multimode lasers*. They can neither be focused to a diffraction-limited spot nor coupled efficiently to a single-mode fiber. Maximum power typically increases sublinearly with laser width. Nevertheless, multimode lasers can be used for many applications such as pumping solid-state lasers and external-drum thermal printing. Total WPE can be about 60%.

The next level up is the laser bar, which is an array of 10–50 side-by-side multimode lasers integrated into a single chip. Standard bar dimensions are about 1.0 cm wide and 1–2 mm long. A single bar can emit 20–60 W of CW power. The most common application for bars is pumping solid-state lasers; the most common pump wavelengths are 785, 792, 808, 915, and 940 nm.

9.1.3 Distributed feedback lasers

Many laser applications require a narrow-band emission spectrum and low noise. These are, first of all, every spectroscopic application and optical communication field, where high-quality laser source is a prerequisite for wavelength division/multiplexing (WDM) performance. A simple Fabry–Pérot edge-emitting laser is good for many applications except for the two above mentioned ones. In 1971 H. Kogelnik and C. V. Shank proposed spatially periodic refractive index of the gain medium or gain itself to promote generation at a given mode of a cavity. They coined the notion of a *distributed feed back* (*DFB*) *laser*. This approach has been realized for dye lasers. In the same period, in 1971–1972 Rudolf Kazarinov and Robert Suris at Ioffe Institute (Leningrad, USSR, now St-Petersburg, Russia) proposed a semiconductor laser with a grating on top of the gain medium. Grating can be performed in the form of grooving (Figure 9.3) or by means of a periodically alternating refractive index. Grooves can be made by photolithography using laser beam interference to develop a sub micrometer periodic image on the surface with subsequent etching or ion milling. The front mirror is replaced by an antireflection coating and the rear mirror features high reflectance. Light propagates in a gain medium with a periodically corrugated surface. Owing to multiple diffraction with inversion of light propagation and light reflection/guiding by reflection from the bottom interface, spectrally selective conditions for gain develop, resulting in narrow-band single-mode lasing. The first semiconductor DFB lasers used optical pumping (Nakamura 1973). Implementation of this idea for injection lasers had met the problem of enhanced surface recombination in the corrugated surface, and an additional carrier confinement layer was introduced to separate areas of current flow and light diffraction. The full theory of a DFB laser is based on a consistent analysis of coupled waves in a periodic structure with gain and can be researched in books on the topic. The third Fourier component is usually used, so that the lasing wavelength λ versus the

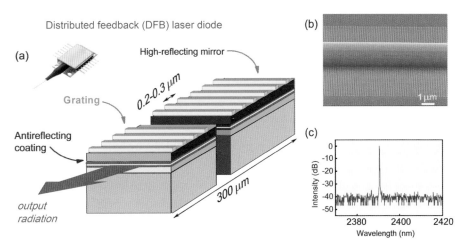

Figure 9.3 A DFB laser diode has a grating on top. Positive feedback provided by diffraction of radiation enables single-mode emission with narrow bandwidth and high signal-to-noise ratio. (a) A general layout. A metal stripe is used on top (not shown) for current injection. (b) Scanning electron micrograph of a real DFB laser diode showing the lateral metal Bragg grating and the stripe on its top. (c) Lasing spectrum of a GaAlInSbAs device. (b) and (c) reprinted with permission from Seufert et al. (2004), copyright Elsevier.

grating period Λ is defined by $\lambda = (2/3)n_g\Lambda$ where n_g is the refractive index of the guiding medium above the gain region. For typical $n_g = 3.5$ and $\lambda = 1.5$ μm, the grating period is about 0.6 μm.

Today, semiconductor single-mode DFB lasers are the essential component of 1.55 μm WDM telecommunication systems. They are also used in many spectroscopy applications, including remote gas sensing. Typically, DFB lasers feature spectral bandwidth well below 1 nm, with high stability and optional minor tunability by means of current density affecting the device temperature in the range 20–30 °C. The typical output power is in the range 1–10 mW, the threshold current being a few tens of milliamps. Standard bipolar laser diodes (using electron–hole recombination in a p–n junction) are available in the market for the range from 760 to 3000 nm. Longer wavelengths become attainable with *cascade lasers*. Optical communication application does not need wide-operation spectral range, but requires extreme accuracy and stability of laser output parameters. Modern DFB lasers for the 1.5 μm communication band offer wavelength accuracy of 0.02 nm with stability better than 0.005 nm per 24 hours; power-level accuracy of 0.5 dB with stability of 0.01 dB per 24 hours; and spectral linewidth about 1 GHz. This type of laser represents a high-precision device weighing about 0.7 kg and costs more than US$2000. It delivers typically 10–20 mW output power.

Quantum cascade lasers represent yet another type of semiconductor quantum well-based laser. The principle of a quantum cascade laser was described in Section 6.7 (Figure 6.14). It is essentially a unipolar device based on resonance tunneling of electrons in an electrically biased quantum well superlattice. Lasing occurs owing to stimulated

BOX 9.1 DISTRIBUTED FEEDBACK AND QUANTUM CASCADE LASERS

Distributed feedback and quantum cascade semiconductor lasers were proposed for the first time by Rudolf Kazarinov and Robert Suris at the Ioffe Institute in Leningrad (now St-Petersburg, Russia) in 1971–1972. Now these lasers are widely used in optical communication and analytical instruments featuring unsurpassed spectral line purity and extreme noise reduction.

The first quantum cascade laser was made in 1994 at Bell Labs by Federico Capasso, Jérôme Faist, and co-workers. In 1998 these scientists shared the Rank Prize in optoelectronics with Kazarinov and Suris. The invention of quantum cascade lasers provided efficient mid-infrared (IR) radiation sources which transformed remote sensing of atmosphere and spectroscopic techniques in chemistry. Nowadays, extension to the terahertz range promises efficient applications in security systems.

Rudolf Kazarinov   Robert Suris   Federico Capasso   Jérôme Faist

inter-subband transitions when electrons tunneled into a neighbor well find themselves at the upper level because of the biased potential profile. The latter experiences considerable bending since the electric field applied is of the order of 100 kV/cm. Though the idea was first put forward in the 1970s (see Box 9.1), the first implementation was reported in 1994 (Faist et al. 1994). The quantum cascade principle extends the operational wavelength of semiconductor lasers toward the mid- and even far-IR, challenging also the sub millimeter (terahertz) range. The actual emergence of quantum cascade lasers has transformed the basic instruments for molecular vibrational spectroscopy and enabled efficient remote sensing of gas pollution in industrial and environmental atmospheres (Zeller et al. 2010). Quantum cascade lasers are commercially available in the range 3–15 μm (see Section 6.8). Similar to traditional lasers discussed above, optical feedback in quantum cascade lasers can be performed either by means of a Fabry–Pérot cavity or by using the DFB concept. Very large output powers have been achieved, as well as narrow linewidths with a very high spectral purity. Nowadays multi-watt output power in CW mode at room temperature with a WPE of more than 10% are seen (Hugi et al. 2015).

In recent years, two main research trends have been highlighted to extend cascade laser operation both to shorter and longer wavelengths. In the first approach, instead of unipolar devices based on superlattices, an *interband cascade laser* has been elaborated. This is a bipolar device making use of electron–hole recombination in a cascaded type II quantum well structure where electrons experience quantum confinement but holes do not. This type of laser may even substitute quantum cascade devices in the range 3–6 μm in the future, owing to lower threshold drive powers, a critical parameter for spectroscopic portable and field devices where energy-saving issues may dominate output power parameters (Vurgaftman et al. 2015). The second approach is driven by persistent need for reliable long-wave radiation sources for millimeter and sub millimeter wavelength range corresponding to the terahertz domain. Emergence of terahertz sources of electromagnetic radiation will transform security systems, allowing for non-perturbing inspection of humans and goods instead of dangerous x-ray exposure. Extension of operating range to 0.25 mm has been reported. Here, however, the room-temperature operation becomes a challenge (Williams 2007). There are reports of operating temperatures up to 200 K, whereas high power at the 1 W level is affordable at liquid helium temperatures only. Though being acceptable for special research instruments, e. g., astronomy, low-temperature devices are not appropriate for routine use in practical applications.

9.1.4 VCSELs: vertical cavity surface-emitting lasers

This type of semiconductor laser was briefly introduced in Chapter 6 (Section 6.5). An optical layout of a VCSEL and a real device image are presented in Figure 9.4. The evident advantages of a surface-emitting laser versus the edge-emitting one are: (1) the possibility of a whole laser being grown epitaxially on a wafer in a single fabrication process; (2) laser beam circular symmetry and higher spatial quality, meaning lower beam divergence; (3) easy options for dicing into single devices, linear or two-dimensional arrays; and (4) on-chip testing in the production process. The VCSEL design was proposed first in 1977

VCSEL: Vertical Cavity Surface Emitting Laser

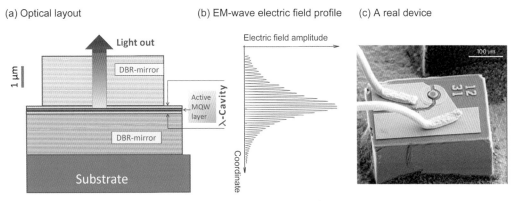

Figure 9.4 VCSEL: vertical cavity surface-emitting laser. (a) Optical layout; (b) the corresponding electric field profile of electromagnetic radiation; (c) a real device. Reprinted with permission from Kapon and Sirbu (2009), copyright Macmillan Publishers Ltd.

by K. Iga, and the first experimental VCSEL was made by him alongside colleagues in 1979. Since then, great progress has been made in research and development, resulting in a multitude of commercial single devices, arrays, and complicated VCSEL-based systems in the market. The first commercial VCSELs appeared in 1990. In 2015, VCSELs took second place in production volume among all types of semiconductor laser, after the Fabry–Pérot-type edge-emitting lasers for use in optical disk drives for data storage. They are the most versatile laser diodes, have enormous market growth potential, and will most probably soon become number one, taking into account that CD and DVD data storage is losing market share to solid-state flash memory devices.

The main applications of VCSELs are optical data transmission, storage, laser printers, and computer mice. For example, owing to VCSELs 2400 dpi (dots per inch) laser printers have been developed using 4×8 VCSEL arrays that allow projecting 32 beams at a time to the photoconductor in the exposure process. In optical communication, InP-based VCSELs are mainly used, whereas in laser printers GaAlAs-based lasers are the most important, emitting at 780 nm.

Consider the possible bandwidth advantages of VCSELs versus edge-emitting counterparts. Recalling the general relation for bandwidth (Eq. (9.12)), one can suppose at first glance that, because of the extremely short cavity (3 μm, i.e., 100 times less than for edge-emitting devices), photon lifetime can become much less than 1 ps. However, this is not the case since low gain coefficients in VCSELs result in a high Q-factor cavity with perfect mirrors (i.e., the cavity is shorter but the number of round trips becomes higher) and the net photon lifetime in VCSEL cavities remains close to 1 ps, i.e., the same as in an edge-emitting laser. Nevertheless, VCSELs do offer higher bandwidths (>10 GHz) owing to operation at higher currents ($J \gg J_0$).

VCSELs cover the wide spectral range from 360 nm to 1600 nm, with the only break in the green range (500–550 nm). These lasers are available from 700 nm to 1600 nm as commercial devices based on GaInAs and InP, and are the subject of experiments on extension to the shorter wavelengths (AlGaN) and toward longer wavelengths using GaSb.

Though the very idea of a surface-emitting semiconductor laser looks very clear and straightforward, its experimental implementation is by no means easy. The vertical design requires a number of contradictive and tough conditions to be met. Optical issues come from the low gain value across a few quantum well or quantum dot layers (a few cm^{-1}). Electric issues come from the current flow management. Growth issues arise from the lattice-matching requirements in a complicated multilayer "sandwich." To summarize:

1. Very short gain length (less than 1 μm) necessitates an extreme cavity Q-factor, i.e., DBR mirrors should reflect more than 99%.

2. The whole device throughout should be conductive and optically transparent at the same time, i.e., the cavity should be made between electrical contacts and mirrors should be involved in the electrical circuit – even minor absorptive losses can cancel lasing because of the low gain values.

3. Epitaxial growth needs good lattice matching from the substrate to the top of the whole device.

To meet lattice-matching requirements, the layers in DBRs should not differ much in the chemical content, and thus refractive index will change only a little and approximately 20–30 periods are necessary to acquire the desirable 99% or even higher reflectance. To ensure the current flow throughout a structure, the appropriate doping of constituent layers should be carefully performed. Finally, in the cavity formed between DBRs, the central, pretty thin, area containing the multiple quantum well (or quantum dot) gain layers should be covered from both sides by additional transparent but conductive spacers to secure the optical cavity length to fit the integer number of half-waves with respect to the desirable lasing wavelength. Finally, the whole system should be designed in a way that electric field amplitude of the standing wave in the cavity should have its maximum (antinode) exactly in the gain area where multiple quantum well (or quantum dot) layers are located, as shown in Figure 9.4(b). One can see that VCSELs can be very tiny. A VCSEL's height is actually defined by the DBRs' thicknesses (1–2 μm each), whereas the base size can measure several wavelengths to ensure low diffraction losses. Vertical arrangement of lightwave propagation gives an accurate output beam with spherical symmetry and lower divergence compared to edge-emitting laser diodes.

The commercial success of VCSELs is essentially the result of advances in technology, design, and modeling to meet the competitive requirements and to account for the complex interplay of optical, electrical, and thermal processes in multilayer semiconductor/metal/dielectric structures. The diameter of commercial VCSELs starts from a few micrometers to get the lowest threshold currents below 0.1 mA and can be as large as 0.1 mm for high output powers beyond 100 mW. However, single-mode operation, the most valuable VCSEL advantage, can be performed up to a few milliwatts, and higher output power operation can be performed in the multimode regime only with loss in laser beam quality.

When optical layout is tentatively defined, the primary issue is to ensure the current flow and to define its surface spreading in the optimal way. Figure 9.5 shows the two most realistic VCSEL designs in which current flow is taken into account. To have higher efficiency, current should not spread far away from the lasing area and therefore a certain *current aperture* is necessary. There are several solutions suggested:

1. Ring- or circular-shaped top electrode. The current flows in the vicinity of the ring electrode, whereas light passes through the center window. This type of electrode is shown in both parts of Figure 9.5. This is easy to fabricate but is not enough for an efficient device since the current cannot be confined completely within a small area due to diffusion.

2. Proton bombardment (Figure 9.5(a)) of the outer portion of the top DBR mirror. An insulating layer is made by proton irradiation to limit the current spreading toward the surrounding area. The process is rather simple, and is used commercially.

3. Buried mesa, including the active region, with a wide-gap semiconductor to limit the current. The refractive index can be small in the surrounding region, resulting in an index-guiding structure so that current and optical confinements are performed simultaneously. However, the process is rather complicated.

4. Air-posts can be developed by etching but nonradiative recombination at the outer wall may cause device performance to deteriorate.

5. Selective oxidation of AlAs and $Al_xGa_{1-x}As$ layers with x close to 1 to Al_xO_y oxide (Figure 9.5(b)). This is very popular in GaAs-based lasers but not suitable for InP-based devices.

Because of the very short cavity, expansion of the optical aperture to get higher output power by means of a larger emitting area readily switches a VSCEL laser into the multimode regime and its principal advantage of higher beam quality vanishes. A few technical solutions have been proposed to keep to single-mode operation for a large emitting surface (Larsson 2011). First, simple increase of the cavity length has been proposed and implemented. In this approach, a several micrometer-long GaAs spacer has been grown epitaxially between the bottom mirror and the active gain layer. Current and optical confinements have been performed by oxide ring(s), as is shown in Figure 9.5(b). A 5 mW barrier has been passed in the CW single-mode operation at 980 nm. Similar results have been obtained with an additional metal aperture on top of the output DBR mirror for spatial filtering of the higher-order modes. Another solution is using the additional doping of the top DBR mirror to generate mode-selective cavity losses through reduced DBR reflectivity. In a similar way, an additional shallow surface relief etched in the top DBR for a spatial variation of the cavity loss through a spatial variation of the phase of the reflection at the surface has also been proven to be useful. The above approaches allow for the 5 mW level to be used in single-mode operation in the ranges 850–980 nm. In Section 9.4 we will show that the photonic crystal paradigm offers yet another approach to scale up VCSELs' power by means of a two-dimensional periodic refractive index lattice instead of the bottom DBR enabling efficient large-square feedback without the top DBR up to multi-watt levels while staying with single-mode operation.

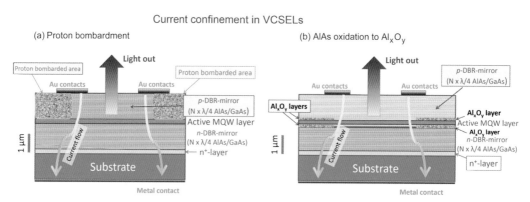

Figure 9.5 Two methods of current confinement in VCSELs: (a) proton bombardment of the outer area; (b) oxidation of $Al_xGa_{1-x}As$ layers to Al_xO_y oxide.

For applications in optical communication, the modulation rate is critically important. For the 1.3 μm optical communication band, both GaAs-based and InP-based structures have been used, whereas for the 1.5 μm band only InP-based structures are appropriate. A 10 GHz direct modulation frequency and 10 GB/s data rates have been demonstrated for both communication bands over distances of a few kilometers. For shortwave communication rages (850, 980, 1060 nm), 20 GB/s have been achieved. To minimize the response time, additional oxide layers are introduced thus reducing the overall capacity of the device.

To understand the multiple technical problems in VCSEL fabrication, let us consider in more detail a multistep process for an Si-integrated 840 nm VCSEL suggested by E. P. Haglund et al. (2015). Its design is presented in Figure 9.6.

The laser consists of the two half-parts fabricated by the two different approaches. The most critical part is developed by means of epitaxial lattice-matched growth of a p-type DBR structure, the gain structure, and the n-type contact layer, all on a monocrystalline

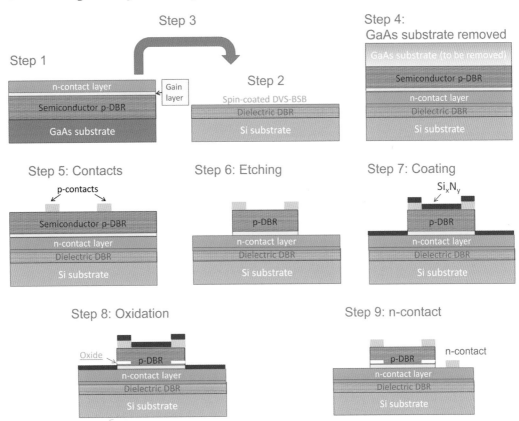

Figure 9.6 Silicon-integrated vertical cavity surface-emitting laser as suggested by E. P. Haglund et al. (2015). Note that the p-contact is outside and the n-contact is inside the cavity. See text for detail.

GaAs wafer (step 1). The second half-part is fabricated by means of non-epitaxial deposition, i.e., without lattice-matched monocrystals, of a dielectric DBR on an Si substrate. Then the two parts are put together (step 3) using the preliminary developed ultrathin divinylsiloxanebis-benzocyclobutene (DVS-BCB) adhesive bonding. Then, metal p-contacts and mesas are formed by the known technique (steps 5 and 6). Afterwards, the surface is protected by an SiN coating (step 7) to allow for the lower layer of AlAs in the DBR mirror to be oxidized to get the current aperture (step 8). Finally, the n-contact metal plate is deposited on the n-contact semiconductor layer. Remarkably, in this design the lower n-contact is inside whereas the p-contact is outside the cavity. The laser is designed for 840 nm and delivers 1.6 mW of optical power at 6 mA current with the 9 μm current oxide aperture. This example clearly shows that basic principles and approaches are well settled, but that every specific laser design needs several interlaced factors to be accounted for to enable technically feasible development of electrical and optical paths in an efficient manner.

Nowadays, researchers are challenged by the problem of wavelength extension toward the 400–600 nm range using GaN-based devices. It is very important for applications in high-resolution printing, high-density optical data storage, head-up displays, backlighting, and chemical/biological sensing. There are serious obstacles that are not inherent in AlGaAs-based VCSELs. With III–N compounds it is not possible to make perfectly reflecting and highly conductive DBR mirrors.

Figure 9.7 presents a technological approach suggested by G. Cosendey et al. (2012) from the École Polytechnique Fédérale de Lausanne for a blue AlInN-based VCSEL. The major technical solutions are:

1. intracavity contacts since DBR mirrors are not conductive;

2. indium tin oxide (ITO) continuous film under the top metal ring contact to manage current flow;

3. thick cavity to inhibit side modes;

4. optical aperture in addition to current aperture;

5. oxide layers in the top DBR instead of III–V compounds;

6. light field management to ensure maximal intensity in the quantum well gain area and minimal intensity in the lossy ITO layer.

In more detail, the device was grown on a free-standing 2-inch GaN substrate using an Aixtron MOCVD reactor. The bottom mirror consisted of 41.5 pairs of $Al_{0.8}In_{0.2}N/GaN$ layers. The use of such an epitaxial bottom DBR eliminates the requirement to remove the cavity from the substrate, which makes the process flow easy. The pair $Al_{0.8}In_{0.2}N/GaN$ features good lattice matching (see Figure 3.3 in Chapter 3). On top of this DBR, a p–i–n-GaN diode structure was grown with the total thickness corresponding to a 7 λ cavity designed for a wavelength of 420 nm. An active region made of five $In_{0.1}Ga_{0.9}N$ (5 nm)/$In_{0.01}Ga_{0.99}N$ (5 nm) quantum wells was centered at an antinode of the optical field. Then a

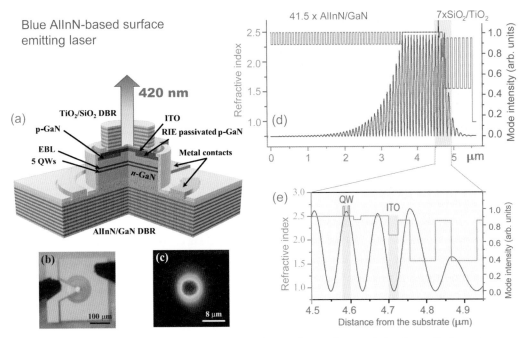

Figure 9.7 A blue monolithic AlInN-based vertical cavity surface-emitting laser diode on free-standing GaN substrate designed and fabricated at the École Polytechnique Fédérale de Lausanne in 2012. (a) The design. (b) Photo of a device under CW current excitation. (c) Near-field real-space image in the linear scale of a device under pulsed excitation with the current aperture diameter of 8 μm. DBR stands for distributed Bragg reflector, ITO stands for indium tin oxide, RIE stands for reactive ion etching, EBL stands for electron blocking layer. Reprinted from Cosendey et al. (2012), with the permission of AIP Publishing. (d,e) Calculated electric field distribution superimposed with the refractive index profile. Note antinode at the quantum wells (QW) position and the node within the ITO layer. Courtesy of W. Nakwaski.

20 nm-thick $Al_{0.2}Ga_{0.8}N$ electron blocking layer (EBL) was grown. The n-GaN (Si doped, 940 nm-thick) and p-GaN (Mg doped, 120 nm-thick) layer thicknesses were grown. Then standard microelectronics techniques were used: mesas were etched by Cl-based ICP RIE (inductively coupled plasma reactive ion etching). The current confinement ring was made by CHF_3/Ar RIE plasma treatment for p-GaN surface passivation. A small circular window in the center of the mesas was protected with photoresist for current injection. An ITO current-spreading layer was then sputtered on top of the mesa. Its thickness was set to be a quarter-wave of the cavity mode to serve as the first half-pair of the top DBR. Metal films were then deposited to form the p-(Ni/Au) and n-(Ti/Al/Ti/Au) contacts. Finally, the top dielectric DBR (seven TiO_2/SiO_2 bilayers) was deposited by e-beam evaporation. Owing to the simple deposition techniques and high reliability, TiO_2/SiO_2 periodic layers are typically used to make commercial DBR mirrors for solid-state or gas lasers. Their refractive index contrast (see Table 4.1) enables 99% reflectivity with just seven periods. Regretfully, III nitrides are not suitable for current-spreading because of low conductivity.

Much better is the conductive semi-transparent material ITO, but even its conductivity is insufficient for current apertures greater than 10 μm. ITO also has noticeable optical loss in the visible spectrum, which is acceptable in liquid crystal display (LCD) structures and in LEDs, but becomes critical in the case of laser design because of higher losses from multiple lightwave round trips in a cavity.

There was also an approach to use both bottom and top mirrors based on purely dielectric rather than epitaxial semiconductor layers. In this case, though mirrors contain typically less than ten periods instead of 30–40 periods in the case of epitaxially grown DBRs, the mirror lift-off from the substrate is involved which disturbs the accuracy of the structure thickness and brings a problem of perfect control of light distribution along the structure.

In spite of the progress in III nitride laser developments, a gap exists in the green range, approximately 500–550 nm, where no laser diode is currently available. Attempts to extend lasing from the blue to the green meets a number of obstacles – e.g., the group from Nichia found that shifting laser parameters to go from the blue to 500 nm have resulted in a drastic ten-fold increase of the threshold current (Kasahara et al. 2011).

9.2 EPITAXIAL QUANTUM DOT LASERS

9.2.1 Edge-emitting laser diodes: commercial devices

Epitaxial quantum dot lasers designed as an edge-emitting device with a Fabry–Pérot cavity formed by crystal facets and delivering radiation at 1.3 μm have recently appeared as commercial products. These advances became possible as a result of the practical needs and progress in self-organized quantum dot growth in strained submonolayer heterostructures owing to research by many teams across the world. In optical communication networks, the bands used for data transfer are defined by both fiber transparency regions and available lasers (and amplifiers for long-distance communication). The main band, C-band, from 1530 nm to 1565 nm corresponds to the lowest fiber losses and is used for long-distance lines using dense wavelength division/multiplexing (DWDM) to increase data transfer rates. The important band is referred to as the O-band and ranges from 1260 nm to 1360 nm. It is widely used for FTTH PON (fiber to the home passive optical network) and LAN (local area network) setups, where one expensive laser is used to deliver data that are further split into up to 32 fiber lines to individual links.

Since the pioneering paper by Y. Arakawa and H. Sakaki (1982), discussed in detail in Chapter 6 (Section 6.6), there has been extensive research activity in the area of practical epitaxial quantum dot laser diodes. Many groups focused at InAs–GaAs heterostructure with the reasonable lattice parameters to obtain a self-organized growth regime of quantum dots in submonolayer epitaxy with the pyramid-like or dome-like nanocrystals of a few nanometers in height and about 10 nm base length (Figure 9.8).

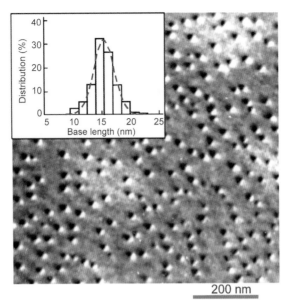

Figure 9.8 Plan view transmission electron microscopy image of InAs quantum dots in a GaAs matrix. The surface density of the array is $3.7 \cdot 10^{10}$ cm^{-2}. The insert shows a histogram of base size distribution centered at 15 nm. Reprinted from Zhukov and Kovsh (2008), copyright Turpion Ltd.

Because of the narrow, delta-function-like electron and hole density of states in an ideal quantum dot, the threshold current density for optical gain (transparency current density) can be much lower than that for quantum well lasers. Additionally, as Y. Arakawa and H. Sakaki suggested in 1982, a narrow density of states spectrum inhibits carrier spreading over energy with temperature, thus temperature dependence of threshold current should ideally vanish. However, implementation of these basic principles in a real laser device became possible within the InAs/GaAs epitaxial system only nearly two decades later owing to efforts toward technological feasibility, material band gap requirements, and electron–hole recombination optimization.

The first problem was achieving the target wavelength range. At first glance, the low band gap energy of InAs (E_g = 0.345 eV, which corresponds to electromagnetic radiation wavelength λ_g = 3.54 μm; see Chapter 2, Table 2.3) provides plenty of opportunities to reach the 1.3 μm (0.95 eV) emission range. However, very small electron effective mass ($0.02m_0$) for typically feasible self-organized InAs dots on a GaAs substrate gives rise to electron ground state in the c-band >0.7 eV and does not allow obtaining durable structures with the first optical transition for wavelength longer than 1.2 μm. Longer wavelengths need bigger dot size, which is not feasible because of a complex interplay of a number of processes in the strain-based dot growth regime.

To go to longer wavelengths, a different matrix was proposed with a narrower band gap than that of GaAs, but with the requirement for self-organized growth due to lattice mismatching safely met. Recall Figure 2.3, which shows that lower potential wells make the energy levels move to lower values. It was found that regrowth of an array of InAs quantum dots with an $In_xGa_{1-x}As$ layer (x = 0.1–0.2) with a thickness of 4–12 nm allows shifting the emission maximum from 1.1 to 1.34 μm. The approach of the optical transition energy control through the matrix parameters rather than through quantum dot material composition and growth regimes is very favorable in a sense of tunability, while keeping dot size/shape/concentration – and therefore the gain medium parameters – constant.

The second problem is inefficient hole population of the quantum dot ground state. Because of the large effective mass of holes ($m_h/m_e > 10$), the number of hole states in a quantum dot is greater than that of the electron states and hole states are more closely spaced in energy than the electron states. This results in a situation in which the injected electrons mainly occupy the ground state within the conduction band, but the injected holes are thermally spread among the closely spaced hole states instead of the ground state. This thermal broadening of the injected holes would reduce the quantum dot ground state gain and increase temperature sensitivity of the threshold current. The use of p-type modulation doping was suggested (Takahashi and Arakawa 1988) to improve properties of InAs quantum dot lasers in terms of temperature insensitivity of the threshold current, and was confirmed experimentally (Shchekin and Deppe 2002).

For a single quantum dot monolayer, the maximal optical gain was found to be no more than a few cm^{-1}, and with the reasonable strip length about 1 mm very low losses should be secured, including multilayer mirrors instead of simple crystal facets. To overcome this limit, multi-quantum-dot-layer heterostructures have been suggested, using up to ten InAs quantum dot layers (Zhukov and Kovsh 2008 and references therein).

Figure 9.9 presents the design of a real experimental InAs quantum dot edge-emitting laser diode with the cavity formed by the two plane-parallel crystal facets, its internal structure, energy diagram, and the result of optical gain calculation for the ten-layer active region. Gain spectra were calculated for the injection current range of 10–30 mA corresponding to the electron surface density per quantum dot layer $n_D = 3.2 \cdot 10^{10}$–$8.3 \cdot 10^{10}$ cm^{-2}. A single quantum dot has been modeled as a disk with a 14 nm radius and a 2 nm height. The calculated wavelengths of the three peaks of quantum dot optical transitions are 1266 nm ($hv_1 = E_{C1} - E_{HH1}$), 1166 nm ($hv_2 = E_{C2} - E_{HH2}$), and 1090 nm ($hv_3 = E_{C3} - E_{HH3}$). The calculated optical gain spectra for different currents were found to perfectly match the experimentally measured data. This example is given to emphasize the complicated multiparametric issues that exist in quantum dot laser design and development.

Commercial quantum dot lasers for the 1.3 μm range are developed by QD Laser, Inc. (Japan). A representative example is given in Figure 9.10. At $T = 25$ °C the device in a

p- doped quantum dot laser

(a)

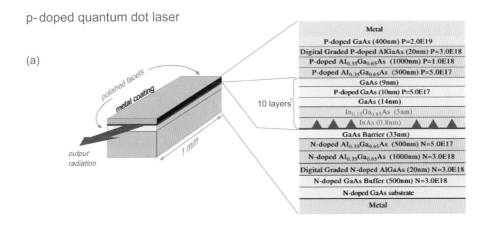

(b)

(c)

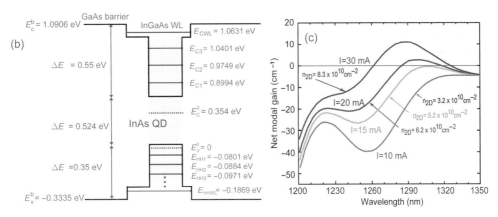

Figure 9.9 Design of InAs p-doped quantum dot strip edge-emitting 1.3 μm laser evaluated by Kim and Chang (2006): (a) design; (b) energy-level diagram; (c) calculated gain spectra for different injection currents.

standard TO-56 case delivers 16 mW in CW mode at 1305 nm wavelength, 60 mA current, and 1.5 V voltage. At $T = 85\,°C$ the same output power needs about 70 mA current, i.e., more than three-fold temperature change results in less than 15% current increase. The typical threshold current at $T = 25\,°C$ is $I_{th} = 13$ mA, i.e., the laser operates at about $4I_{th}$.

9.2.2 Vertical cavity surface-emitting lasers (VCSEL): extensive research

VCSELs have a number of evident advantages that were discussed in detail in Chapter 6 (Section 6.5). First, the surface-emitting design enables compliance with the standard chip-on-wafer production flow in microelectronics and *in situ* test options. Second, multilayer highly reflective mirrors (DBRs) can be fabricated in a single epitaxial running process instead of using natural reflectance from crystal facets. Third, the vertical

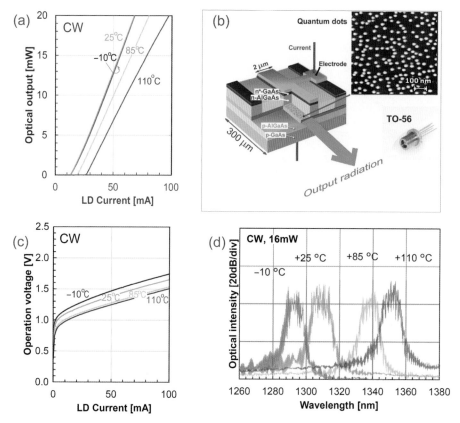

Figure 9.10 Design and parameters of a commercial TO-56 quantum dot laser for application in optical communication systems. (a) Power–current characteristics; (b) layout; (c) voltage–current characteristics; (d) emission spectra. QD Laser, Inc. (2015), reproduced with permission.

design gives a much better radiation pattern than edge-emitting analogs. However, the low gain coefficients even for ten-layer epitaxial quantum dot structures needs the reflection coefficient of mirrors to be close to 100%, which in turn requires many periods of alternating layers since refraction indices are close for semiconductor materials with small deviation chemical composition, and higher difference in composition is not possible because the structure must provide the current flow through with DBRs involved in the electric circuit. One can see that the clear physical principles and ideas can only be implemented owing to a number of technological solutions and compromises. Quantum dot VCSELs are the subject of active research in many laboratories and companies.

Collaboration of researchers from the Ioffe Institute in St-Petersburg (Russia) and the Technical University of Berlin (Germany) pioneered the development of the first ever quantum dot VCSEL for 1.3 μm (Lott et al. 2000) and demonstrated a number of records

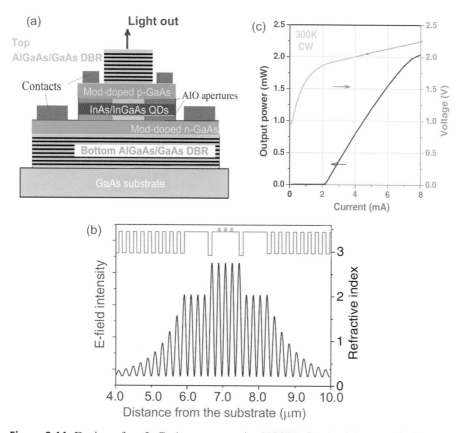

Figure 9.11 Design of an InGaAs quantum dot VCSEL for the 1.3 μm optical communication range. (a) The general layout; (b) electric field amplitude and refractive index spatial profile; (c) output power and voltage versus current. Adapted from Ustinov et al. (2005), with permission from John Wiley and Sons. © 2005 WILEY-VCH Verlag GmbH & Co. KGaA, Weinheim.

in optimized structure design for the best performance. Figure 9.11 gives a representative example of the design suggested for efficient CW lasing operation.

Design and fabrication of efficient quantum dot-based VCSELs meets a number of challenging issues.

1. *Controlled growth of multiple quantum dot layers.* To get ten successive quantum dot layers buried in InGaAs, the spacing layer thickness should be carefully optimized to ensure that conditions for self-organized quantum dot formation in strained heterostructures are met in successive layers. A thickness of 33 nm was found to give the best results in this context.

2. *Gain peak and microcavity mode matching.* Quantum dots, when compared to quantum wells, feature much narrower gain spectrum because of the narrow density of states distribution. This can result in lower threshold current, but this advantage can be implemented

in a device only when perfect matching is realized for the microcavity mode wavelength and quantum dot gain spectrum. Matching needs accurate structure design and precise fabrication.

3. *High-reflective lossless mirrors*. Extremely low gain value across quantum dot layers necessitates perfect mirrors with negligible losses and high, nearly 99% reflectance. This is aggravated by relatively small refraction index difference of technologically feasible, lattice-matched, and transparent in the 1.3 μm range pairs of semiconductor materials, and eventually up to 30-period periodic structures (DBRs) are necessary to meet the 99% reflectance condition. The situation is compounded by the current flow condition. If DBRs form a part of the current flow path, they should consist of doped semiconductors to ensure reasonable free carrier concentration. This in turn gives rise to optical dissipation losses because of free carrier absorption, especially in the case of p-doped materials. To overcome this obstacle, DBRs should be placed outside the current path circuit, i.e., design of intracavity electrical contacts becomes critical.

The above approaches are embodied in the device presented in Figure 9.11. This VCSEL structure has intracavity contacts, p- and n-doped conductive layers, two selectively oxidized current apertures, top and bottom AlGaAs/GaAs DBRs, and the active region. The quantum dot active region consists of three sets of triple-stacked InAs/InGaAs quantum dots. Quantum dot layers are placed within the 2λ-thick undoped GaAs layer and surrounded by n- and p-doped $Al_{0.98}Ga_{0.02}As$ quarter-wave aperture layers which were later selectively oxidized to form both current and waveguide apertures. These aperture layers are followed by 1.75λ-thick intracavity contact/current-spreading layers which are followed by 29 top pairs and 35 bottom pairs of undoped $Al_{0.9}Ga_{0.1}As$/GaAs DBRs. The 1.75λ-thick intracavity contact layer is p-doped with Be to $10^{18}\,cm^{-3}$ and includes two $\lambda/16$-thick Be spikes doped to $10^{19}\,cm^{-3}$ centered at the two standing wave nodes closest to the top DBR. Similarly, the lower n-doped 1.75λ-thick intracavity contact layer is doped with Si to $1.5\cdot10^{18}\,cm^{-3}$ and includes two $\lambda/16$-thick Si spikes doped to $4\cdot10^{18}\,cm^{-3}$ centered at the two standing wave nodes closest to the bottom DBR. The structure is engineered to get each set of triple-stacked InAs/InGaAs quantum dots exactly at the antinodes of the electric field intensity (Figure 9.11(b)). At room temperature the device delivers 1 mW of radiation power at 4 mA current and 2 V voltage, thus featuring WPE above 10%. The maximal output is 2 mW and the threshold current is 2 mA.

9.3   COLLOIDAL LASERS

Colloidal quantum dots with size smaller than exciton Bohr radius of the parent bulk crystal allow for wide-range absorption and emission spectra tuning, thus offering a means for tunable colloidal lasers. Consider an ensemble of quantum dots with strong confinement of electrons and holes dispersed in some transparent material. Let it be excited by the light whose frequency corresponds to interband transitions much higher than the absorption

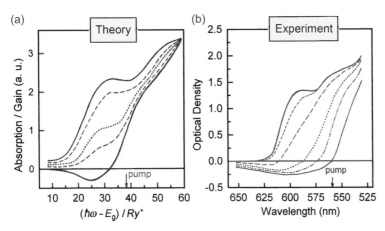

Figure 9.12 Calculated (a) and observed (b) absorption spectra of CdSe nanocrystals (Hu et al. 1996). Upper curves in each panel are the linear absorption spectra. The other curves correspond to successively growing (theory) population and (experiment) excitation intensity, from top to bottom. The mean nanocrystal radius is $a = 2.5$ nm, the Gaussian size distribution in the calculation was given as $0.1a$. Reprinted from Hu et al. (1996) with permission from Elsevier.

onset (Figure 9.12). Every dot can have a single electron–hole pair, two electron–hole pairs, or three or even more. When an ensemble is optically excited, the average number of electron–hole pairs per dot grows from zero upward. The number of electron–hole pairs per dot can be described by the Poisson distribution function.

At low intensities of external light sources, the average number of electron–hole pairs per dot is very close to 0, i.e., only a very small portion of dots has one electron–hole pair, and an even smaller amount can have two, and even fewer gain three pairs. The absorption spectrum then features the shape defined by the theory developed without electron–hole pairs existing in dots taken into account. With growing excitation level (higher intensity of light) the average number of electron–hole pairs per dot grows up to one pair per dot. This situation corresponds to the absorption saturation, as was discussed in Chapter 6 (Section 6.1, Figure 6.1). For further increase in excitation intensity, the average number of electron–hole pairs per dot will approach two. Under this condition, optical gain develops, similar to population inversion discussed for a three-level system in Chapter 6 (Section 6.1, Figure 6.2). Absorption coefficient then becomes negative, the transmission coefficient of a slab becomes higher than 1, and optical density becomes negative. This is seen in Figure 9.12 for both the theory and the experiment. Notably, optical gain develops always for frequencies lower than the optical excitation frequency. At the excitation frequency, zero absorption coefficient (absorption saturation) is an extreme.

The above presentation of optical properties of semiconductor quantum dots in the strong confinement regime was elaborated theoretically and verified experimentally in the 1990s (see, e.g., books by Bányai and Koch 1993; Gaponenko 1998). The first experimental evidence on lasing for glasses containing semiconductor quantum dots (CdSe) was reported by Vandyshev et al. (1991) from the Moscow State University for liquid nitrogen temperature (T = 80 K), at 640 nm under pumping with the second harmonic of an Nd:YAG laser (532 nm), and has been followed up with extensive research. Successful observation of optical gain has been reported by various groups for InAs, PbS, and PbSe nanocrystals in a glass and in a polymer matrix.

However, the implementation of this simple optical gain idea for quantum dot lasers faced a serious obstacle. When two or more electron–hole pairs exist in the same dot, the strong wave functions overlap and a lack of translational symmetry (lifts on momentum conservation) enables an efficient Auger process, resulting in fast nonradiative recombination. This effect was discussed in Chapter 8 (Section 8.2, Figure 8.8). Auger recombination shunts stimulate the emission process and additionally promote fast photodegradation. To combat the Auger process, V. Klimov from Los Alamos National Lab (USA) suggested using the type II core–shell quantum dots rather than type I ones (Figure 9.13).

In type II core–shell quantum dots, only one type of charge carrier, either an electron or a hole, has a potential well (see also Figure 3.22). Then an electron and a hole wave functions and the corresponding probability densities separate in space, resulting

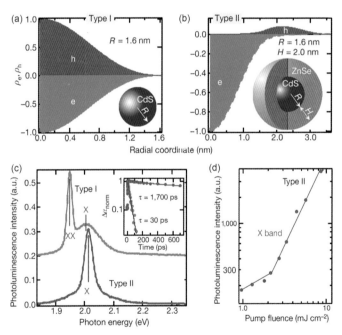

Figure 9.13 Single-exciton optical gain in semiconductor nanocrystals. (a) Electron and hole wave functions (probability density) in a generic CdS quantum dot (type I bands alignment). (b) Electron and hole wave functions (probability density) in a core–shell CdS/ZnSe quantum dot (type II alignment). (c,d) Photoluminescence at high excitation, X denotes single exciton, XX denotes biexciton. Reprinted by permission from Macmillan Publishers Ltd.; Klimov et al. (2007).

in charge separation (Figure 9.13(b)) and a local electric field develops upon optical excitation. This field gives rise to a small energy shift of the first transition energy so that a single-exciton emission band moves slightly to lower energies as compared to the single photon absorption band. Thus a familiar three-level system develops dynamically upon excitation, enabling efficient stimulated emission and optical gain in a quantum dot ensemble with a single electron–hole pair per dot. The Auger process does not develop; as a result, the type II dots feature much slower recombination rates (inset in Figure 9.13(c)) and stimulated emission for the single-exciton resonant emission band (Figure 9.13(c,d)).

The idea of single-exciton optical gain in type II colloidal dots has been successfully implemented by several groups, including not only experiments on luminescence line-narrowing owing to dominating stimulated emission, but also lasing when dots are placed in a cavity. However, perfect mirrors are necessary because of the relatively low gain for dots

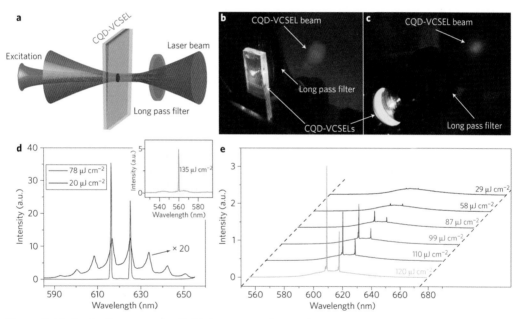

Figure 9.14 Optically pumped colloidal quantum dot vertical cavity surface-emitting laser (CQD-VCSEL) in the red and green ranges. (a) Schematic of a vertically pump CQD-VCSEL with a long pass filter to remove any residual pump excitation beam. CQD gain medium is placed inside a wedge cavity for a variable cavity length. The wedge angle is $1.2 \cdot 10^{-3}$ rad, and two DBRs have reflectivity higher than 99%. (b,c) Photographic images of red and green CQD-VCSELs showing spatially well-defined output beams, which are collinear with the pump beam. (d) Spectra from a red CQD-VCSEL structure below and above threshold. Inset: single-mode lasing for a green CQD-VCSEL from a shorter cavity. From the linewidth of laser emission, the quality factor of the cavity was estimated to be 1300. (e) Emergence of laser modes from spontaneous emission in a CQD-VCSEL when increasing pump power. Reprinted by permission from Macmillan Publishers Ltd.; Dang et al. (2012).

dispersed in a thin polymer film. Figure 9.14 gives a representative examples of green and red lasing in a planar cavity with type II quantum dots under optical pumping.

Colloidal lasers made of semiconductor nanocrystals have been the most promising, particularly because of their solution-processing capability, which makes them applicable essentially to any arbitrary substrate. One specific demonstration is an all-colloidal laser (Guzelturk et al. 2015). The basic idea of all-colloidal lasers is to integrate colloidal gain intimately into the colloidal cavity (Figure 9.15). For a proof-of-concept demonstration, in the work of Guzelturk et al., the cavity was made of a pair of Bragg reflectors, each of which contained alternating layers of titania and silica, all spin-coated. The gain material of semiconductor nanocrystals, which is sandwiched between the colloidal Bragg reflectors, was also introduced via spin-coating. As a result, this whole structure is truly all-solution processed. Although this all-colloidal laser structure provides high performance, it is still limited in performance owing to the fundamental barriers intrinsic to colloidal quantum dots. These colloidal systems suffer relatively large optical gain thresholds, which are typically several hundreds of $\mu J/cm^2$ under pulse excitation. This is due to the small optical gain coefficients that colloidal quantum dots can provide, which are typically in the range of low hundreds of cm^{-1}. The root cause of this limited performance is their ultrashort gain lifetimes, commonly shorter than 50 ps, which is constrained by Auger recombination dominant under high optical fluence. Also, their relatively small absorption cross-section makes pumping less effective. All of these factors combined make practical lasers made of colloidal quantum dots technically challenging. To address these limitations, another interesting family of colloidal materials holds promise: colloidal quantum wells.

Colloidal quantum wells, also usually referred to as semiconductor nanoplatelets, represent yet another type of promising active media for colloidal lasers. Because of translational symmetry, Auger recombination is not an issue for nanoplatelets, contrary to quantum dots. By 2017, a number of groups had reported on successful observation of

All-colloidal laser

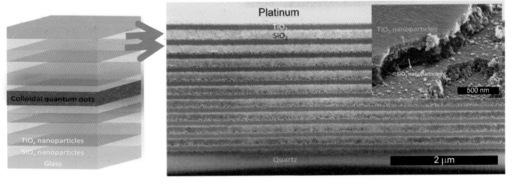

Figure 9.15 All-colloidal laser made of only solution-processed films incorporating colloidal gain into the colloidal cavity: structure schematics and cross-sectional scanning electron microscopy image of the fabricated colloidal reflector. The scale bar is 2 μm.

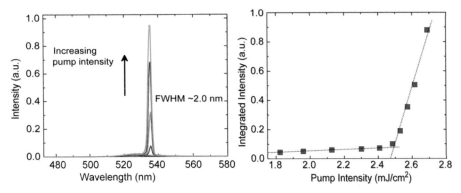

Figure 9.16 All-colloidal lasers of nanoplatelets: emission spectra and integrated emission intensity as a function of pump intensity.

narrow-band stimulated emission for nanoplatelets, which is a prerequisite for lasing to offer high-gain performance. First of all, they exhibit reduced optical gain thresholds in the ranges of low tens of $\mu J/cm^2$ under pulsed excitation, enabled by large optical gain coefficients of the order of thousands of cm^{-1} and accompanied by long gain lifetimes of about 150 ps. As one of the first demonstrations of lasing in core/crown nanoplatelets (Guzelturk et al. 2014), Figure 9.16 presents the characteristic spectral narrowing with increasing pump intensity, reaching a full-width half-maximum of 2.0 nm at room temperature, along with the characteristic integrated emission intensity versus pump intensity.

9.4 LASERS WITH PHOTONIC CRYSTALS

For decades (since 1971) DBRs developed on dielectric substrates have been used routinely to form cavities in solid-state and gas lasers. During recent decades, DBR structures have been integrated in commercial surface-emitting semiconductor lasers. Formally speaking, DBRs can be treated as application of photonic crystals (PCs) to laser design. However, since DBR mirrors were developed a few decades prior to the photonic crystal concept was elaborated, using DBR is not actually an application of the PC concept to lasers. In this section, we consider the options that two- and three-dimensional PCs offer for laser design and fabrication. Photonic crystals can be used to make ultrasmall high Q-factor cavities and to form multidirectional DFB. In the first case, a laser can become smaller, going down to the ultimate volume limit of less than 1 μm^3 and sub-microampere threshold current. This is very important for application in photonic and optoelectronic circuitry in the context of growing demands for higher integration levels in optical communication, optical interconnect, and data storage systems. In the second case, *vice versa*, application of photonic crystals enables square scaling of vertically emitting lasers without loss in output beam quality, promising higher power level in commercial devices.

9.4.1 Photonic crystal nanocavity semiconductor lasers

In this approach, a photonic crystal serves as a matrix in which a small defect (a cavity) can gain extremely high Q-factor owing to strong confinement of lightwaves therein (see Chapter 4; Section 4.6, Figure 4.29). This is a straightforward idea for photonic crystal application to lasers which has many followers but still remains at the proof-of-concept stage. It was first put forward and implemented for a semiconductor quantum well laser with optical pumping in 1999 by O. Painter and colleagues at Caltech, USA. An example of this design is shown in Figure 9.17. Note that although the cavity itself is in the sub-micrometer range, as is the mode confinement volume, the actual physical volume of such a laser measures tens of cubic micrometers since at least several periods of photonic crystal lattice are necessary for a cavity to develop.

In what follows, we shall consider further progress toward nanocavity photonic crystal-based lasers. The main efforts have been directed toward implementing the starting PC-nanocavity idea for electrically pumped devices. Figure 9.18 presents the result reported by the team from Stanford University and Berkeley National Laboratory on an ultralow-threshold 2D PC semiconductor diode laser. The nanocavity laser is electrically pumped by a lateral p–i–n junction (Figure 9.18(a)). The intrinsic region is 400 nm wide in the cavity region, extending to a width of 5 μm to the sides of the cavity. This design

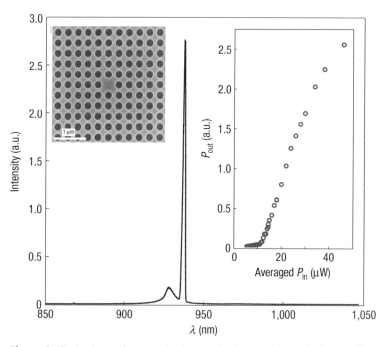

Figure 9.17 A photonic crystal microcavity laser with optical pumping: the emission spectrum, output power versus pump power, and the design. Reprinted by permission from Macmillan Publishers Ltd.; Altug et al. (2006).

enables efficient current flow through the cavity region. A modified three-hole defect photonic crystal cavity design has been used (Figure 9.18(b)). Implantation of high-energy ions causes some lattice damage that will lead to a reduction in gain; it is critical that the p- and n-regions are precisely aligned to the photonic crystal cavity to avoid damaging the active region. A fabrication procedure has been developed in which ion implantation is used through silicon nitride masks patterned by electron-beam lithography to achieve an alignment accuracy of 30 nm. The gain material for the laser comprises three layers of high-density (300 μm^{-2}) InAs quantum dots. The laser features extremely low-threshold current (below 100 nA), but operates at low temperatures (150 K) only.

A nanocavity with high Q-factor can also be fabricated using a *three-dimensional photonic crystal* (3DPC) as a matrix. In this case, the full 3D confinement of lightwaves is ensured by the PC design owing to high refractive index of typical semiconductor materials ($n > 3$). At the same time, a PC structure must be conductive to ensure carrier injection into the active region of such a laser. The 3DPC laser is currently at the proof-of-concept stage. 3DPC with controllable structure integrated with the gain medium needs multistep submicron lithography/etching/alignment processing combined with epitaxy to grow quantum well or quantum dot layers. In the earlier experiments in the 1990s, colloidal crystals impregnated with laser dyes were used. However, this approach cannot be considered seriously in the context of technological implementations because colloidal crystals do not feature an omnidirectional band gap in the IR–visible–UV range as a result of the

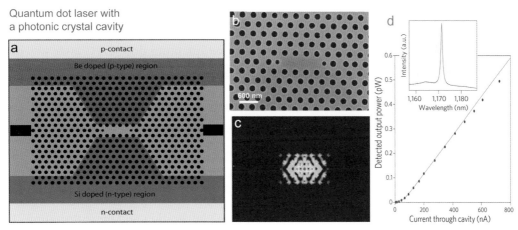

Figure 9.18 Design of the electrically pumped two-dimensional photonic crystal nanocavity InAs/ GaAs quantum dot laser. (a) Schematic diagram of the structure. The p-type (n-type) doping region is indicated in red (blue). The intrinsic region width is narrow in the cavity region to direct current flow to the active region of the laser. A trench is added to the sides of the cavity to reduce leakage current. (b) The fabricated modified three-hole defect photonic crystal cavity structure. (c) Theoretical simulation of the E-field of the cavity mode in such a structure. (d) Output power versus current through the cavity (leakage beyond the cavity subtracted) and (inset) the spectrum of emitted radiation. Reprinted by permission from Macmillan Publishers Ltd.; Ellis et al. (2011).

low refraction index contrast, nor do organic dyes exhibit the necessary photostability for serious laser applications beyond research setup. Though being arduous and complicated, but still controllable, the multistep woodpile photonic crystal fabrication approach has been successfully implemented in the research reported by the group of Y. Arakawa from the University of Tokyo (Figure 9.19). Because of the high refractive index (n = 3.37 at 1.15 µm), GaAs woodpile structures enable a full 3D photonic band gap in the near-IR (see Chapter 4, Section 4.3). The group from Tokyo University suggested this intricate multistep approach to get a microcavity 3D PC laser using an inner InAs quantum dot multilayer insert as a gain medium. They grew a multilayer InAs quantum dot heterostructure on top of a 3D woodpile GaAs/air photonic crystal slab and then added sequential PC layers while monitoring the Q-factor, emission spectrum, and output versus input optical power behavior.

In more detail: 150 nm-thick GaAs layers used for the fabrication of the 3DPC were grown on a 1 µm-thick $Al_{0.7}Ga_{0.3}As$ sacrificial layer using metal–organic chemical vapor deposition (MOCVD). Line-and-space patterns were formed by means of electron-beam lithography and then transferred through the GaAs slab by inductively coupled plasma reactive ion etching. The sacrificial layer was removed by a wet-etching process with hydrogen fluoride (HF) solution to form air-bridge structures. An active layer was prepared in the same way as the GaAs layers, but contained three sublayer stacked InAs quantum dot layers in which the middle quantum dot layer was at the center of the slab. These GaAs layers were stacked in a 3D structure by means of a micromanipulation system installed in an SEM chamber. The nanocavity was designed so that the resonance of the lasing mode was located as close as possible to the center wavelength of the photonic crystal band gap.

In this way, the first ever optically pumped quantum dot 3DPC laser was developed. The 3DPC cavity Q-factor was found to grow monotonically with the number of PC layers over the quantum dots gain insert. However, the output power did not bear witness to monotonic behavior. Lasing was not observed for four PC overlayers; it featured maximum output intensity value for six overlayers and then went down for subsequent PC layers attachment (8 and 12 layers).

9.4.2 How small can a laser be?

Modern optoelectronics needs more and more integration and miniaturization. The physical limit for electron circuitry size is the de Broglie wavelength of charge carriers, which typically measures a few nanometers (see Chapter 3). In this range, electron and hole energies become discrete and size-dependent, and at the same time multiple tunneling processes develop. Therefore, the notion of *nanoelectronics* implies manipulation of conductivity in quantum-confined semiconductor structures. These quantum effects may form a natural physical restriction to the known Moore's law (an electronic element's size continuously

Quantum dot laser with a 3D photonic crystal cavity

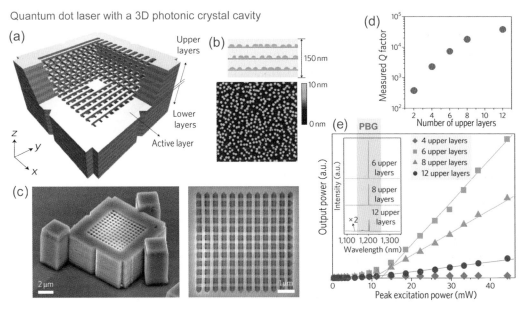

Figure 9.19 A woodpile optically pumped 3DPC semiconductor laser developed at the University of Tokyo. (a) Schematic of a fabricated 3DPC structure. A portion of the upper layers is removed to show the cross-section of the stacked structure and to reveal the cavity structure. (b) Illustration of a cross-section of an active layer showing three-layer stacked quantum dots (upper). Also shown (lower) is a $1 \times 1 \ \mu m^2$ planar atomic force microscope image of InAs quantum dots on a GaAs substrate. (c) SEM images of the fabricated 25-layer woodpile structure shown in bird's eye (left) and top (right) views. (d) Dependence of the measured Q-factor upon the number of the upper layers. (e) The measured output power versus excitation power for different numbers of upper layers. The inset shows emission spectra for 6, 8, and 12 upper layers. PBG is photonic band gap. Lasing has not been observed for four upper layers. Reprinted by permission from Macmillan Publishers Ltd.; Tandaechanurat et al. (2011).

reduces with time on a semi-logarithmic size/time plot). In this context, it is reasonable to discuss the physical restrictions defining the ultimate limits for a laser size.

For lightwave confinement, the wavelength represents the physical limit. Here, wavelength in a medium merits λ/n, with $n \approx 3$ being the refractive index of a typical semiconductor. It appears to be in the range of 200–500 nm, depending on the operating spectral range. Thus there is a definite and well-recognized mismatch between the physically possible electronic circuitry miniaturization and its optical counterpart. Efforts toward integrated photonics over the last few decades are not as impressive as electronic integrated circuitry. In this connection, it should be emphasized that the minimal laser size a_{min} is defined by the condition $a_{min} = \lambda/(2n)$ to ensure the standing wave at the laser wavelength in a cavity. In this case, the round trip length of a cavity equals the desirable laser wavelength. This size is to be complemented by at least a few periods of alternating refractive index materials to form mirrors, either planar ones (DBRs) or more complicated photonic crystal structures. Therefore, the actual physical restriction for laser size is in the range of at least 1 μm or even more. If a cavity is planar, then its cross-sectional size should be at least several wavelengths to minimize diffraction losses. In a two-dimensional photonic crystal cavity in which vertical confinement is supposed to come from the total reflection, the high-index reflecting layers should again have thickness greater than wavelength to secure the minimal evanescent leakage of electromagnetic energy. Thus, the value of a few micrometers for the side length and approximately 20–30 μm^3 for laser volume should be set as the reasonable limit, depending on material refractive index and operation wavelength.

9.4.3 Photonic crystal surface-emitting lasers

In 1998–1999 researchers from Bell Laboratories (M. Berggeren et al.) and Kyoto University (M. Imada et al.) suggested application of a two-dimensional photonic crystal (2DPC) beneath a gain layer to perform a new type of optical feedback by means of multiple wave coupling in the PC plane. It was later classified as *multidirectional distributed feedback* and the relevant theory has been developed (Imada et al. 2002). The group from Kyoto (M. Imada, S. Noda, and co-workers), starting from the first experiments, considered mainly diode lasers with multidirectional feedback by means of the 2DPC design and suggested using it for electrically pumped surface-emitting lasers. In what follows we consider the recent impressive implementation of this approach, which seems to constitute a far-reaching roadmap toward high-power, multi-watt range, single-mode, surface-emitting diode lasers. Development of high-power VCSELs promises dramatic extension of their applications. The current applications of VCSELs are restricted to communications and interconnections by their relatively low power, in the milliwatt range. The principal problem in increasing their output power to the watt level is maintaining single-mode operation upon scaling the emitting surface area. This issue prevents their application in high-power fields such as material processing, laser medicine, and nonlinear optics, despite their advantageous properties of circular beams, wafer-based fabrication/testing/dicing, and suitability for two-dimensional integration. Researchers from Hamamatsu Photonics, in collaboration with the group of S. Noda (Kyoto University), reported on the 2DPC-based VCSEL design and demonstrated its far-reaching possibilities in terms of multiple enhancements of individual surface-emitting laser output power to the multi-watt scale. This new type of watt-class high-power surface-emitting laser demonstrates single-mode CW operation at room temperature owing to a 2DPC beneath the gain layer. The two-dimensional band edge resonant effect of a PC formed by MOCVD enables a 1000 times broader coherent-oscillation area compared to VCSELs, which results in a high beam quality of $M^2 \leq 1.1$ and a smaller focal spot by two orders of magnitude.

   A sketch of lightwave propagation, interference, and diffraction (Figure 9.20(d,e)) provides a rationale for the feedback formed by a 2DPC structure located beneath the multiple quantum well gain layer. The laser is essentially based on the band edge effect of a photonic crystal. For wavelengths close to the band edge, the group velocity of light tends to zero (see Chapter 4, Section 4.2), the lightwaves propagating in various 2D directions

BOX 9.2 THREE PERSONS SHAPING MODERN SEMICONDUCTOR LASERS

In 1977 **Ken-ichi Iga** from Tokyo Institute of Technology invented a surface-emitting laser that, unlike the traditional edge-emitting counterpart, offers unmatched fabrication versatility and beam quality but needs precise design and fabrication because of low gain in a few quantum well layers. These lasers acquired the acronym VCSEL and comprise today the fastest growing sector in the laser diodes market.

Yasuhiko Arakawa, together with H. Sasaki, predicted in 1982 temperature-independent threshold current for a quantum dot laser owing to discrete energy spectra of electrons and holes. Commercial quantum dot lasers are already on the market and breathtaking news from laboratories promise a bright future for them.

Susumu Noda from Kyoto University has an impressive list of records on manipulation with light using ingeniously fabricated semiconductor photonic crystals that pave the way to ultimate nanolasers. In 1999 he suggested a new type of feedback with a 2DPC beneath the gain medium which offers the possibility of power upscaling without beam quality loss.

are coupled with each other, and a 2D standing wave or 2D cavity mode develops in the PC plane (Figure 9.20(d)). Furthermore, the in-plane waves experience diffraction toward the vertical (z) direction because of the first-order Bragg diffraction condition (Figure 9.20(e)). This vertically diffracted wave constitutes the output radiation of the laser under consideration. Standing waves in-plane and in the vertical direction combined with the gain medium enable positive feedback, resulting in lasing. The full analytical description can be made based on the coupled waves theory developed first for DFB lasers. To efficiently utilize the downward-emitted electromagnetic radiation, it is necessary to have phase-matching of the back-reflected and upward-radiated waves such that constructive interference occurs when they are coupled into free space. Otherwise, the efficiency might fall considerably. Accordingly, the thickness of the p-GaAs contact layer (just above the p-side electrode) was adjusted to tune the phase difference between the upward-radiated and back-reflected waves. Thus, the feedback provided by a PC layer eliminates both the bottom and the top distributed Bragg mirrors inherent in the standard VCSEL design.

To summarize, PC lasers have passed through the proof-of-concept stage but still remain the subject of extensive research rather than being ready for commercial prototypes. Most

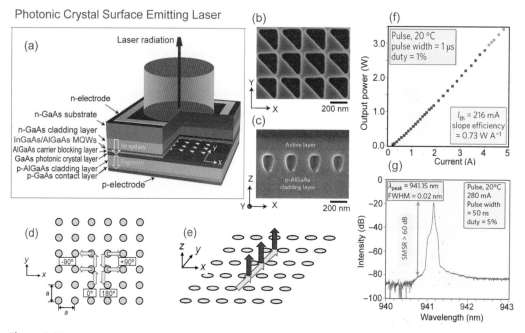

Figure 9.20 A multi-watt PCSEL (photonic crystal surface-emitting laser) developed by Hamamatsu Photonics in cooperation with Kyoto University. (a) Schematic of the PCSEL structure. Arrows indicate the growth direction of the first epitaxial and regrowth structures. (b) Top view SEM image of the fabricated square-lattice PC before burial by MOCVD regrowth. The lattice constant is 287 nm. (c) Cross-sectional SEM image of PC air holes in the *x*-direction after burial. (d,e) A sketch of lightwave propagation, interference, and diffraction. (f,g) Room-temperature lasing characteristics under pulsed conditions: output power versus current and lasing spectrum. Reprinted with permission from Hirose et al. (2014), copyright Macmillan Publishers Ltd.

probably, PCSELs will offer commercial solutions for high-power multi-watt single-mode laser diodes, whereas devices based on photonic crystal cavities will pave the way toward the ultimate size limit of a laser, which measures a few micrometers.

9.5 Q-SWITCHING AND MODE-LOCKING WITH QUANTUM DOTS AND QUANTUM WELL STRUCTURES

9.5.1 Q-switching with a saturable absorber

In Chapter 6 absorption saturation was considered in a simple two-level system resulting in rising optical transmission with increasing input light intensity owing to equalization of the ground and excited state populations (see Figure 6.1 and comments on it). This phenomenon is routinely used in solid-state lasers to obtain giant nanosecond output pulses by means of dynamical switching of the cavity Q-factor from low to high value during active medium pumping on a submicrosecond timescale (Figure 9.21).

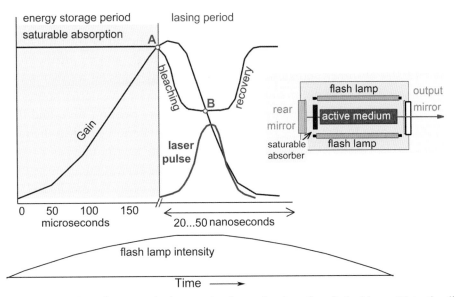

Figure 9.21 Time diagram of a laser pulse formation in a Q-switched laser. Note the different timescales for energy period and lasing period. The *A* point is the first threshold and the *B* point is the second threshold. See text for detail.

Actually, the saturable absorber placed inside the laser cavity "shuts" and "opens" the rear mirror. At the initial (energy storage) stage, high losses from the saturable absorber prevent earlier lasing and enable very high gain of the active medium to be attained. At a certain level defined by the pump intensity and saturable absorber properties, radiation intensity in a cavity becomes high enough to initiate bleaching of the absorber (the first threshold point *A*). The further processes take the form of an avalanche to give fast absorber bleaching and rapid decay of population inversion. The laser appears immediately overpumped well beyond the threshold defined by the cavity drop losses and the high population inversion quickly decays, resulting in a single nanosecond pulse at the laser output. Although gain falls, the intensity in the cavity still rises until the second threshold point *B*, where reduced gain now equals the reduced cavity loss. This point defines the approximate position of the output pulse and the pulse temporal width is defined by a combination of the absorber excited state lifetime, its high-to-low transmission ratio, and the population inversion level. The efficient Q-switching in terms of high extraction efficiency of the energy stored in the gain medium to the output giant pulse is defined by the condition (Siegman 1986),

$$\delta_{g0} \frac{T}{\tau_a} \frac{\alpha_{g0} L_g}{\alpha_{a0} L_a} \frac{\sigma_a}{\sigma_g} > 1, \qquad (9.16)$$

where $\delta_{g0} = dI / I$ is the dimensionless coefficient of fractional intensity I gain (dI/I) in the cavity during the time equal to the single round trip period T; τ_a is the saturable absorber lifetime defining its recovery time; α_{g0} is the maximal optical gain (absorption) coefficient of the gain medium; α_{a0} is the maximal absorption coefficient of the saturable absorber; $L_g (L_a)$ is the geometrical gain medium (absorber) length; σ_a and σ_g are the cross-sections of the saturable absorber and gain medium, respectively. One can see that in this relation the first factor defines the pump rate of the gain medium by an external source. The pump rate characterizes the rise in intensity I inside the cavity per unit time, and the value of δ_{g0} is chosen here to define intensity gain for the time period equal to the round trip T of the cavity. For example, in the case of a 15 cm cavity length, one has $T = 1$ ns. Pump rate is defined by the external light source and since the typical pump pulse time is a few microseconds, an incremental increase of intensity in the cavity for the time of the order of 1 ns will be much less than 1. The second factor is the ratio of this time and the absorber lifetime τ_a. The third factor is the ratio of gain and absorption increments αL defining maximal gain $G = \exp(\alpha_{g0} L_g)$ and maximal loss $I_0 / I = \exp(\alpha_{a0} L_a)$ when radiation propagates through the gain medium and through the saturable absorber. One can see that efficient Q-switching requires the pump rate to be fast enough, depending on gain medium and saturable absorption parameters. Recalling the definition of absorption cross-section as a factor coupling concentration, N, and absorption coefficient , $\alpha = \sigma N$, the values α / σ for the gain medium and the absorber in Eq. (9.16) can be replaced by the relevant concentrations to get a more simple relation,

$$\delta_{g0} \frac{T}{\tau_a} \frac{N_g L_g}{N_a L_a} > 1, \quad T = \frac{2L}{c}. \tag{9.17}$$

Here, N_g and N_a are concentration of absorber and gain medium, respectively. Now it becomes clear that the product of gain medium concentration and its length must be higher than the same value of the absorber, the absorber should be as fast as possible (small τ_a), whereas the resonator length should be as long as possible (large T) so that for a given pump rate δ_{g0}, higher power will be injected into the gain medium during the time interval T.

9.5.2 Saturable absorber as laser pulse width compressor and intensity dynamic range expander

Absorption saturation results in low transmission of a saturable absorber plate or film for low intensities and high transmission for high intensities. If a laser pulse width essentially exceeds the absorber excited state lifetime, then the transmission of such an absorber will follow the instantaneous radiation intensity and respond accordingly. That means that low-intensity pulse wings will be absorbed more than the higher-intensity portion at the pulse maximum. The laser pulse will become shorter, but at the expense of losing a portion of its energy. Calculations show that for a simple two-level system, absorption saturation can give 40% pulse compression at maximum for a Gaussian pulse, its intensity at the output being approximately four times lower (Siegman 1986). This is shown in Figure 9.22.

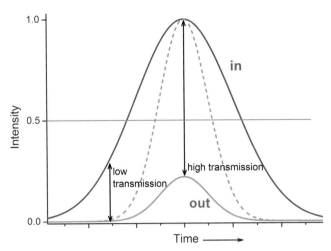

Figure 9.22 Compression of a Gaussian laser pulse by a saturable absorber: 40% compression occurs at the cost of 25% transmitted power. The dashed line shows the output pulse normalized to the incident pulse peak intensity.

At the same time, the dynamic range of the input signal expands. This is clear from Figure 9.22, in which lower transmission for lower intensity and higher transmission for higher intensity increases the maximum/minimum ratio of any incoming signal, i.e., the signal dynamic range expands.

9.5.3 Mode-locking with a saturable absorber

The absorption saturation resulting in Q-switching of a laser cavity and dynamic range expanding can be used to perform the special laser operation regime known as mode-locking. Every cavity has an infinite number of longitudinal modes with different wavelengths satisfying a condition $N\lambda_N / 2 = L$, $N = 1, 2, 3...$ (see, e.g., Figure 2.1). On the frequency scale these modes are separated by a distance $c/2L$ (c being the speed of light in a vacuum), as shown in Figure 9.23. The gain spectrum always has finite width and lasing becomes possible for those modes only where gain exceeds losses. These are shown by thick red lines, whereas the rest are shown as thin pink lines. If gain is kept low enough with respect to saturable absorption losses, there might be a situation in which the two threshold conditions discussed in the comments regarding Figure 9.21 (points A and B) may be met for a single mode only. Then a short pulse can be generated after certain round trips in the cavity whose time width will be defined by multiple compression events when passing through the absorber and gain events when passing through the gain medium. The lower limit for the pulse duration is set by the frequency–time relation due to Fourier transform restrictions. Roughly, the product of pulse length $\Delta\tau$ and its spectral width $\Delta\omega$ should by approximately 1. Pulses whose temporal width corresponds to the Fourier transform limit are referred to as *bandwidth-limited pulses*. This operating regime of a laser is termed *mode-locking*. For typical solid-state laser media, pulses of about 10 ps can be generated. In the steady-state regime, output pulses form a train with spacing equal to the round trip in the cavity, i.e., $2L/c$.

The buildup of an ultrashort pulse in a cavity can be explained in a simple form as follows. The mode for which gain exceeds losses will get the so-called "net gain window" (shaded areas in Figure 9.23(b)) and, starting first from the noise present in the cavity, will experience multiple compression/gain round trip cycles until finally a short output pulse will be generated. The dynamic range expansion featured by a saturable absorber will successively diminish other signals in the cavity. In an ideal case of the very fast absorber and constant gain value, a steady-state regime is maintained, as shown in Figure 9.23(b). The pulse shape obeys a simple relation (Svelto 1998),

$$I(t) = I_{\max}(t)\,\mathrm{sech}^2(t/\Delta\tau). \tag{9.18}$$

Remarkably, mode-locking and ultrashort light pulse generation can be performed also with a "slow" absorber resulting in a generated pulse width much shorter than the absorber lifetime. This becomes possible if alternating gain value is maintained in a way that the short "gain window" is defined by correlated dynamics of both saturable absorption and

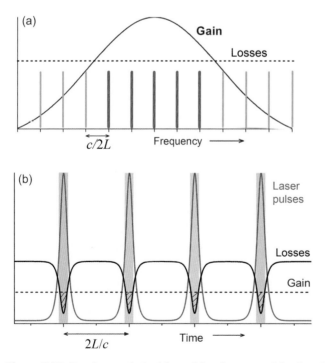

Figure 9.23 Passive mode-locking with a fast saturable absorber. (a) Modes in a cavity superimposed with gain spectrum. (b) Time diagram for output pulse train and losses.

saturable gain, as is shown in Figure 9.24. Though this regime looks very tricky, it has been successfully performed for many solid-state lasers.

9.5.4 Glasses doped with quantum dots as efficient saturable absorbers

Glasses doped with semiconductor nanocrystals (quantum dots) have been commercially used as optical cutoff red, orange, and yellow filters (see Chapter 3; Section 3.5, Figure 3.16) for many decades. In the first years after invention of lasers in the 1960s, red glasses were shown to be suitable for Q-switching of a ruby laser generated at 694 nm. Since then, quantum confinement phenomena and absorption saturation in glasses containing CdSSe quantum dots have been thoroughly examined (see Gaponenko 1998, 2010 for details). An example of genuine absorption saturation in quantum dot-doped glass was given in Section 6.1 (Figure 6.1). Glasses with quantum dots feature a number of properties that make them good Q-switching and mode-locking elements in solid-state lasers. The main advantages as compared to other saturable media are:

1. fast recovery time of bleaching in the nano- and picosecond range;

2. high ratio of the saturable absorption coefficient to the nonsaturable background;

3. high photostability;

4. tunable spectral range in which nonlinear response is available by means of size-dependent absorption spectra;

5. large bandwidth (typically tens of nanometers) under conditions of monochromatic pumping.

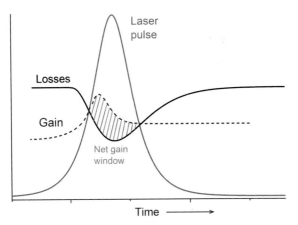

Figure 9.24 Development of a narrow net gain window owing to gain saturation enables short-pulse generation with a slow saturable absorber.

Fast recovery time, high saturable-to-non saturable absorption contrast, good photostability, and tunability options are necessary in all applications. Broad-band bleaching is crucial for mode-locking to obtain ultrashort light pulses.

Most modern solid-state lasers generate radiation pulses in the near-IR, including lasers for material processing and optical communications (see Figure 6.5). To get absorption edge and absorption saturation in this range, quantum dots of narrow-band semiconductor materials developed on a glass matrix should be fabricated. PbS- and PbSe-doped glasses with size-dependent optical absorption in the IR are now available. Narrow-band semiconductors feature low electron effective mass (see Table 2.3) and therefore a strong confinement regime can be readily performed in a few nanometer-size crystallites offering easy tunability of the absorption spectrum from 1 to 2 μm. The representative example of size-dependent absorption spectra for PbS-doped glasses is given in Figure 9.25.

PbS-doped glasses have been successfully used for Q-switching and passive mode-locking for a number of solid-state IR lasers (Figure 9.26). Yb-, Nd-, and Cr-based lasers have been successfully mode-locked to give ultashort pulses from a few picoseconds to 150 ps; Nd-, Er-, Tm-, and Ho-based lasers have been shown to be efficiently Q-switched to give single pulses of a few tens of nanoseconds.

Figure 9.27 represents application of a PbS-doped glass saturable absorber for a Cr^{4+}-YAG laser mode-locking resulting in generation of 10 ps $sech^2$-shaped pulses in the principal optical communication wavelength range around 1.5 μm. The repetition rate was 235 MHz, the average output power being 35 mW for the CW pump power of 6 W. The pump source was a Yb-doped fiber laser. This mode-locked laser was tunable from

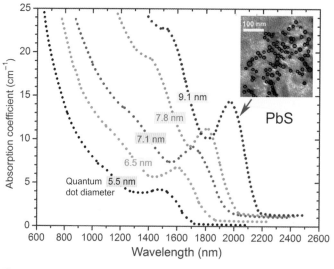

Figure 9.25 Optical absorption spectra of PbS-doped glasses containing quantum dots of different mean size. Reprinted with permission from Loiko et al. (2012), copyright Elsevier.

1460 nm to 1550 nm. As shown in the figure, tuning was accomplished by a fused-silica prism. For the observed spectral width of 0.3 nm at a center wavelength of 1509.5 nm, the implied duration–bandwidth product was 0.4 The minimal duration–bandwidth product for sech²-shaped pulses defined by the Fourier transform limit is 0.315.

9.5.5 SESAM: semiconductor saturable absorption mirror

Passive Q-switching has been successfully used in solid-state laser design for a long time, since the late 1960s, whereas passive mode-locking has been performed for dye lasers (now obsolete) only because of problems with Q-switching/mode-locking interplay resulting in dramatic instabilities. The solution was found only in the 1990s when Ursula Keller and co-workers at the Swiss Federal Institute of Technology (ETHZ, Zurich) suggested a device which she later called the semiconductor saturable absorber mirror (SESAM) (Jung et al. 1995). It consists of an ultrathin semiconductor layer (quantum well in the first reports and also quantum dot layer suggested later on) featuring ultrafast absorption saturation which is epitaxially grown on top of a DBR mirror. A quantum well or a quantum dot layer experiences absorption saturation at relatively low power density with fast relaxation time. The recovery of saturated absorption typically has a short, subpicosecond period (evolution of electron–hole energy distribution function to quasi-Fermi-like profile in wells and hole intraband relaxation in dots) followed by a longer one, of the order of 10^{-12} to 10^{-11} s, including cooling down of the electron–hole system to the lattice temperature and electron–hole recombination. The latter shows the tendency to shorter times in thinner layers as compared to micrometer-thick epilayers. The SESAM invention represents a genuine example of interdisciplinary development based on laser physics (Q-switching and mode-locking concepts) and solid-state laser technology complemented

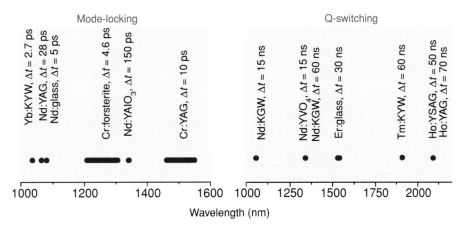

Figure 9.26 A sketch with summary of mode-locking and Q-switching performed for a number of solid-state lasers using PbS quantum-dot-doped glasses as fast saturable absorbers (Malyarevich et al. 2008 and references therein).

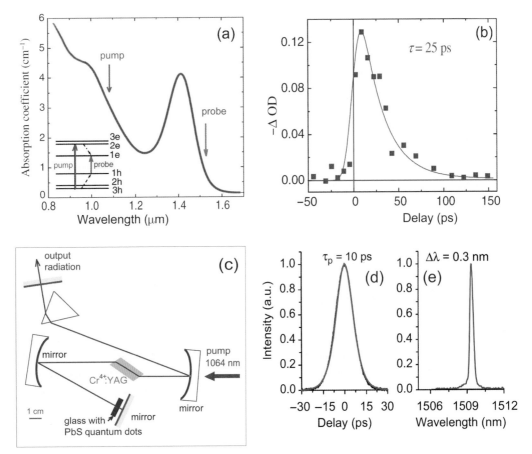

Figure 9.27 Passive mode-locking with silicate glass doped with PbS quantum dots. (a) Room-temperature absorption spectrum. The inset shows the energy-level diagram. (b) Kinetics of bleaching relaxation for PbS-doped glass at 1.524 μm after the pump at 1.08 μm (squares) and result of the best fit (line); ΔOD is the change in optical density upon excitation. (c) An optically pumped solid-state Cr^{4+}:YAG laser with PbS quantum dots used in the mode-locking fast saturable absorber. (d) Intensity autocorrelation, the red curve is the $sech^2$-fit. (e) Spectrum of mode-locked pulses. Adapted from Lagatsky et al. (2004), copyright Elsevier.

by semiconductor physics (ultrafast optical nonlinearities) and semiconductor epitaxial technology.

The design of a SESAM is shown in Figure 9.28. A quantum well layer is grown on top of a Bragg mirror and, typically, special measures are performed to enhance the electric field amplitude at the saturable layer position to ensure the lowest possible saturation fluxes. The initial reflection is set at approximately 95–97% and then after absorption saturation reflectance rises to 99% or even close to 100%. This intensity- and time-dependent minor change in a laser cavity loss defines the interplay of gain–loss processes to give well-defined pulse trains in picosecond and subpicosecond time ranges. An example of a laser setup with a SESAM instead of a rear mirror was presented in Figure 6.7.

The invention of SESAM had a strong impact on ultrafast solid-state lasers design, development, and commercial production. Nowadays most ultrafast solid-state lasers are based on passive rather than active mode-locking, enabled by SESAM. The pulse energy of ultrafast solid-state lasers was raised by several orders of magnitude, to the 10 μJ level; the pulse duration ranges from subpicosecond values to tens of picoseconds; and the average power in a few cases exceeds 100 W. The optical-to-optical efficiency reaches 40%, and the pulse repetition rates are in the tens of megahertz range. The field is experiencing lots of activity aimed at shorter pulses, higher power, and higher repetition rate. For example, Lagatsky et al. (2010) have reported on bandwidth-limited sub-100 fs pulses for a Cr^{4+}:forsterite (1.28 μm wavelength) mode-locked laser using a SESAM with InAs quantum dots. Multiple quantum dot layers were grown on top of a Bragg reflector with a number of intermediate spacers and external layers forming cavity a few wavelengths long to ensure that every quantum dot layer is located at the antinode of electric field distribution.

Not only solid-state laser design and performance benefited from the invention of SESAM; semiconductor laser R&D activity has also experienced a noticeable effect. Here, the concept of surface-emitting semiconductor lasers with external cavities appears to be very efficient, resulting in the emergence of mode-locked integrated external-cavity lasers and vertical external-cavity lasers. These will be the subject of Section 9.6.

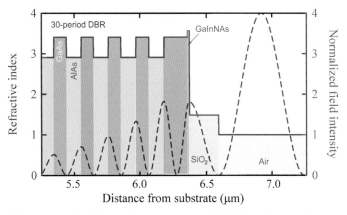

Figure 9.28 A design of a quantum well SESAM for 1314 nm (Nd:YLF laser) with ultrafast recovery enabling mode-locking to generate picosecond pulse train. The blue line shows the refractive index profile and the red line shows the radiation electric field distribution. Adapted from Spühler et al. (2005), with permission of Springer.

9.6 QUANTUM WELL AND QUANTUM DOT LASERS WITH AN EXTERNAL CAVITY (VECSELS)

9.6.1 What is VECSEL?

VECSEL stands for vertical *external cavity* surface-emitting laser. This is a general term for laser design in which an epitaxially grown multilayer semiconductor structure with a bottom reflector, gain section, and antireflective coating is complemented by external mirror(s) and (optionally) additional components. Thus, an epitaxially grown multilayer structure represents a "half-laser" that is to be complemented by an external mirror at least and optionally by passive or active Q-switching/mode-locking elements. It is important that a gain section features relatively high reflectance because of the ultrathin semiconductor gain layer, and therefore it should be necessarily complemented by an antireflection coating to minimize cavity losses. If compared to VCSELs, this approach enables going from a microchip-type electrically driven laser of milliwatt range with only possible electrical control of regimes, to a variety of designs enabling optical, electro-optical, and electrical control options for various regimes, as well as higher power levels up to multi-watt CW operation. VECSELs are typically optically pumped, though electrical pumping is also used. VECSEL design is also often referred to as a *semiconductor disk laser* (SDL).

The core VECSEL component is the *gain chip* (Figure 9.29), which includes the bottom DBR attached to a metal heat sink; the gain section on top of the DBR is covered with a number of antireflective layers. In the simplest version, the gain chip is optically pumped from an external source (typically a laser) and an output coupler forms the cavity together with the bottom mirror as shown in the bottom of Figure 9.29. The main VECSEL challenge is power scaling by means of square increase to develop multi-watt lasers for instrumentation, medicine, and other applications. Besides power scaling, a distant output mirror allows Q-switching and mode-locking components to be inserted. Typical VECSELs have length of about 10 cm.

Since the first demonstration of the VECSEL in 1997, CW output powers for optically pumped devices have been increased up to approximately 20 W in the fundamental mode and single-frequency operation, and exceeds 100 W in multimode operation. The optically pumped VECSEL can be considered an efficient light converter from semiconductor diodes' incoherent broad-band radiation to a high-quality Gaussian monochromatic beam with an optical-to-optical efficiency exceeding 50% in the best experiments.

The spectral range available extends from 244 nm to 6 μm; however, there are certain breaks in the spectrum and definite technical problems. One can see from Figure 9.30 that all UV and visible emission cases are realized by means of higher harmonic generation (second for the visible and third and fourth for UV), with two exceptions: a red laser at 680 nm based on InGaP quantum well structure and a 390 nm laser based on GaN

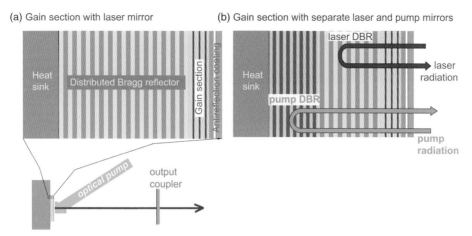

Figure 9.29 Sketch of VECSEL with optical pumping. In the simplest version there is only one mirror (DBR) to form the laser cavity (a); in the more complicated version there are two distinct DBRs – one for the laser cavity and another one for pump radiation (b).

quantum wells. However, the latter features pretty low CW power (below 10 mW). The dominant design relies on quantum wells, with the only exception for InAs quantum dot-based activity by the groups of O. Okhotnikov and D. Bimberg.

For optically pumped VECSELs, the most challenging applications can be laser-based projector TV sets, and for this red–green–blue sources are important, with desirable emission wavelengths of about 620, 530, and 460 nm to get the wider gamut. The latter range is also important for medical applications in oncology (photodynamic therapy). IR lasers can become efficient components for molecular analysis equipment.

9.6.2 CW VECSEL devices with optical and electrical pumping

Consider the two representative examples of recently reported highly efficient CW-emitting VECSELs with optical and electrical pumping. Butkus et al. (2009) proposed the design of a 1032 nm CW quantum dot VECSEL with a disk shape of a gain chip (a disk laser) with optical pumping delivering 4 W of CW radiation when pumped by a 20 W, 808 nm CW external laser. The device follows the so-called *V-geometry*, with an intermediate curved mirror to improve the beam quality. The distinctive feature is the natural-diamond top heat spreader using the liquid capillary bonding technique for appropriate heat management in addition to a standard copper bottom heat sink. The gain chip was designed as follows. The bottom DBR consisted of 29.5 pairs of GaAs/$Al_{0.9}Ga_{0.1}As$ layers grown for the designed wavelength of 1040 nm. The reflectivity was calculated to be 99.99%. The active region, grown on top of the DBR, was $7.5 \cdot \lambda/2$ long and consisted of five groups of seven quantum dot layers positioned at the antinodes of the E-field standing

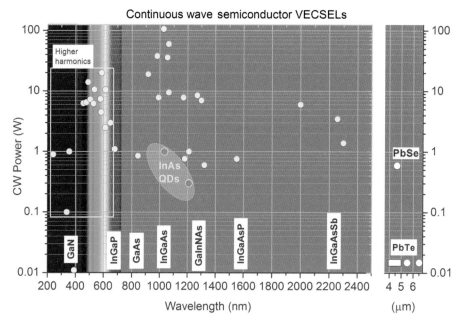

Figure 9.30 Semiconductor disk lasers (VECSELs) with optical pumping: wavelengths, CW power, and materials involved. All wavelengths shorter than 670 nm are generated by higher harmonics (shown by a frame) except for the GaN laser at 390 nm. All lasers are based on quantum wells except for the two cases of InAs quantum dot structures marked by an oval area. Sources: Germann et al. (2008a, 2008b), Calvez et al. (2009), Rahim et al. (2010), Tilma et al. (2015).

wave. The quantum dot layers with a dot density of $3 \cdot 10^{10}$ cm^{-2} were separated by 10 nm GaAs spacers to minimize material defects. Between the groups of quantum dot layers, GaAs spacers were made featuring simultaneously high transparency for the laser design wavelength and high absorbance for the 808 nm pump light. A 15 nm-thick $Al_{0.9}Ga_{0.1}As$ cap layer was used to prevent surface recombination of the excited carriers. Pump spot diameter was 120 μm.

Researchers from Princeton Optronics (Zhao et al. 2014) have developed an experimental electrically pumped VECSEL delivering 4.7 W at 531 nm with a 20 W wall-plug supply. A 7 cm long device is based on a 2×2 mm^2 quantum well gain chip generating up to 8 W IR radiation at 1062 nm under 20 W electrical pumping, i.e., the gain chip represents a quantum well laser with 40% WPE. The gain chip is mounted on a 1×1 cm^2 copper plate with thermoelectrical cooling. The device further includes intracavity frequency doubling based on a lithium niobate crystal with output facets covered with a multilayer structure featuring high reflection for the fundamental 1062 nm and high transmission for the second harmonic at 531 nm. The crystal is 7 mm long and 1 mm thick. Further components inside the external cavity are a Brewster plate (polarization selection), a Fabry–Pérot etalon (spectrum narrowing), and a 15 mm focusing lens. Two comments are worthwhile for this advanced device.

First, the gain chip itself should be explained. It consisted of a MOCVD grown n-type GaAs substrate, the bottom and the top DBRs, and an active gain section in between. The gain section consisted of strained InGaAs/GaAs multiple quantum wells sandwiched between n-type spacers/DBR and p-type spacers/DBR layers, both of which are made of highly doped GaAs/AlGaAs layers for electrical conduction. The p-type bottom DBR has a high reflectivity (>99%) and served as the end-mirror of the laser cavity. On the other side of the multiple quantum wells, the n-type DBR has a lower reflectivity and forms an internal cavity with the p-type bottom DBR. The emission area of the device was also controlled by oxide apertures with 0.4 mm diameter.

Second, the frequency-doubling crystal also needs a comment. Frequency doubling occurs owing to the nonlinear I^2 polarization response of crystals without a center of inversion in their structure. However, the high nonlinear susceptibility is not the only necessary condition for frequency doubling. Another condition requires that the fundamental and the second harmonic radiations feature phase-matching to ensure coherent interplay of both types of radiation upon propagation in a crystal. To meet the phase-matching conditions, the lithium niobate crystal used in the device under consideration has been periodically poled (re-oriented) along the light beam propagation direction on a micrometer scale by means of special high-voltage treatment upon fabrication.

9.6.3 Ultrafast VECSELs

Ultrafast VECSELs are at the research stage, with the main studies being performed at ETHZ, Zurich (Switzerland) by U. Keller and co-workers; at the University of Southampton (UK) by the group of A. C. Tropper; at Tempere University (Finland) by the group of O. Okhotnikov; and at Stuttgart University (Germany) by P. Michler's group. For ultrashort pulse generation, the semiconductor laser is mode-locked with a SESAM. The device usually obeys a V-shape geometry (Figure 9.31), with the total cavity round trip of the order of 10 cm enabling pulse trains with spacing below 1 ns and pulse repetition rate exceeding 1 GHz.

Both the active section of a VECSEL and the core of a SESAM can be fabricated using either single/multiple quantum well layer(s) or single/multiple quantum dot layer(s), the latter being fabricated by self-organized epitaxial growth under condition of lattice mismatch. Experimentally, all possible combinations have been reported for gain/mode-locker nanostructures, i.e., QW/QW, QW/QD, QD/QW, and QD/QD. The data reported in the period 2001–2016 are presented in Table 9.1 and are additionally summarized in Figure 9.32. Record parameters (shortest pulse duration, highest CW power, shortest and longest wavelengths, are marked with bold letters). The data are grouped according to the nanostructures used and then sorted in chronological order within every group.

Most experimental efforts concentrate on InGaAs structures in the gain section enabling lasing between 950 nm and 1050 nm, the gain section being quantum well and quantum dot layers, and SESAM being quantum well- or quantum dot-based, but an

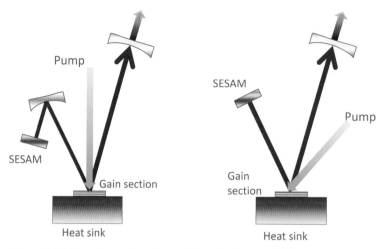

Figure 9.31 Sample realizations of V-cavity in passively mode-locked VECSELs.

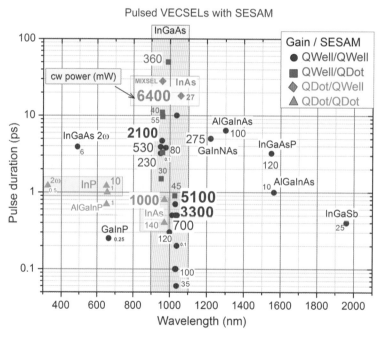

Figure 9.32 Summary of VECSELs mode-locked with a SESAM and optically pumped reported from 2001 to 2016. Different symbols (circle, square, diamond, triangle) correspond to different combinations of quantum wells and quantum dots in gain/SESAM sections (see the legend). Numbers show the reported CW power, the digit size correlating with the power value.

all-quantum dot device (quantum dot gain and quantum dot SESAM) for InGaAs has not been reported. Based on InGaAs quantum well structures, the maximal power (CW) of up to 5.1 W from one chip (with pulse duration 680 fs; University of Arizona, Scheller et al. 2012) has been obtained. InGaAs mode-locked VECSELs feature pulse duration from

Table 9.1 **VECSEL mode-locked with SESAM (semiconductor saturable absorber mirror)**

Gain material	Mode-locker material	Wavelength (nm)	Duration (ps)	CW power (mW)	Research team (first author)	Year
Quantum well gain section/quantum well SESAM						
InGaAs	InGaAs	950	3.2	230	Zurich (Haring)	2001
InGaAs	InGaAs	1040	0.5	100	Southampton (Gamache)	2002
InGaAs	InGaAs	950	3.9	530	Zurich (Haring)	2002
InGaAs	InGaAs	1032	0.5	700	Southampton (Hoogland)	2005
InGaAs	InGaAs	978	3.8	80	Kaiserslautern (Casel)	2005
	Frequency doubling	489	3.9	6		
InGaAs	InGaAs	957	4.7	2100	Zurich (Aschwanden)	2005
InGaAsP	GaInNAs	1550	3.2	120	Chalmers (Lindberg)	2005
InGaAs	InGaAs	1036	10	-	Tempere (Saarinen)	2007
GaInNAs	GaInNAs	1220	5	275	Tempere (Rautiainen)	2008
InGaAs	InGaAs	1037	**0.06**	35	Southampton (Quarterman)	2009
InGaAs	InGaAs	999	0.3	120	Southampton (Wilcox)	2010
AlGaInAs	AlGaInAs	1300	6.4	100	Tempere (Rautiainen)	2010
AlGaInAs	GaInNAs	1560	2.3	15	Marcoussis (Khadour)	2010
AlGaInAs	InGaAsNSb	1564	1	10	Marcoussis (Zhao)	2011
InGaSb	InGaSb	**1960**	0.4	25	Tempere (Härkönen)	2011

Table 9.1 **(Cont.)**

Gain material	Mode-locker material	Wavelength (nm)	Duration (ps)	CW power (mW)	Research team (first author)	Year
InGaAs	InGaAs	1030	0.7	5100	Tucson (Scheller)	2012
GaInP	GaInP	664	0.25	0.5	Stuttgart (Bek)	2013
InGaAs	InGaAs	1013	0.5	3300	Southampton (Wilcox)	2013
InGaAs	InGaAs	1038	0.2	0.1	Zurich (Zaugg)	2014
InGaAs	InGaAs	1030	0.1	3	Zurich (Zaugg)	2014
InGaAs	InGaAs	1034	0.1	100	Zurich (Waldburger)	2016
Quantum well gain section/quantum dot SESAM						
InGaAs	InAs	960	9.7	55	Zurich (Lorenser)	2004
InGaAs	InAs	960	3.3	0.1	Zurich (Lorenser)	2006
InGaAs	InGaAs	989	50	360	Birmingham (Butkus)	2008
InGaAs	InAs	1027	0.9	45	Southampton (Wilcox)	2009
InGaAs	InAs	953	1.5	30	Zurich (Hoffmann)	2010
InGaAs	InAs	958	11	40	Zurich (Zaugg)	2012
Quantum dot gain section/quantum dot SESAM						
InAs	InAs	970	0.8	1000	Zurich (Hoffmann)	2011
InAs	InAs	970	0.4	140	Zurich (Hoffmann)	2011
InP	InP	655	1	1	Stuttgart (Bek)	2014
AlGaInP	AlGaInP	651	0.7	1	Stuttgart (Bek)	2015
InP	InP	650	1.2	**10**	Stuttgart (Bek)	2015
Frequency doubling		325	1.2	0.5	Stuttgart (Bek)	2015

Table 9.1 **(Cont.)**

Gain material	Mode-locker material	Wavelength (nm)	Duration (ps)	CW power (mW)	Research team (first author)	Year
			Quantum dot gain section/quantum well SESAM			
InAs	InGaAs	1059	18	27	Zurich (Hoffman)	2008
InAs	InGaAs	959	28	**6400**	Zurich (Rudin) MIXSEL	2010

60 fs to 50 ps. Here, the shortest ever reported passively mode-locked VECSEL with 60 fs (35 mW CW power) pulses (Quarterman et al. 2009) should be noted. Longer wavelengths are obtained with ternary or quaternary Ga–Group V compounds. In all cases both gain and SESAM sections are based on quantum wells.

Only two groups have reported on passively mode-locked VECSELs for the visible spectrum. Casel et al. (2005) at Kaiserslautern University obtained 489 nm by intracavity second harmonic generation in an InGaAs device. The Stuttgart group (Bek et al. 2015), with InP-based structures, reported on successful red lasing (around 650 nm) and even demonstrated intracavity second harmonic generation to get 325 nm, which is the shortest wavelength ever reported for mode-locked VECSELs.

Noteworthy is that all quantum dot systems look very promising. First, an InP-all-quantum-dot device exhibits 10 mW CW radiation power in the red (664 nm) without frequency doubling, as in the case of InGaAs. Second, an InAs-based all-quantum-dot laser has shown outstanding parameters: 1 W at 970 nm with 800 fs pulse duration (Hoffman et al. 2011). These numbers obtained just in the first experiments with quantum dots in ultrafast lasers correspond to the best achievements with InGaAs quantum well VECSELs. Therefore, all-quantum-dot InAs mode-locked VECSEL can become a competitive platform for commercial devices in this range of wavelengths and durations. Last but not least, the record CW power for ultrafast surface-emitting laser (6.4 W, 28 ps, 959 nm) has been achieved with an InAs quantum dot gain section using the MIXSEL design (Rudin et al. 2010). This approach will be discussed in detail in the following subsection.

Progress toward longer wavelengths (1.2–2 μm) needs quaternary compounds. This is an area in which the following groups are active: Tempere University (O. Okhotnikov and co-workers), Chalmers University (Lindberg et al.), and CNRS Lab at Marcoussis (J.-L. Oudar et al.).

A representative experiment on a subpicosecond laser is given in Figure 9.33. The V-shaped cavity has an overall length of 18 cm; the outcoupling mirror, with a reflectivity of 99.7%, served as a folding mirror, which resulted in two outcoupled beams. The inset in Figure 9.33(c) shows the appearance of side pulses for a larger scan range of 15 ps. The distance between these pulses is 8.95 ps, which is the round trip time expected from

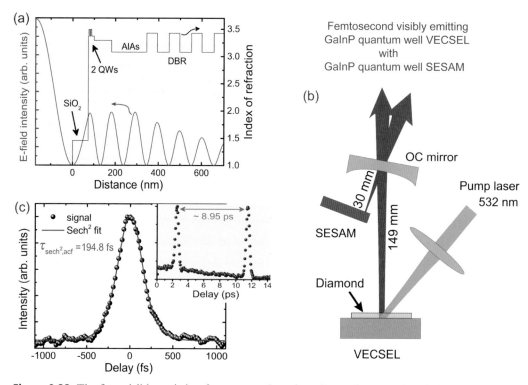

Figure 9.33 The first visibly emitting femtosecond semiconductor laser. Red-emitting (663 nm) quantum well GaInP-based femtosecond semiconductor laser passively mode-locked with a GaInP quantum well SESAM (Stuttgart University, P. Michler and co-workers). (a) Index of refraction and electric field intensity in the SESAM simulated with the transfer matrix method. Two quantum wells are located close to the surface of the near-resonant semiconductor structure, which is coated with a quasi-λ/4 SiO$_2$ layer. (b) Experimental setup of the mode-locked VECSEL. The V-shaped cavity is used for tight focusing of the laser mode onto the absorber. (c) Autocorrelation trace with 5 ps scan range showing a full-width half-maximum pulse duration of 222 fs. Inset: autocorrelation measurement with a scan range of 15 ps. Due to the intracavity diamond heat spreader, side pulses appear with a spacing of 8.95 ps. Reprinted with permission from Bek et al. (2013), copyright AIP.

the subcavity of the diamond heat spreader acting as a Fabry–Pérot etalon. Thus, further improvement of the heat management components is desirable to ensure pulse trains defined by the laser cavity round trip time only.

9.6.4 MIXSEL: mode-locked integrated external-cavity surface-emitting laser

With modern advanced epitaxial techniques it is possible to integrate in a single chip all components of a mode-locked semiconductor vertically emitting laser except for an external-cavity mirror (an output coupler). In 2007, U. Keller and co-workers (Maas et al.

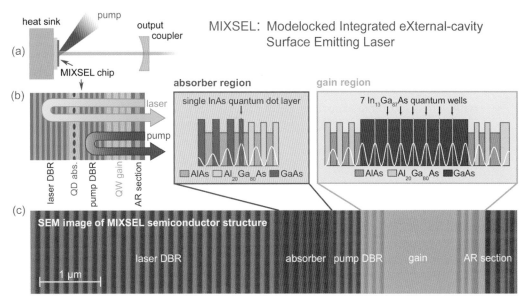

Figure 9.34 MIXSEL: the concept and the design. (a) The scheme of the device; (b) design of a MIXSEL chip; (c) scanning electron microscopy (SEM) image of a real device pumped at 808 nm, which generates 28 ps pulses at 959 nm. See text for detail. Reprinted from Rudin et al. (2010), with permission from OSA Publishing.

2007) suggested the MIXSEL: **m**ode-locked **i**ntegrated e**x**ternal-cavity **s**urface-**e**mitting **l**aser (Figure 9.34).

The MIXSEL device comprises an optically pumped semiconductor chip with a built-in mode-locking component and an external output coupler to form a laser cavity. The chip includes a bottom laser mirror (DBR), a fast saturable absorber layer (a self-organized quantum dot layer in the case under consideration), a multiple quantum well gain region, and an antireflective layer on top. One more principal component is the additional DBR structure representing a mirror for pump radiation, but transparent at laser wavelength. It is grown between the saturable absorber and the optical gain sections of the chip. It prevents bleaching of the saturable absorber by the pump light and at the same time increases absorption of the pump light by the gain medium. The laser DBR reflects the laser light and forms the laser cavity, together with the external output coupler. The white oscillations in the upper sketches in Figure 9.34 represent the square of the electric field of the laser light. Note that the quantum dot layers as well as the quantum well layers are placed in antinodes to ensure lowest possible lasing and bleaching thresholds. The laser chip has total thickness of 8 μm, 5 × 5 mm square, and the external output coupler is placed at a distance of a few centimeters. Heat management has been successfully performed by means of wafer removal and mounting the 8 μm-thick MIXSEL structure directly onto a CVD (chemical vapor deposition) diamond heat spreader. To make this possible, the whole structure was grown in reverse order, i.e., the antireflective coating was grown first. The chip was cleaved from the wafer, metalized with Ti/Pt/In/Au, and soldered onto the diamond plate metalized with Ti/Pt/Au using a fluxless indium soldering process in a vacuum. The laser presented here operates at 959 nm wavelength and features outstanding parameters: 6 W average power, 28 ps pulses at 2.5 GHz repetition rate. The pulse shape perfectly obeys the sech^2-shape predicted for passive mode-locking with a fast saturable absorber. Pumping has been performed with a fiber-coupled laser diode array radiation (37 W at 808 nm) focused to a 0.2 mm spot. A Peltier element was used to keep the chip at a constant temperature of –15 °C. The MIXSEL concept does open the way for femtosecond optically pumped semiconductor lasers. In 2015, the group of U. Keller reported on a 250 fs pulse MIXSEL with 10 GHz rate in the 1 μm wavelength range (Mangold et al. 2015).

To summarize, VECSELs, or SDLs, are the subject of extensive research. Most of the devices reported use optical pumping from an LED or a laser featuring outstanding optical-to-optical efficiency up to 50%. The first studies toward electrically pumped VECSELs have succeeded with nearly 20% WPE. Most of the structures used are quantum wells, though rare experiments with quantum dot structures suggest their high potential for the future. A quantum well or a quantum dot gain section can be combined with a SESAM to get pico- and femtosecond pulses down to sub-100 fs by means of passive mode-locking. Here, the integration of gain and mode-locking components into a single epitaxially grown structure (MIXSEL) have demonstrated the CW power record of 6.4 W.

9.7 CHALLENGES AND OUTLOOK

9.7.1 Will VCSELs completely replace edge-emitting lasers?

VCSELs represent the fastest growing sector of the market. Unlike edge-emitting lasers, VCSELs offer much better beam quality and control, though generating high-power radiation from a single chip simply by enlarging the aperture of gain area remains an issue since inhibition of side modes becomes questionable for a large aperture. Probably, integrating a gain section with a 2DPC structure paves the way toward powerful VCSELs to replace edge-emitting devices in massive applications. The only sector in which edge-emitting lasers will remain unsurpassed is edge-emitting DFB lasers since this type of laser features unparalleled mode purity and narrow bandwidth with fine tunability, which are important for optical communication systems.

9.7.2 Ultimate size of a laser?

At first glance, the ultimate size of a laser, which is defined by the diffraction limit (a few wavelengths) in the cross-sectional plane and a wavelength in the lasing direction, will never fall to the scale available for electronic circuitry since light wavelengths exceeds electron wavelengths by two orders of magnitude. Therefore, despite the desire to integrate photonic and electronic components, the photonic and electronic subsystems in these arrangements seem to inevitably have different scales. Nevertheless, there is some light at the end of the tunnel in the sense that new solutions toward small lasers are being sought. One possible trend is to use strong light confinement in plasmonic structures. Owing to localized surface plasmon polariton generation, light wavelengths reduce drastically and light confinement within the volume whose size is considerably smaller than the wavelength in a vacuum becomes feasible. One more trend is based on light confinement in subwavelength semiconductor nanorods. This approach offers the promise of going beyond the diffraction limit in the cross-sectional area of a laser. However, both trends are at the initial research level only and no prediction can be made on their potential future applications.

9.7.3 Will quantum dot lasers replace quantum well lasers?

Quantum dot lasers, when compared to quantum well lasers, feature lower threshold current with less dependence on temperature, which is favorable for most applications. However, fabrication of multilayer quantum dot structures in a controllable manner for desirable wavelengths has many technological problems. Controlling size and keeping the lattice mismatching reasonable at the same time as avoiding complex interplay phenomena in a multilayer strain-driven system of self-organized dots is not easy in the same structure and within the same technological run. Quantum dot lasers are believed to compete with and possibly to substitute quantum well counterparts, but due to growth problems this process may occur for selected applications and selected wavelengths only. Quantum well lasers are still considered to be the mainstream within the semiconductor laser industry.

9.7.4 Poor heat conductivity of an epitaxial DBR is the bottleneck for higher output of surface-emitting lasers in many cases

Heat management forms the bottleneck in laser performance for most devices. In spite of many attractive features of surface-emitting laser design – first-class beam quality and fabrication versatility in terms of wafer testing and array scalability – this approach needs the multilayer DBR to appear as a thermal interface between the gain medium and a heat spreader. Lattice-matching conditions need many periods of alternating composition to be used since close chemical composition results in close refractive indices of the materials used. Managing heat conductance in multilayer semiconductor structures calls for new ideas, probably along with refractive index engineering.

9.7.5 Narrow-band Bragg mirror reflectance forms a bandwidth limit for ultrafast lasers

Close values of refractive indices of semiconductor materials coupled to get a DBR in surface-emitting semiconductor lasers define the maximal attainable spectral width at which high reflection develops. A wider reflectance spectrum needs more difference in refractive indices, which in turn should occur along with lattice matching. Since the lattice matching condition necessitates coupling materials with close refractive indices, the resulting narrow-band DBR mirrors form a limit for cavity bandwidth and therefore become an obstacle toward shorter pulses in the subpicosecond range in many cases. New ideas are necessary to obtain broader reflection spectra with a reasonable number of layers involved.

9.7.6 The green gap in III-nitride-based lasers

GaN-based lasers, similarly to GaN-based LEDs, cannot be designed for any wavelength in spite of the fact that AlInGaN quaternary solutions offer UV–visible–IR band gap when using AlGaN to InN structures. Growth issues and lattice-matching conditions in solid solution development both need further study. Many researchers consider a possible contribution from InN dots to optical properties of InGaN quantum well structures. InN dots are supposed to develop at high In content in GaN owing to poor In solubility. Therefore the existing green gap in GaN-based semiconductor lasers is still waiting for new ideas and approaches. Probably, quantum dot structures rather than quantum well structures will become the mainstream here. A recent report by Weng et al. (2016) on successful development of a room-temperature CW InGaN quantum dot laser at 560 nm with 0.6 mA threshold current density looks promising.

Colloidal quantum dots (nanocrystals) and quantum wells (nanoplatelets) are extensively examined as potential gain materials for lasers. Although at present only optically pumped lasing has been reported, electrically driven devices can also be foreseen, considering that colloidal LEDs have been successfully demonstrated by many groups to date. Otherwise, even in the case that optical pumping remains the most efficient power source for colloidal lasers, these can be viewed as cheap spectrum transformers for epitaxial lasers to extend their operating range. In the case of successful colloidal laser diode development, the whole process of their fabrication can become free of the expensive epitaxial processes and then cheap bottom-up laser fabrication may appear in the laser industry.

9.7.7 Silicon-based lasers

Development of Si-based or at least Si-compatible lasers is important for integrating photonic components with silicon microelectronics. The most important and also most challenging is direct epitaxial growth of lasers on a silicon wafer. There is active research in this field using both quantum well (InGaAs) and quantum dot (InAs) structures with direct injection of carriers grown in an Si wafer (Liu et al. 2015). Comparison of quantum dot versus quantum well lasers grown on silicon clearly demonstrates the quantum dot advantage. Room-temperature, low-threshold operation has been demonstrated with InAs quantum dot lasers directly grown on silicon. Quantum dot-based lasers directly grown on silicon are interesting candidates as light sources for silicon photonic interconnects.

Table 9.2 summarizes the existing, emerging, and forthcoming applications of nanostructures in lasers with separately listed electron confinement and lightwave confinement physical phenomena involved.

Table 9.2 **Nanostructures with electron and lightwave confinement in lasers**

Nanostructure	Function	Status
Electron confinement		
Quantum well epitaxial	Gain medium in diode lasers	Mature commercial products
Quantum well epitaxial	Saturable absorber mirror	New commercial product, active R&D
Quantum well epitaxial superlattices	Active medium in quantum cascade lasers	Mature commercial products
Quantum well colloidal	Gain medium	First reports on lasing with optical pump, emerging research field
Quantum dot epitaxial	Gain medium in diode lasers	New commercial products, extensive research
Quantum dots in glasses	Saturable absorber	Extensive R&D
Quantum dot colloidal	Gain medium	First reports on lasing with optical pump, emerging research field
Lightwave confinement		
Multilayer structures	Mirrors (DBR)	Commercial, the component of semiconductors and solid-state lasers
Gratings	Distributed feedback	Commercial, the component of distributed feedback lasers
Photonic crystal nanocavities	Active component of lasers	Extensive research
Photonic crystal 2D gratings	Multidirectional distributed feedback	Extensive research

Conclusion

- Nanophotonic solutions, including quantum confinement of electrons/holes in semiconductors and lightwave confinement/control in multilayer structures, photonic crystals, and nanocavities are essential for the current production and future progress in the fields of semiconductor lasers and compact solid-state lasers, with the latter being optionally efficiently pumped by semiconductor lasers or LEDs.

- Quantum well laser diodes dominate in the semiconductor laser industry, ranging from near-UV to middle-IR wavelengths, whereas commercial quantum dot laser diodes fill the range around 1.3 μm only. Edge-emitting lasers still constitute the major semiconductor laser sector in the market. Edge-emitting lasers can be made using distributed feedback with a grating that can be viewed as a nanophotonic component because of purposeful control over lightwaves and evanescent wave propagation, diffraction, and interference in a periodic structure.

- Vertical cavity surface-emitting lasers (VCSELs), owing to a number of fabrication advantages and higher beam quality, represent the fastest growing (30% revenue growth per annum) portion of the semiconductor laser industry. Commercial VCSELs are based on multiple quantum wells, though extensive research toward efficient quantum dot VCSELs promises to reach the commercial stage soon. Commercial VCSELs essentially include multi-period epitaxial DBRs, i.e., lightwave control in a subwavelength periodic medium, often referred to as a one-dimensional photonic crystal.

- Two- and three-dimensional photonic crystals combined with a semiconductor active gain medium (quantum wells or quantum dots) promise downsizing of lasers toward the ultimate limit of the order of 10 μm³ to meet the growing demand for higher integration level in optical communication and data storage components.

- A two-dimensional photonic crystal developed beneath the active gain region in a VCSEL enables multidirectional distributed feedback and promises fabrication of large-square single-mode multi-watt class VCSELs, the top mirror being eliminated in this case. This will probably pave the way for output power of VCSEL device being scaled up while keeping high beam quality owing to single-mode operation.

- Colloidal quantum dot lasers with optical pumping have been demonstrated and first experimental evidence of optical gain in colloidal quantum well lasers has been reported.

- Glasses doped with semiconductor quantum dots represent efficient saturable absorbers suitable for Q-switching and mode-locking of solid-state lasers to get nano- and pico-second pulses, respectively. Active research with multiple examples of efficient application to commercial solid-state lasers promises fast commercialization of quantum dot-based Q-switchers and mode-lockers in the near future, including compact vertical external-cavity surface-emitting lasers (VECSELs) based on quantum well and quantum dot structures.

- SESAM (semiconductor saturable absorber mirror), a semiconductor epitaxial DBR combined with a thin semiconductor layer exhibiting saturable absorption, is making its first steps on the market as a Q-switching and mode-locking component of semiconductor VECSELs and solid-state compact lasers pumped with laser diodes or LEDs. SESAM enables generation of ultrashort pulses in the pico- and subpicosecond range.

Problems

9.1 Why can power not be enhanced by simple size scaling of a classical Fabry–Pérot laser without reducing the beam quality?

9.2 Explain the main advantages of surface-emitting lasers versus edge-emitting lasers. Explain the challenges in surface-emitting laser design.

9.3 Explain why the shorter cavity in a VCSEL versus an edge-emitting diode does not result in shorter relaxation time (bigger bandwidth). What can be done to make VCSEL response faster?

9.4 Compare semiconductor epitaxial and oxide non-epitaxial distributed Bragg reflectors. Why do the latter need fewer periods than the former? Evaluate the geometrical thickness of the two types of mirrors. Does a smaller number of periods lead to much thinner mirrors?

9.5 Explain the advantages and problems of quantum dot lasers.

9.6 What can a photonic crystal paradigm bring to semiconductor laser design?

9.7 What is the ultimate size limit for a laser?

9.8 Recall all cases of electron confinement (quantum size effects) and light confinement (photonic crystals and microcavities) implementations in semiconductor and solid-state laser design. Emphasize where both confinements are combined. Consider ideas and approaches that have been commercialized, are close to the commercial stage, or are still at the stage of research or proof-of-principle.

Further reading

Alferov, Z. I. (1998). The history and future of semiconductor heterostructures. *Semiconductors*, **32**, 1–14.

Carroll, J. E., Whiteaway, J., and Plumb, D. (1998). *Distributed Feedback Semiconductor Lasers*, vol. **10**. IET.

Chow, W. W., and Jahnke, F. (2013). On the physics of semiconductor quantum dots for applications in lasers and quantum optics. *Prog Quantum Electron*, **37**, 109–184.

Coleman, J., Young, J., and Garg, A. (2011). Semiconductor quantum dot lasers: a tutorial. *J Lightwave Technol*, **29**, 499–510.

Gmachl, C., Capasso, F., Sivco, D. L., and Cho, A. Y. (2001). Recent progress in quantum cascade lasers and applications. *Rep Prog Phys*, **64**, 1533–1601.

Iga, K. (2000). Surface-emitting laser: its birth and generation of new optoelectronics field. *IEEE J Sel Top Quantum Electron*, **6**, 1201–1215.

Kazarinov, R. F., and Suris, R. A. (1971). Possibility of amplification of electromagnetic waves in a semiconductor with a superlattice. *Sov Phys Semicond*, **5**, 707–709.

Kazarinov, R. F., and Suris, R. A. (1972/1973). Injection heterojunction laser with a diffraction grating on its contact surface. *Sov Phys Semicond*, **6**, 1184.

Keller, U. (2010). Ultrafast solid-state laser oscillators: a success story for the last 20 years with no end in sight. *Appl Phys B: Lasers Opt*, **100**, 15–28.

Ledentsov, N. N., Ustinov, V. M., Shchukin, V. A., et al. (1998). Quantum dot heterostructures: fabrication, properties, lasers (review). *Semiconductors*, **32**, 343–365.

Liu, J. M. (2009). *Photonic Devices*. Cambridge University Press.

Michalzik, R. (ed.) (2013). *VCSELs: Fundamentals, Technology and Applications of Vertical-Cavity Surface-Emitting Lasers*. Springer.

Morthier, G., and Vankwikelberge, P. (2013). *Handbook of Distributed Feedback Laser Diodes*. Artech House.

Ning, C.-Z. (2010). Semiconductor nanolasers (a tutorial). *Phys Status Solidi B*, **247**, 774–788.

Okhotnikov, O. G. (ed.) (2010). *Semiconductor Disk Lasers: Physics and Technology*. Wiley-VCH.

Rafailov, E. U. (2014). *The Physics and Engineering of Compact Quantum Dot-Based Lasers for Biophotonics*. Wiley-VCH.

Rafailov, E. U., Cataluna, M. A., and Avrutin, E. A. (2011). *Ultrafast Lasers Based on Quantum Dot Structures: Physics and Devices*. John Wiley & Sons.

Svelto, O. (1998). *Principles of Lasers*. Springer-Verlag.

Ustinov, V. M., Zhukov, A. E., Egorov, A. Y., and Maleev, N. A. (2003). *Quantum Dot Lasers*. Oxford University Press.

Zhukov, A. E., and Kovsh, A. R. (2008). Quantum dot diode lasers for optical communication systems. *Quantum Electron*, **38**, 409–423.

References

Altug, H., Englund, D., and Vuckovic, E. (2006). Ultrafast photonic crystal nanocavity laser. *Nature Physics*, **2**, 484–488.

Arakawa, Y., and Sakaki, H. (1982). Multidimensional quantum well laser and temperature dependence of its threshold current. *Appl Phys Lett*, **40**, 939–941.

Bányai, L., and Koch, S. W. (1993). *Semiconductor Quantum Dots*. World Scientific Publishers.

Bek, R., Kahle, H., Schwarzbäck, T., Jetter, M., and Michler, P. (2013). Mode-locked red-emitting semiconductor disk laser with sub-250 fs pulses. *Appl Phys Lett*, **103**(24), 242101.

Bek, R., Baumgärtner, S., Sauter, F., et al. (2015). Intra-cavity frequency-doubled mode-locked semiconductor disk laser at 325 nm. *Opt Express*, **23**, 19947–19953.

Butkus, M., Wilcox, K. G., Rautiainen, J., et al. (2009). High-power quantum-dot-based semiconductor disk laser. *Opt Lett*, **34**, 1672–1674.

Calvez, S., Hastie, J. E., Guina, M., Okhotnikov, O. G., and Dawson, M. D. (2009). Semiconductor disk lasers for the generation of visible and ultraviolet radiation. *Laser Photonics Rev*, **3**(5), 407–434.

Casel, O., Woll, D., Tremont, M. A., et al. (2005). Blue 489-nm picosecond pulses generated by intracavity frequency doubling in a passively mode-locked optically pumped semiconductor disk laser. *Applied Phys B*, **81**, 443–446.

Cosendey, G., Castiglia, A., Rossbach, G., Carlin, J. F., and Grandjean, N. (2012). Blue monolithic AlInN-based vertical cavity surface emitting laser diode on free-standing GaN substrate. *Appl Phys Lett*, **101**, 151113.

Dang, C., Lee, J., Breen, C., et al. (2012). Red, green and blue lasing enabled by single-exciton gain in colloidal quantum dot films. *Nat Nanotechnol*, **7**, 335–339.

Ellis, B., Mayer, M. A., Shambat, G., et al. (2011). Ultralow-threshold electrically pumped quantum-dot photonic-crystal nanocavity laser. *Nat Photonics*, **5**, 297–300.

Faist, J., Capasso, F., Sivco, D. L., et al. (1994). Quantum cascade laser. *Science*, **264**(5158), 553–556.

Gaponenko, S. V. (1998). *Optical Properties of Semiconductor Nanocrystals*. Cambridge University Press.

Gaponenko, S. V. (2010). *Introduction to Nanophotonics*. Cambridge University Press.

Germann, T. D., Strittmatter, A., Pohl, J., et al. (2008). High-power semiconductor disk laser based on InAs/GaAs submonolayer quantum dots. *Appl Phys Lett*, **92**, 101123.

Germann, T. D., Strittmatter, A., Pohl, J., et al. (2008). Temperature-stable operation of a quantum dot semiconductor disk laser. *Appl Phys Lett*, **93**, 051104.

Guzelturk, B., Kelestemur, Y., Olutas, M., Delikanli, S., and Demir, H. V. (2014). Amplified spontaneous emission and lasing in colloidal nanoplatelets. *ACS Nano*, **8**, 6599–6605.

Guzelturk, B., Kelestemur, Y., Gungor, K., et al. (2015). Stable and low-threshold optical gain in CdSe/CdS quantum dots: an all-colloidal frequency up-converted laser. *Adv Mater*, **27**, 2741–2746.

Haglund, E. P., Kumari, S., Westbergh, P., et al. (2015). Silicon-integrated short-wavelength hybrid-cavity VCSEL. *Opt Express*, **23**, 33634–33640.

Hirose, K., Liang, Y., Kurosaka, Y., et al. (2014). Watt-class high-power, high-beam-quality photonic-crystal lasers. *Nat Photonics*, **8**, 406–411.

Hoffmann, M., Sieber, O. D., Wittwer, V. J., et al. (2011). Femtosecond high-power quantum dot vertical external cavity surface emitting laser. *Opt Express*, **19**, 8108–8116.

Hu, Y. Z., Koch, S. W., and Peyghambarian, N. (1996). Strongly confined semiconductor quantum dots: pair excitations and optical properties. *J Luminescence*, **70**, 185–202.

Hugi, A., Maulini, R., and Faist, J. (2010). External cavity quantum cascade laser. *Semicond Sci Technol*, **25**, 083001.

Iga, K. (2008). Vertical-cavity surface-emitting laser: its conception and evolution. *Jpn J Appl Phys*, **47**, 1–11.

Imada, M., Chutinan, A., Noda, S., and Mochizuki, M. (2002). Multidirectionally distributed feedback photonic crystal lasers. *Physical Review B*, **65**, 195306.

Jung, I. D., Brovelli, L. R., Kamp, M., Keller, U., and Moser, M. (1995). Scaling of the antiresonant Fabry–Perot saturable absorber design toward a thin saturable absorber. *Opt Lett*, **20**(14), 1559–1561.

Kapon, E., and Sirbu, A. (2009). Long-wavelength VCSELs: power-efficient answer. *Nat Photonics*, **3**, 27–29.

Kasahara, D., Morita, D., Kosugi, T., et al. (2011). Demonstration of blue and green GaN-based vertical-cavity surface-emitting lasers by current injection at room temperature. *Appl Phys Express*, **4**, 072103.

Kim, J., and Chuang, S. L. (2006). Theoretical and experimental study of optical gain, refractive index change, and linewidth enhancement factor of p-doped quantum-dot lasers. *IEEE J Quantum Electron*, **42**, 942–952.

Klimov, V. I., Ivanov, S. A., Nanda, J., et al. (2007). Single-exciton optical gain in semiconductor nanocrystals. *Nature*, **447**, 441–446.

Kogelnik, H., and Shank, C. V. (1971). Stimulated emission in a periodic structure. *Appl Phys Lett*, **18**, 152–154.

Lagatsky, A. A., Leburn, C. G., Brown, C. T. A., et al. (2004). Passive mode-locking of a Cr⁴⁺: YAG laser by PbS quantum-dot-doped glass saturable absorber. *Optics Commun*, **241**, 449–454.

Lagatsky, A. A., Leburn, C. G., Brown, C. T. A., et al. (2010). Ultrashort-pulse lasers passively mode locked by quantum-dot-based saturable absorbers. *Prog Quantum Electron*, **34**, 1–45.

Larsson, A. (2011). Advances in VCSELs for communication and sensing. *IEEE J Sel Top Quantum Electron*, **17**,1552–1567.

Liu, A. Y., Srinivasan, S., Norman, J., Gossard, A. C., and Bowers, J. E. (2015). Quantum dot lasers for silicon photonics. *Photonics Res*, **3**, B1–B9.

Loiko, P. A., Rachkovskaya, G. E., Zacharevich, G. B., et al. (2012). Optical properties of novel PbS and PbSe quantum-dot-doped alumino-alkali-silicate glasses. *J Non-Cryst Solids*, **358**, 1840–1845.

Lott, J. A., Ledentsov, N. N., Ustinov, V. M., et al. (2000). *Electron Lett*, **36**, 1384–1386.

Maas, D. J. H. C., Bellancourt, A.-R., Rudin, B., et al. (2007). Vertical integration of ultrafast semiconductor lasers. *Appl Phys B*, **88**, 493–497.

Malyarevich, A. M., Yumashev, K. V., and Lipovskii, A. A. (2008). Semiconductor-doped glass saturable absorbers for near-infrared solid-state lasers. *J Appl Phys*, **103**(8), 4–14.

Mangold, M., Golling, M., Gini, E., Tilma, B. W., and Keller, U. (2015). Sub-300-femtosecond operation from a MIXSEL. *Opt Express*, **23**(17), 22043–22059.

Nakamura, M., Yariv, A., Yen, H. W., Somekh, S., and Garvin, H. L. (1973). Optically pumped GaAs surface laser with corrugation feedback. *Appl Phys Lett*, **22**(10), 515–516.

Painter, O., Lee, R. K., Scherer, A., et al. (1999). Two-dimensional photonic band-gap defect mode laser. *Science*, **284**, 1819–1821.

QD Laser, Inc. (2015). Technical Data Laser Diode QLF131 F-P16. www.qdlaser.com (accessed May 1, 2017).

Quarterman, A. H., Wilcox, K. G., Apostolopoulos, V., et al. (2009). A passively mode-locked external-cavity semiconductor laser emitting 60-fs pulses. *Nat Photonics*, **3**, 729–731.

Rahim, M., Khiar, A., Felder, F., et al. (2010). 5-μm vertical external-cavity surface-emitting laser (VECSEL) for spectroscopic applications. *Appl Phys B: Lasers Opt*, **100**, 261–264.

Rudin, B., Wittwer, V. J., Maas, D. J. H. C., et al. (2010). High-power MIXSEL: an integrated ultrafast semiconductor laser with 6.4 W average power. *Opt Express*, **18**, 27582–27588.

Scheller, M., Wang, T. L., Kunert, B., et al. (2012). Passively modelocked VECSEL emitting 682 fs pulses with 5.1 W of average output power. *Electron Lett*, **48**, 588–589.

Seufert, J., Fischer, M., Legge, M., et al. (2004). DFB laser diodes in the wavelength range from 760 nm to 2.5 μm. *Spectrochim Acta, Part A*, **60**, 3243–3247.

Shchekin, O. B., and Deppe, D. G. (2002). Low-threshold high-T_0 1.3-μm InAs quantum-dot lasers due to p-type modulation doping of the active region. *IEEE Photon Technol Lett*, **14**, 1231–1233.

Siegman, A. E. (1986). *Lasers*. University Science Books.

Spühler, G. J., Weingarten, K. J., Grange, R., et al. (2005). Semiconductor saturable absorber mirror structures with low saturation fluence. *Applied Physics B*, **81**, 27–32.

Svelto, O. (1998). *Principles of Lasers*. Springer-Verlag.

Takahashi, T., and Arakawa, Y. (1988). Theoretical analysis of gain and dynamic properties of quantum well box lasers. *Optoelectron Dev Technol*, **3**, 155–162.

Tandaechanurat, A., Ishida, S., Guimard, D., et al. (2011). Lasing oscillation in a three-dimensional photonic crystal nanocavity with a complete bandgap. *Nat Photonics*, **5**, 91–94.

Tilma, B. W., Mangold, M., Zaugg, C. A., et al. (2015). Recent advances in ultrafast semiconductor disk lasers. *Light Sci Appl*, **4**, e310.

Ustinov, V. M., Maleev, N. A., Kovsh, A. R., and Zhukov, A. E. (2005). Quantum dot VCSELs. *Physica Status Solidi (a)*, **202**, 396–402.

Vandyshev, Y. V., Dneprovskii, V. S., Klimov, V. I., and Okorokov, D. K. (1991). Laser generation in semiconductor quasi-zero-dimensional structure on a transition between size quantization levels. *JETP Lett*, **54**, 441–444.

Vurgaftman, I., Weih, R., Kamp, M., et al. (2015). Interband cascade lasers. *J Phys D: Appl Phys*, **48**, 123001.

Weng, G., Mei, Y., Liu, J., et al. (2016). Low threshold continuous-wave lasing of yellow–green InGaN-QD vertical-cavity surface-emitting lasers. *Opt Express*, **24**, 15546–15553.

Williams, B. S. (2007). Terahertz quantum-cascade lasers. *Nat Photonics*, **1**, 517–525.

Zeller, W., Naehle, L., Fuchs, P., et al. (2010). DFB lasers between 760 nm and 16 μm for sensing applications. *Sensors*, **10**, 2492–2510.

Zhao, P., Xu, B., van Leeuwen, R., et al. (2014). Compact 4.7 W, 18.3% wall-plug efficiency green laser based on an electrically pumped VECSEL using intracavity frequency doubling. *Opt Lett*, **39**, 4766–4768.

Zhukov, A. E., and Kovsh, A. R. (2008). Quantum dot diode lasers for optical communication systems. *Quantum Electron*, **38**, 409–423.

10 Photonic circuitry

The photonic crystal paradigm offers a new generation of compact waveguides, splitters, and demultiplexers. Coupling of waveguides with microcavities enables further uses, including electro-optical and all-optical switching. These ideas and their experimental implementations are the subject of this chapter. Problems on the way toward efficient integration of photonic and electronic devices and issues related to compatibility with the major platforms (InP platform photonics and Si platform) are discussed as well. Because of the complicated calculations involved in photonic circuitry modeling and complex interplay of many physical and technological issues, the discussion is kept at the conceptual level to highlight the basic principal approaches, physical phenomena involved, and ideas, without diving into detail.

10.1 PHOTONIC CRYSTAL WAVEGUIDES

A linear defect in a photonic crystal (PC) forms a waveguide that supports propagation of waves through a defect channel owing to multiple scattering and interference of waves whose frequency falls into a photonic band gap of a PC under consideration. The most instructive and useful is waveguiding in a two-dimensional photonic crystal (2DPC) since it can be made using templated etching through a submicron mask similar to the submicroelectronic chip production scheme. Guiding electromagnetic waves in a 2DPC gives rise to the notion of the *photonic crystal circuitry*. The basic ideas are explained in Figure 10.1. Wave confinement and guiding within the figure plane occurs owing to periodic structure and defects therein, and wave confinement in the vertical plane occurs by means of Fresnel reflection, which for typical semiconductor materials (Si, GaAs, InP) measures about 30% at normal incidence and then sharply grows with angle of incidence, reaching the total reflection regime for angles approximately >15° (see Figure 4.5 for reflection data). A "point-like" single-rod defect then forms a cavity (Figure 10.1(b)), and a linear defect that can be viewed as a coupled point-like defect forms a guiding channel (Figure 10.1(a)). Linear waveguides can be coupled to a cavity or multiple cavities when certain topological conditions are met for a given wave mode, as is shown in Figure 10.1(c).

It is of principal importance that propagation of lightwaves through a linear defect in PCs occurs owing to complicated scattering and interference processes. It is therefore

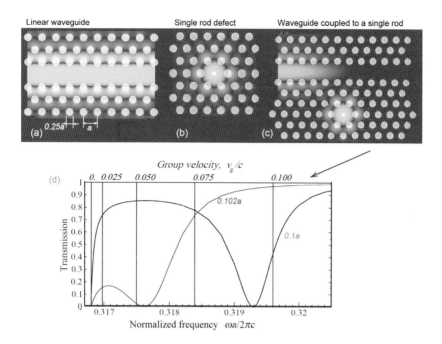

Figure 10.1 Principles of photonic circuitry based on a planar 2DPC consisting of Si rods in air, the crystal period is a, and the rod diameter is $0.25a$. (a) A linear defect consisting of missing rods forms a waveguide enabling efficient light guiding though a portion of electromagnetic energy spreads into the photonic crystal. (b) A "point-like" defect (a rod with radius $r_{def} = 0.1a$) forms a high-Q microcavity. (c) A photonic crystal waveguide coupled to a microcavity formed by a single rod. The field structure is shown for the frequency corresponding to the peak in the reflection spectrum (dip in transmission). (d) Transmission function versus normalized frequency for a structure shown in panel (c); different curves correspond to different defect rod radius r_{def}. The band edge is located at dimensionless frequency 0.3168. Here the group velocity for light in the waveguide tends to zero. Transmission features strong dependence on frequency as well as on the defect rod radius r_{def}. Courtesy of S. F. Mingaleev.

essentially frequency- and polarization-dependent. Moreover, guiding becomes possible at the expense of drastic slowing down. The group velocity then features strong frequency dependence and can be many times lower than in a vacuum or in a solid material in which a photonic crystal structure has been developed.

An example of slowing down and complicated frequency dependence for group velocity is shown in Figure 10.1(d) for a linear waveguide coupled to a cavity. One can see that (1) cavity size drastically affects transmittivity of a waveguide, and (2) both group velocity and transmittivity tend to zero for frequencies close to the band edge. For higher frequencies, group velocity monotonically rises with frequency but still remains a small fraction of c.

waveguiding channel

$D_2 < D < D_1$

Figure 10.2 An optimized photonic crystal waveguide design enabling high transmission with lower group velocity dispersion. The group velocity is $v_{g} = c/34$. See text for detail.

For practical applications it is important to design guiding structures with low losses and frequency-independent group velocity. A possible approach has been suggested by Frandsen et al. (2006), and is shown in Figure 10.2 for a PC based on cylindrical air holes etched in silicon. The technology is based on e-beam lithography and ion etching. The rows nearest to the guiding channel have smaller and bigger diameters. The structure fabricated according to this design has shown losses of 5 dB/mm and nearly constant group velocity of $\varepsilon_g = c/34$ within the 11 nm wavelength band.

A traditional waveguide (an optical fiber) operates using total internal reflection and therefore does not allow strong bending because of leakage when the total internal reflection condition is not perfectly met. A PC waveguide allows for bending in a PC plane since the total reflection condition is not necessary here (Figure 10.3(a)). However, simple bending of a linear guiding defect gives rise to big losses at every bend. Calculated field distribution is shown in Figure 10.3(c); the loss per bend value may exceed 10 dB (Figure 10.3(e), blue curve). Bending losses can be minimized by means of adaptive optimization of scatterers' topology, as is shown in Figure 10.3(b,d). Then losses per bend can be made below 1 dB (the red curve in Figure 10.3(e)). This ingenious approach has become a standard trend in waveguide design for micro- and nanophotonic circuitry. It became feasible owing to development of efficient computational techniques for light propagation in PC structures.

Three-dimensional guiding still remains challenging for experimental implementation. Development of a controllable 3D photonic crystal (3DPC) structure by multiple etching/ alignment procedures (see Chapter 4, Figure 4.21 and comments therein) is complicated and needs very accurate processing. The development of a 2D waveguide in a 3DPC is even more laborious. An example of this waveguide fabrication is given in Figure 10.4. One can see that transmission through a waveguide is higher by three orders of magnitude compared to the original PC, but still remains far from unity.

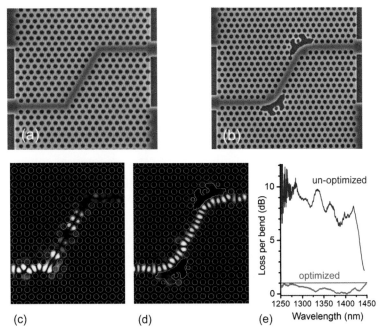

Figure 10.3 (a,b) Fabricated 2DPC structures with a waveguide using a silicon wafer and submicron e-beam lithography and ion etching technique. The hole diameter is 275 nm. (a) corresponds to a generic design, whereas (b) represents the optimized design of bends areas. The contrast and brightness of the images have been adjusted for clarity. (c,d) Calculated steady-state field distribution for the fundamental photonic band gap using the two-dimensional finite difference time-domain method for (c) generic and (d) optimized design. (e) Measured loss per bend for the unoptimized 60° bends and the topology-optimized 60° bends. Both spectra have been normalized to transmission through a straight PC waveguide of the same length. A blue horizontal line marks a bend loss of 1 dB. Adapted from Frandsen et al. (2004), with permission from OSA Publishing.

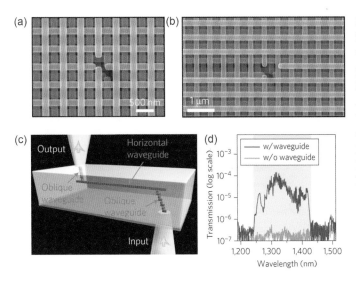

Figure 10.4 3D waveguiding in a silicon PC. (a) An oblique waveguide. (b) A connecting point of an oblique and a horizontal waveguide. (c) A sketch of light guiding through the two oblique guides connected by a linear waveguide. Reprinted by permission from Macmillan Publishers Ltd.; Ishizaki et al. (2013).

Progress in 2DPC waveguiding with low losses per bend design and fabrication allows for more complicated PC circuitry components to be implemented, namely beam splitters, demultiplexers, and interferometers. A Y-splitter has been demonstrated at the Technical University of Denmark (Borel et al. 2005). This element of the optical circuitry delivers each of the two wavelengths into the desirable guiding channel. Two beams with different wavelengths λ_1, λ_2 can be discriminated and sent to individual channels #1 and #2, provided the spectral spacing is approximately $|\lambda_1 - \lambda_2| > 50$ nm.

The optical components discussed in this section demonstrate the emerging field of *silicon photonics*. Optical transparency of silicon in the strategic optical communication spectral range of 1.3–1.5 μm can be combined with the unprecedented advances in silicon submicron-scale technology to develop a family of silicon photonic components. The above nanostructures are made of silicon using *silicon-on-insulator (SOI) technology* combined with high-resolution e-beam lithography. SOI nano-imprint technology has been developed in recent years, promising a cheap route toward commercial planar nanostructures. Imprinting is performed by means of reactive ion etching using a master stamp structure made by e-beam lithography.

Another example of basic photonic circuitry components is a *directional coupler* (Figure 10.5). This consists of two waveguides that are made so close to each other that the fields can couple in a section of a certain length. Then, periodic exchange of power takes place from one guide to the other. A portion of lightwave energy can be tapped to the neighboring channel. A directional coupler with optional control of the refractive index of constituent materials by means of electric field effects or optical nonlinearities can be used for (electro)optical modulation and switching.

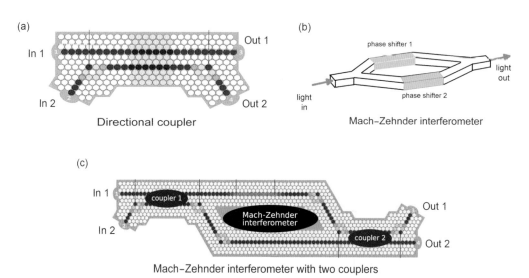

Figure 10.5 Photonic circuitry components: a directional coupler, a Mach–Zehnder interferometer, and a combination of an interferometer with two couplers. Adapted from Hermann et al. (2008), with permission from OSA Publishing.

Optical modulation and switching can also be performed using a Mach–Zehnder interferometer (Figure 10.5(b)). In this device the two Y-splitters are connected via two phase shifters. Depending on phase-matching/mismatching of the two beams in the output fiber, the output light intensity can be drastically changed. Then, if one of the phase shifters is electrically or optically controlled, the light propagation through a device can be modulated or switched. A PC version of a Mach–Zehnder interferometer is presented in Figure 10.5(c). It is an important building block in emerging photonic integrated circuitry.

10.2 COUPLING LIGHTWAVES THROUGH TUNNELING

In Section 4.5 we discussed examples of lightwave tunneling. Tunneling occurs by means of an evanescent wave penetrating through a "barrier," the latter of which can be formed by a thin mirror. Evanescence occurs in a metal or in a PC slab featuring a photonic band gap for the lightwave in question. For our consideration of photonic circuitry ideas, it is important that development of slow light in PC waveguides and high-Q values in PC cavities favorably enhance tunneling efficiency (i.e., coupling strength) in photonic band gap substructures like waveguides and cavities. This idea becomes clear when resonant tunneling is recalled (see Figure 4.28). A cavity between two mirrors promotes light tunneling throughout the whole structure. In a similar way, slowing light propagation in a PC waveguide enhances its penetration by evanescence to a closely spaced parallel guiding channel in a directional coupler (Figure 10.5). For a waveguide going near a cavity, efficient coupling occurs owing to tunneling enhanced by a high Q-factor value of a cavity in the case that the resonance between guiding frequency and a cavity is met. Such coupling occurs not only within the PC structures but in "solid" fiber waveguides and cavities as well. An example was given in Figure 4.30. In a similar way, coupling of lightwaves can be performed for two planar microcavities; between two microcavities made in the form of microdisks or microrings; or between a linear waveguide and any cavity, including a planar one, a PC one, a microsphere, a microdisk, or a microring. For a microsphere, a microdisk, and a microring, coupling to the so-called whispering gallery mode(s) is efficient owing to its high Q-factor. Coupling efficiency is strongly dependent on distance and resonant conditions. Therefore, if one of the components – say, a cavity in a "cavity–waveguide" system – can be tuned by changing the optical path length nd (with n being the refractive index and d being the geometrical length (electrically, optically, thermally), then coupling efficiency can be purposefully altered to enable light propagation control. This consideration, though simplified, offers a reasonable intuitive insight to PC circuitry ideas.

10.3 OPTICAL SWITCHING

10.3.1 Kramers–Kronig relations

Optical switching needs a combination of a physical process that can result in changing the optical parameter(s) of the material (absorption or refraction) and a device that converts small changes of the material parameter into sharp changes of the transmitted light intensity. Dissipative losses should be minimized to avoid the need for an optical amplifier after a switch.

Recalling the general notion of the complex refractive index (see Eq. (4.60)),

$$\tilde{n}(\omega) = n + i\kappa, \tag{10.1}$$

it is important to emphasize that the real n and imaginary κ parts of the complex refractive index are related to each other by the universal relation. In other words, in a medium in which electromagnetic wave velocity depends on its frequency, electromagnetic energy will necessarily be absorbed, and, vice versa, every absorbing medium will feature frequency-dependent velocity of electromagnetic waves. Knowing the absorption spectrum, one can calculate the refraction spectrum and vice versa. This important property is a consequence of the causality principle which states that a medium response to electromagnetic perturbation cannot occur prior to perturbation. The universal relations for imaginary and real parts of a material response function were derived by H. A. Kramers and R. Kronig in 1926 and are referred to as *Kramers–Kronig relations*. These relations read:

$$n(\omega') = 1 + \frac{2}{\pi} P \int_0^\infty \frac{\omega \kappa(\omega)}{\omega^2 - \omega'^2} d\omega, \tag{10.2}$$

$$\kappa(\omega') = -\frac{2\omega'}{\pi} P \int_0^\infty \frac{n(\omega') - 1}{\omega^2 - \omega'^2} d\omega, \tag{10.3}$$

where P in front of the integrals denotes the principal value. For an isolated absorption band with growing frequency, refractive index typically grows (normal dispersion) along with absorption, then falls near the absorption maximum (anomalous dispersion) and then rises again. The relation is more complicated for bulk semiconductors, but the main trend is the same, i.e., refractive index rises up with frequency (see Figure 4.2) in the transparency region and near the fundamental absorption edge. It is instructive to compare data for refractive index spectra (Figure 4.2) with semiconductor crystal band gaps (Figure 2.13) in this context. Therefore, to enable electro-optical switching, one needs to search for the electric field effect on absorption or refraction. To enable all-optical switching, one needs to search for the phenomena resulting in absorptive or refractive changes upon optical excitation. Dissipative losses are usually not welcome and therefore the research is mainly focused on looking for refractive index control while absorptive changes are typically used as a quick reference or as a preliminary indicator for desirable refractive index changes. Once the mechanism of refractive index control is settled, then a device can be designed based on tuning/detuning of resonant transmission through a Fabry–Pérot interferometer, a Mach–Zehnder interferometer, a waveguide coupled to a microcavity, a microring, or a microdisk resonator.

10.3.2 Electric field effect on optical properties of semiconductor nanostructures

Consider possible effects of the electric field on optical absorption/refraction of semiconductor nanostructures. An external electric field does change the probability of electron transitions since it adds to the periodic potential of a crystal lattice and disturbs electron–hole

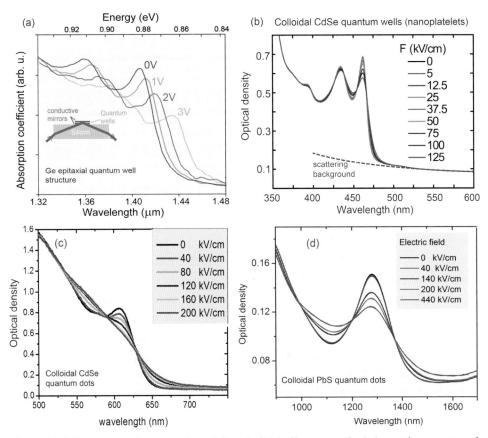

Figure 10.6 Representative examples of electric field effect on optical absorption spectra of quantum well and quantum dots. (a) Ge quantum wells in an Si/Ge epitaxial nanostructure. The total thickness of the multiple quantum well layer is 0.26 μm. Reprinted by permission from Macmillan Publishers Ltd.; Kuo et al. (2005). (b) Colloidal CdSe quantum wells (nanoplatelets) in a polymer film. Adapted with permission from Achtstein et al. (2014). Copyright 2014 American Chemical Society. (c) Colloidal CdSe and (d) PbS quantum dots in polymers.

Coulomb interaction. A quick estimate of the necessary electric field strength at which optical absorption will be modified can be made based on exciton Rydberg energy Ry^* divided by exciton Bohr radius a_B or quantum well (quantum dot) size a. For typical semiconductor materials (Ry^* is of the order of 10^{-2} eV and a is of the order of 10 nm), one has $E = Ry^*/(ea)$ ≈ 10 kV/cm. In a thin-film structure this value can be maintained with a few volts of external voltage. Electric field effect on optical absorption (and refraction, owing to the universal Kramers–Kronig relations) is pronounced in semiconductor quantum wells and quantum dots, and is often referred to as the quantum-confined Stark effect to emphasize the similarity between excitons and atoms in electric fields. The first evidence for an enhanced electric field effect in quantum well structures dates back to the 1980s with the work of D. A. B. Miller and co-workers. In the 1990s, the strong electric field effect on optical absorption was also revealed for quantum dots. Representative examples are given in Figure 10.6. A direct gap in a germanium multiple quantum well structure developed in Si_xGe_{1-x} heterostructures gives rise to absorption in the spectral range close to the commercial optical communication band (1.5 μm) and simultaneously enables CMOS (complementary metal-oxide-semiconductor) compatibility owing to the silicon platform. Si_xGe_{1-x} heterostructures exhibit strong electric field dependence (Figure 10.6(a)) and have been used to demonstrate electro-optical switching on a silicon wafer for oblique incidence, as shown in the inset. A submicrometer-thick multiple Ge/Si_xGe_{1-x} quantum well structure serves as an active medium of an interferometer formed by the two parallel conductive mirrors. Modulation of absorption by an external electric field modulates the interferometer reflectance.

Recently, a strong electric field effect on optical absorption has been found for colloidal quantum well structures – nanoplatelets (Figure 10.6(b)). A comparative study using the same synthesis route and the same experimental setup has revealed that colloidal nanoplatelets show higher electro-optic response as compared to colloidal nanorods and colloidal quantum dots. This opens the way to colloidal quantum well electro-optical components in the case that the desirable spectral ranges used in commercial communication circuitry will be attained. This is principally possible using PbS or PbSe as the basic materials for colloidal quantum well structures.

Colloidal quantum dots have been the subject of extensive electro-optical experiments since the 1990s. Under the external field the signature of the discrete optical absorption band can be fully smeared in the result of absorption band broadening without noticeable spectral shift (Figure 10.6(c)). In the same manner, PbS colloidal dots exhibit electro-optical response near 1.3 μm (Figure 10.6(d)), which can also be shifted to 1.5 μm when necessary. However, there is no efficient demonstration of colloidal quantum dot application for switching. One of the major obstacles is a complex interplay of photo- and electro-induced phenomena in colloidal quantum dot thin-film structures (surface charging, carrier trapping, partial photoionization promoted by high electric field). These phenomena are poorly controlled and deteriorate the time response of electro-optical switches. In this context, colloidal quantum wells and nanoplatelets represent a new generation of nanostructures which can probably combine advantages of epitaxial quantum wells but will not bring about the side effects inherent in colloidal quantum dots.

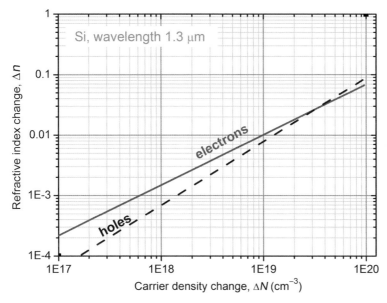

Figure 10.7 Calculated refractive index change for common semiconductors versus carrier density (Eqs. (10.4) and (10.5)). Results for free electrons and free holes are given. The wavelength in every case meets the condition $E_g - \hbar\omega = 0.17$ eV.

10.3.3 Electro-optical switching based on carrier density effects on optical properties

Increasing carrier density by doping of a semiconductor is known to change optical absorption (and refraction) properties since otherwise free electron (hole) states in the conduction (valence) band become populated and the rate of optical transition changes. In a similar way, optical properties of a semiconductor material can be altered dynamically by carrier injection in a slightly doped semiconductor or carrier depletion in a heavily doped semiconductor. Researchers avoid using current-modulated absorption because of undesirable insertion losses (these result in lower signal amplitude). Currently the main efforts are focused on refractive index modulation to perform noticeable phase shift, which is then combined with a phase-sensitive arrangement like a Fabry–Pérot interferometer, a Mach–Zehnder interferometer, a directional PC coupler, or a waveguide coupled to a microring or microdisk cavity. In every case, a small change in refractive index of an active material gives rise to strong transmission modulation of a device. The desired goal is to get a strong refractive index change with a high response rate and negligible heating. The latter results in slowly relaxing refractive index change because of the temperature-dependent band gap. Band gap shrinkage with temperature results in absorption edge red shift and (via Kramers–Kronig relations) in positive change of refractive index. The critical points here are: (1) small refractive index changes necessitate long paths; (2) time response is essentially defined by electron–hole recombination, which cannot be efficiently controlled; and (3) current flow gives rise to dissipation losses and heating which in turn changes refractive index but features very slow relaxation time.

The carrier injection/depletion approach dominates current research activity toward optical switching components in photonic circuitry. Regretfully, refractive index n of any material cannot be altered noticeably either by external electric field or by carrier injection/depletion. For external electric field (Stark effect), typically $\Delta n/n \approx 10^{-3}$ represents the ultimate limit. For carrier injection/depletion, refractive index modulation is even smaller to prevent undesirable contributions from current-induced heating. The changes in Δn with carrier density N for GaAs, InP, and Si show the same order for the magnitude at $N \approx 10^{18}$–10^{19} cm^{-3} (Figure 10.7), which represent the reasonable injection/depletion levels to avoid side heating-related effects. In the case of silicon, refractive index change Δn and absorption coefficient change $\Delta\alpha$ can be approximated by empirical relations (Soref and Bennet 1986),

$$\lambda = 1.55 \ \mu m$$
$$\Delta n = -8.8 \cdot 10^{-22} \Delta N (\text{cm}^{-3}) - 8.5 \cdot 10^{-18} \Delta P^{0.8} (\text{cm}^{-3}),$$
$$\Delta\alpha (\text{cm}^{-1}) = 8.5 \cdot 10^{-18} \Delta N (\text{cm}^{-3}) + 6.0 \cdot 10^{-18} \Delta P (\text{cm}^{-3}),$$

(10.4)

$$\lambda = 1.3 \ \mu m$$
$$\Delta n = -6.2 \cdot 10^{-22} \Delta N (\text{cm}^{-3}) - 6.0 \cdot 10^{-18} \Delta P^{0.8} (\text{cm}^{-3}),$$
$$\Delta\alpha (\text{cm}^{-1}) = 6.0 \cdot 10^{-18} \Delta N (\text{cm}^{-3}) + 4.0 \cdot 10^{-18} \Delta P (\text{cm}^{-3}),$$

(10.5)

which gives only $\Delta n < 10^{-3}$ for $\Delta N \approx 10^{17}$ cm^{-3}. The important figure of merit is L_π, the path length where the phase shift equal to π develops. The small changes in refraction necessitate millimeter-scale path length L_π in the Mach–Zehnder arrangement to get phase shift of π since the condition $\Delta n L_\pi = \lambda/2$ is to be met to get maximal change in transmitted light intensity. The other important parameter is the product $V_\pi L_\pi$, with V_π being an operation voltage that results in phase shift equal to π. Many authors have reported the values 1 V·cm $< V_\pi L_\pi <$ 10 V·cm for Mach–Zehnder electro-optical switches. Table 10.1

Table 10.1 **Representative examples of optical modulator parameters based on Mach–Zehnder interferometers**

Device length (mm)*	1.35	1	2.4	2
$V_\pi L_\pi$ (V·cm)	11	2.8	2.4	2.4
Insertion loss at maximum transmission (dB)	15	3.7	4.3	4.1
Speed (Gbit/s)	40	50	30	50
Reference	Gardes et al. 2011	Thomson et al. 2012	Chen et al. 2011	Dong et al. 2012

*This is the phase-shifter length, the full length of an interferometer is several times greater.

gives a few representative examples of switching based on carrier injection/depletion in Mach–Zehnder devices.

Another reasonable phase-sensitive arrangement is a linear waveguide coupled to a high Q-factor whispering gallery mode inherent in a microring or a microdisk. For cavity-based

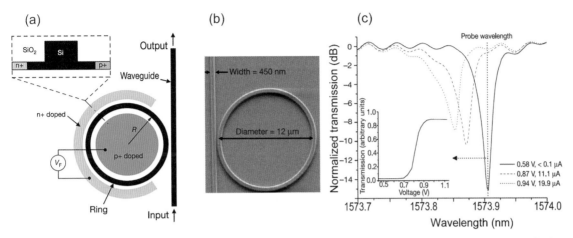

Figure 10.8 Silicon electro-optical modulator based on a waveguide coupled to an electrically controlled ring resonator. (a) Schematic layout. The inset shows the cross-section of the ring. R, radius of ring. V_F, voltage applied on the modulator. (b) Top view scanning electron microscope image. (c) DC measurement of the ring resonator. The main panel shows the transmission spectra of the ring resonator at the bias voltages of 0.58 V, 0.87 V, and 0.94 V, respectively. The vertical dashed line marks the position of the probe wavelength used in the transfer function and dynamic modulation measurements. The inset shows the transfer function of the modulator for light with a wavelength of 1573.9 nm. Reprinted by permission from Macmillan Publishers Ltd.; Xu et al. (2005).

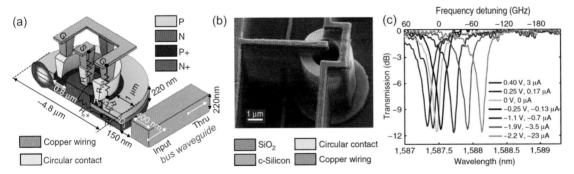

Figure 10.9 Silicon vertical junction microdisk electro-optical modulator. (a) A sketch of the electro-optical silicon microdisk modulator, showing the cross-section, size, metal connections, and the optical mode overlapped with the vertical p–n junction. The gray regions illustrate the depletion region in the p–n junction within the microdisk and the undoped bus waveguide adjacent to the microdisk. (b) A scanning electron microscopy image of the modulator, revealed by dry etching the SiO_2 around the modulator to show the metal interconnect, circular contact, silicon bus waveguide and the microdisk. The signal pad, connected by short wires, is shown on the left side of the image. (c) The measured transmission spectra of the resonant modulator at 26.5 °C and applied DC bias voltages ranging from 0.4 to –2.2 V. Reprinted from Timurdogan et al. (2014).

design, phase shift, which is necessary for maximal modulation, is of the order of π/Q i.e., it is approximately Q times smaller than that in the Mach–Zehnder design. A representative example for an Si microring-based modulator is given in Figure 10.8. The Q-factor for a 12 μm ring resonator was about $4 \cdot 10^4$, the operating voltage of the order of 1 V and the data rate was estimated to be of the order of 1 Gbit/s. This design offers much more compact devices compared to Mach–Zehnder interferometers. A vertical p–n junction microring cavity coupled to a silicon bus waveguide (Figure 10.9) has been shown to feature ultralow (1 fJ) power consumption per bit with a data rate of 25 Gbits/s, the operating voltage being in the range of 1 V. It is based on carrier depletion in the p–n junction area by external voltage (Figure 10.9).

10.3.4 All-optical switching: controlling light with light

We saw in Chapter 6 that at high radiation intensity the high population of excited states of matter results in absorption saturation. This effect is pronounced in semiconductors and in semiconductor nanostructures. Its manifestation in quantum dots is extensively used in laser Q-switching and mode-locking. Its manifestation in quantum wells or thin semiconductor layers resulted in development of SESAM (semiconductor saturable absorption mirror) devices for laser mode-locking. These applications were discussed in detail in Chapter 9 (Section 9.5). Intensity-dependent optical properties of matter and the related phenomena form the subject of *nonlinear optics*, contrary to traditional linear optics, in which material response is independent of radiation intensity.

Absorption saturation can be used for all-optical switching. However, it appears to be strongly frequency-dependent (needs matching of absorption resonance conditions) and have side effects like heating, incomplete transparency because of excited state absorption (thus leading to high insertion loss of a switch), and finite relaxation time defined by recombination rate that cannot be fully controlled. Owing to Kramers–Kronig relations, absorption saturation will result in refractive index changes that in turn can be used for all-optical switching. Population-induced phenomena were used at early stages of optical switching research in the 1980s. However, population-induced nonlinear response brings about the issue of recombination time and partial heating. Therefore, population-induced nonlinear optical phenomena are typically avoided now when all-optical switching design is considered. Instead, nonlinear susceptibility is used, which is population-independent.

In the case when the medium reacts to a change in the incident electromagnetic field *instantaneously*, the polarizability of the medium can be described in terms of *dielectric susceptibility* χ, defined via a relation $\mathbf{P} = \chi \varepsilon_0 \mathbf{E}$, whence a simple relation between dielectric permittivity and dielectric susceptibility reads $\varepsilon(\omega) = 1 + \chi(\omega)$. Then, polarization vector components can be expanded into a power series of the field amplitude as

$$\frac{1}{\varepsilon_0} P_i = \sum_j \chi_{ij}^{(1)} E_j + \sum_{j,k} \chi_{ijk}^{(2)} E_j E_k + \sum_{i,k,l} \chi_{ijkl}^{(3)} E_j E_k E_l + \dots, \tag{10.6}$$

implying the susceptibility is a tensor value. In a simpler scalar form, polarization versus incident field reads

$$P = \varepsilon_0 \left[\chi^{(1)} E + \chi^{(2)} E^2 + \chi^{(3)} E^3 + ... \right]. \tag{10.7}$$

The first term in this expansion describes the linear response, the second term corresponds to frequency doubling, summation, and subtraction, and the third term describes nonlinear refraction and its consequences, like four-wave mixing and self-focusing/defocusing. Values $\chi^{(i)}$ are referred to as ith order susceptibilities. The second-order $\chi^{(2)}$ susceptibility equals zero in all materials with symmetry with respect to inversion. For an isotropic medium ($\chi^{(2)} = 0$) at reasonably high but not extreme radiation intensity only $\chi^{(1)}$ and $\chi^{(3)}$ are used. Then the approximate expression is valid for refractive index versus radiation intensity I,

$$n = n_0 + n_2 I \tag{10.8}$$

where n_0 is refractive index at low intensity (linear refractive index) and n_2 (dimensionality W^{-1} cm^2) is the nonlinear refraction coefficient. The third-order nonlinearity expressed by a nonzero $\chi^{(3)}$ value and Eq. (10.8) is referred to as *Kerr nonlinearity*. Kerr nonlinearity is instantaneous. Therefore, when it is used in a device that is sensitive to refractive index change, the device response rate is defined only by radiation accumulation/depletion rate.

Nonlinear susceptibility $\chi^{(3)}$ has dimensionality of m^2/V^2 (SI units), but often is reported in electrostatic units (ESUs). The relations between $\chi^{(3)}$ in different units and nonlinear refraction index coefficient n_2 read (Boyd 2008):

$$\chi^{(3)}(\text{m}^2 / \text{V}^2) = 1.40 \times 10^{-8} \chi^{(3)}(\text{ESU}), \tag{10.9}$$

$$n_2(\text{m}^2 / \text{W}) = \frac{3}{4\tilde{n}n_0\varepsilon_0 c} \chi^{(3)}(\text{m}^2 / \text{V}^2) = \frac{283}{\tilde{n}n_0} \chi^{(3)}(\text{m}^2 / \text{V}^2), \tag{10.10}$$

$$n_2(\text{m}^2 / \text{W}) = \frac{12\pi^2}{\tilde{n}n_0 c} 10^7 \chi^{(3)}(\text{ESU}) = \frac{0.0395}{\tilde{n}n_0} \chi^{(3)}(\text{ESU}). \tag{10.11}$$

It is not possible to affect refraction noticeably – e.g., for fused silica $n_2 \approx 3 \cdot 10^{-16}$ cm^2/W, which gives $\Delta n = 10^{-7}$ only at radiation intensity 1 GW/cm^2. For silicon, $n_2 \approx 2 \cdot 10^{-14}$ cm^2/W in the range of wavelengths 1.3–1.5 µm (Lin et al. 2007). Higher values of nonlinearity occur under conditions of high electron–hole population density. In this case, nonlinear refraction results from absorption saturation in accordance with Kramers–Kronig relations. For glasses doped with CdS$_x$Se$_{1-x}$ quantum dots, the values of $n_2 \approx 10^{-11}$ cm^2/W were found at 500 nm (Olbright and Peyghambarian 1986), i.e., $\Delta n = 10^{-3}$ can be obtained at radiation intensity 100 MW/cm^2, which is, however, still a very high intensity level. Population-induced nonlinearity features finite relaxation time defined as the inverse of electron–hole recombination rate. Therefore, higher nonlinearity occurs at the expense of longer decay time and switching time. In this context it is similar to injection-based electro-optical response.

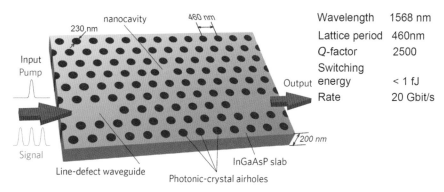

Figure 10.10 InGaAsP PC cavity-based optical switch. Adapted with permission from Macmillan Publishers Ltd.; Nozaki et al. (2010).

A microcavity containing an inner material with nonlinear refraction promotes higher field intensity inside (roughly Q-fold intensity enhancement), but this occurs simultaneously with development of the cavity photon lifetime (approximately QT, $T \approx 1$ fs being the oscillation period) which for $Q > 10^3$ exceeds 1 ps. In the case of population-induced refractive index change, cavity-induced recombination rate enhancement (Purcell factor, see Chapter 5) by the factor of Q moves recombination rate from its typical nanosecond range to the picosecond range. Thus, recombination time can become close to photon cavity lifetime and therefore population-induced nonlinear refraction can be used instead of Kerr nonlinearity, without deterioration of the switching speed.

Figure 10.10 shows an example of a PC-nanocavity coupled to a waveguide whose transmission is altered by external laser pulses by means of population-induced refractive index change. InGaAsP composition was adjusted to obtain a band gap wavelength of 1.47 μm to ensure that the desirable operating wavelength range (1550–1570 nm) is close enough to ensure strong refractive index change owing to band edge proximity and at the same time to keep operating range corresponding to low absorption. As a result, non-linear refractive index coefficient was relatively large, $\Delta n = -8.2 \cdot 10^{-20} \Delta N (\text{cm}^{-3})$, nearly two orders of magnitude higher than for Si in the same spectral range, as expressed by Eq. (10.4). The cavity photon lifetime was 5.4 ps, which is slightly longer than the calculated carrier lifetime including Purcell effect by the cavity. This design features much smaller size and much lower switching energy as compared to PC Mach–Zehnder devices reported by many groups. Nozaki et al. (2010) suggested an instructive estimation of the performance of this type of all-optical switches. The switching energy W_{switch} is defined as the pulse energy at the device input, which produces a nonlinear refraction-induced phase shift equal to the cavity spectral width,

$$\Delta n / n_0 = \Delta \lambda / \lambda_0 = 1/Q \;\Rightarrow\; n_2 I_{\text{switch}} = n_0 / Q \;\Rightarrow\; I_{\text{switch}} = n_0 / (n_2 Q). \qquad (10.12)$$

Now it is important to account for which portion η of energy delivered to the input port U_{in} of the waveguide is tapped to the cavity, U_{cavity}. It reads

$$\eta = \frac{U_{\text{cavity}}}{U_{\text{in}}} = \frac{4\tau_{\text{ph}}^2 A}{\tau_{\text{cpl}}} / Q \; ; \; \tau_{\text{ph}} = \frac{1}{(\tau_{\text{int}})^{-1} + (\tau_{\text{cpl}})^{-1} + A},$$ (10.13)

where τ_{ph} is cavity photon full lifetime, τ_{int} is intrinsic cavity loss lifetime (inverse to the intrinsic loss rate in the cavity), τ_{cpl} is the cavity outcoupling time (inverse to the outcoupling rate to the waveguide), and A is the absorption rate. Then the switching energy U_{switch} at the input can be estimated as $U_{\text{switch}} = I_{\text{switch}} \tau / \eta$, where τ is the laser pulse duration.

Slow light in a PC waveguide enhances nonlinearity of the material used to fabricate the PC structure since slowing down is equivalent to increasing intensity. A number of groups reported on application of the slow light in a PC waveguide for optical switching. However, the size of these devices is rather large, of the order of 1 mm, similar to Mach–Zehnder interferometers with a nonlinear arm.

One more approach to an all-optical switching device is presented in Figure 10.11. Here, a number of microring resonators are coupled and a three-port system is organized

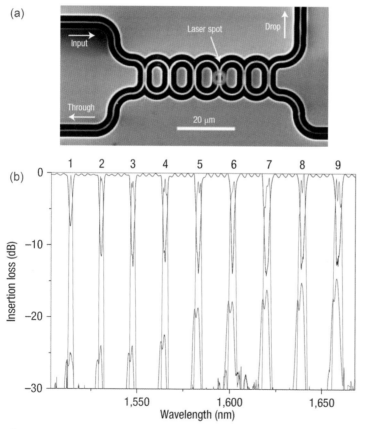

Figure 10.11 All-optical switching using a fifth-order coupled microring silicon resonator. (a) Scanning electron image of the device. (b) Switching operation. The red curve represents drop-port transmission spectrum, the black curve shows through-port transmission, and the blue curve is the drop-port transmission spectrum under optical excitation ("switch-on" state). Reprinted by permission from Macmillan Publishers Ltd.; Vlasov et al. (2008).

using waveguides. The circuit is developed on a silicon wafer. The authors used silicon-on-insulator 200 mm wafers with a 2 mm buried oxide layer and a thin silicon layer of thickness 226 nm on a standard CMOS fabrication line.

Radiation coming from the input port goes either into the "through" port or "drop" port, depending on coupling between microrings. Thus, a device serves as a router rather than a simple switching modulator. Every ring resonator has diameter about 10 μm, i.e., similar to the single microring device shown in Figure 10.8. Appropriate apodization of the coupling coefficients between neighboring high Q-factor microrings results in a flat-top pass band at the drop-port with amplitude ripples <0.5 dB. When laser pulse radiation is applied, pass bands feature fast shift to shorter wavelengths with increasing pump laser power, as is expected for the free carrier plasma-dispersion effect in silicon. The required concentration of laser-induced free carriers is $10^{19}\,cm^{-3}$ when a single ring is illuminated and drops to $10^{18}\,cm^{-3}$ when two rings are excited. The observed switching time of 2 ns is longer than the carrier recombination lifetime (about 0.5 ns), and is attributed to the intrinsic optical response of the switch defined by the cavity design.

10.4 CHALLENGES AND OUTLOOK

In microelectronics, progress in Si-platform-based miniaturization has resulted in unprecedented growth of integration level, with doubling of the number of components per unit area every two years (Moore's law) over several decades. Progress in technology suggests downscaling to 10–20 nm until quantum size phenomena and tunneling alter the components' functionality significantly. Then a new generation of electronic components exploiting size-dependent electronic properties can be foreseen.

There is also a trend toward integrated optical components known as photonic integrated circuitry. Here, the principal building blocks are (1) active components: lasers, optical amplifiers, modulators, routers, detectors; and (2) passive components: waveguides, couplers, splitters, filters, demultiplexers. There has been a pronounced tendency over the last two decades toward higher integration of components within the InP platform. InP is the basic material for lasers used in optical communication and therefore other components should be technologically compatible with InP lasers. The highest number of elements per photonic chip has reached a few hundred. Along with the general trend toward miniaturization, photonic integrated circuitry meets the desire to minimize the number of expensive and power-consuming optical-to-electronic-to-optical conversions by signal processing in the optical components.

Silicon photonics technology has also become a very active trend in research toward photonic integrated circuitry. The silicon platform offers most of the functionalities available with InP except for lasers and amplifiers. However, the Si-based approach promises lower costs because of compatibility with CMOS technology.

The PC paradigm paves a way to efficient integration of multiple photonic components by using 2D periodic structures obtained by templated etching within the engineering rate attainable in modern microelectronics. Most of the examples in this chapter were based on Si technology; however, InP-based structures are also the subject of active research. Successful reports on waveguides, splitters, demultiplexers, and couplers reviewed in this chapter, along with progress in compact lasers using PCs (see Chapter 9), form a solid basis for higher integration levels. Here, modulators seem to be a bottleneck toward higher miniaturization.

In Europe, the two major photonic integration technologies are supported through the two organizations: Joint European Platform for Photonic Integration of Components and Circuits (JePPIX) for InP-based integration technology,[1] and the European Silicon Photonics Alliance (ePIXfab),[2] which promote the science, technology, and application of silicon photonics. These organizations provide open access for research purposes to a relatively mature integration technology: the JePPIX platform to the InP-based integration technology of the COBRA Institute of the Technical University in Eindhoven, and later also the platform technologies of Oclaro and Fraunhofer HHI, the ePIXfab platform to the SOI technology at IMEC and LETI. Access to technologies is provided through multi-project wafer runs, a well-known approach in microelectronics, nowadays being extended to photonics.

Nevertheless, even successful commercialization of PC-based integrated components cannot bridge the existing fundamental gap between electronic and optical length scales. Each of these scales is defined by the relevant wavelength, i.e., the 10 nm scale inherent in electronics and 10^3 nm scale in photonics. Here, nanoplasmonics is believed to pave the way toward subwavelength light manipulation owing to extreme spatial concentration of radiation beyond the diffraction limit.

Conclusion

- Lightwave confinement in PCs and microcavities allows guiding, coupling, splitting, demultiplexing, and switching. Guiding occurs by means of multiple scattering in periodic structures with defects, and coupling is performed essentially by tunneling of the evanescent field. There are a number of successful demonstrations of various photonic components based on these principles, but only at the research level rather than commercial scale. These components combined with small semiconductor lasers integrated with PCs and with semiconductor photodetectors form a generic integration technology in photonics.

- Regretfully, photonics cannot offer today the integrity and scaling comparable to its microelectronic counterparts. There are basic limits related to operational principles of lasers, waveguiding, and optical switching. Optical switching/modulation represents a

---

[1] See the website www.JePPIX.eu (accessed January 7, 2018).
[2] See the website www.epixfab.eu (accessed January 7, 2018).

major bottleneck in photonic component integration and miniaturization. High-speed efficient Mach–Zehnder interferometers have lengths of the order of 1 cm (!). Microdisk-based switches are downsized to the order of 10–20 μm, which is still not a reason to refer to these devices as components of nanophotonics. PC microcavity-based switches are a few micrometers in size, which is the smallest among the suggested designs.

- Since switching/modulator design seems to be the major bottleneck in integrated photonics, new ideas and paradigms are necessary in this area. Nowadays, plasmonics is considered to offer a possible alternative to the traditional light confinement approaches. In plasmonics, development of the specific modes featuring very high wave numbers (and thus very short wavelength) offers a way to subwavelength confinement of light.

- One can speak also about possible development of light sources which do not use optical feedback and thus can operate without a cavity or a grating. Such sources can probably be foreseen based on single quantum dot emitters whose efficiency is enhanced by plasmonics.

Problems

10.1 Explain principles of light guiding in PCs. What is the propagation speed in a PC waveguide compared to a continuous medium?

10.2 Recall physical phenomena that can be used to develop an optical switch.

10.3 Consider advantages and disadvantages of population-induced electro-optical and all-optical switching principles versus population-independent counterparts.

10.4 Explain why a cavity can downsize a phase-sensitive photonic element only at the expense of a slower operating rate.

10.5 Explain why cavity-based switches and modulators need much smaller changes in refractive index compared to Mach–Zehnder devices.

10.6 Explain why a cavity and a PC waveguide enhance optical nonlinearity.

10.7 Explain why enhanced recombination rate of electron–hole pairs in a microcavity does not result in operation rate enhancement.

Further reading

Bozhevolnyi, S. I. (ed.) (2009). *Plasmonic Nanoguides and Circuits*. Pan Stanford Publishing.

Gilardi, G., and Smit, M. K. (2014). Generic InP-based integration technology: present and prospects. *Progr Electromagnetics Res*, **147**, 23–35.

Gramotnev, D. K., and Bozhevolnyi, S. I. (2010). Plasmonics beyond the diffraction limit. *Nat Photonics*, **4**, 83–91.

Jin, C. Y., and Wada, O. (2014). Photonic switching devices based on semiconductor nanostructures. *J Physics D: Applied Physics*, **47**, 133001.

Krauss, T. F. (2008). Why do we need slow light? *Nat Photonics*, **2**, 448–450.

MacDonald, K. F., and Zheludev, N. I. (2010). Active plasmonics: current status. *Laser Photonics Rev*, **4**, 562–567.

Miller, D. A. B. (2009). Device requirements for optical interconnects to silicon chips. *Proc IEEE*, **97**, 1166–1185.

Mingaleev, S. F., Miroshnichenko, A. E., and Kivshar, Yu. S. (2007). Low-threshold bistability of slow light in photonic-crystal waveguides. *Opt Express*, **15**, 12380–12385.

Niemi, T., Frandsen, L. H., Hede, K. K., et al. (2006). Wavelength division de-multiplexing using photonic crystal waveguides. *IEEE Photon Technol Lett*, **11**, 226–228.

Notomi, M. (2010). Manipulating light with strongly modulated photonic crystals. *Rep Prog Phys*, **73**, 096501.

Peiponen, K. E., Vartiainen, E. M., and Asakura, T. (1998). *Dispersion, Complex Analysis and Optical Spectroscopy: Classical Theory*. Springer Science & Business Media.

Priolo, F., Gregorkiewicz, T., Galli, M., and Krauss, T. F. (2014). Silicon nanostructures for photonics and photovoltaics. *Nature Nanotechn*, **9**, 19–32.

Smit, M., van der Tol, J., and Hill, M. (2012). Moore's law in photonics. *Laser Photonics Rev*, **6**, 1–13.

Sorger, V. J., Oulton, R. F., Ma, R.-M., and Zhang, X. (2012). Toward integrated plasmonic circuits. *MRS Bulletin*, **37**, 728–738.

References

Achtstein, A. W., Prudnikau, A. V., Ermolenko, M. V., et al. (2014). Electroabsorption by 0D, 1D, and 2D nanocrystals: a comparative study of CdSe colloidal quantum dots, nanorods, and nanoplatelets. *ACS Nano*, **8**, 7678–7686.

Borel, P. I., Frandsen, L. H., Harpøth, A., et al. (2005). Topology optimised broadband photonic crystal Y-splitter. *Electron Lett*, **41**, 69–71.

Boyd, R. W. (2008). *Nonlinear Optics*. Academic Press.

Chen, L., Doerr, C. R., Dong, P., and Chen, Y. K. (2011). Monolithic silicon chip with 10 modulator channels at 25 Gbps and 100-GHz spacing. *Opt Express*, **19**, B946–B951.

Dong, P., Chen, L., and Chen, Y. K. (2012). High-speed low-voltage single-drive push–pull silicon Mach–Zehnder modulators. *Opt Express*, **20**, 6163–6169.

Frandsen, L. H., Harpøth, A., Borel, P. I., et al. (2004). Broadband photonic crystal waveguide 60°-bend obtained utilizing topology optimization. *Opt Express*, **12**, 5916–5921.

Frandsen, L. H., Lavrinenko, A. V., Fage-Pedersen, J., and Borel, P. I. (2006). Photonic crystal waveguides with semi-slow light and tailored dispersion properties. *Opt Express*, **14**, 9444–9450.

Gardes, F. Y., Thomson, D. J., Emerson, N. G., and Reed, G. T. (2011). 40 Gb/s silicon photonics modulator for TE and TM polarisations. *Opt Express*, **19**, 11804–11814.

Hermann, D., Schillinger, M., Mingaleev, S. F., and Busch, K. (2008). Wannier-function based scattering-matrix formalism for photonic crystal circuitry. *J Opt Soc Amer B*, **25**, 202–209.

Ishizaki, K., Koumura, M., Suzuki, K., Gondaira, K., and Noda, S. (2013). Realization of three-dimensional guiding of photons in photonic crystals. *Nat Photonics*, **7**, 133–137.

Kuo, Y. H., Lee, Y. K., Ge, Y., et al. (2005). Strong quantum-confined Stark effect in germanium quantum-well structures on silicon. *Nature*, **437**, 1334–1336.

Lin, Q., Zhang, J., Piredda, G., et al. (2007). Dispersion of silicon nonlinearities in the near infrared region. *Appl Phys Lett*, **91**, 021111.

Nozaki, K., Tanabe, T., Shinya, A., et al. (2010). Sub-femtojoule all-optical switching using a photonic-crystal nanocavity. *Nat Photonics*, **4**, 477–483.

Olbright, G. R., and Peyghambarian, N. (1986). Interferometric measurement of the nonlinear index of refraction, n_2, of CdS_xSe_{1-x}-doped glasses. *Appl Phys Lett*, **48**, 1184–1186.

Soref, R. A., and Bennett, B. R. (1986). Kramers–Kronig analysis of electro-optical switching in silicon. *Proc SPIE*, **704**, 32–37.

Thomson, D. J., Gardes, F. Y., Fedeli, J. M., et al. (2012). 50-Gb/s silicon optical modulator. *IEEE Photonics Technol Lett*, **24**, 234–236.

Timurdogan, E., Sorace-Agaskar, C. M., Sun, J., et al. (2014). An ultralow power athermal silicon modulator. *Nature Commun*, **5**(4008), 1–11.

Vlasov, Y., Green, W. M., and Xia, F. (2008). High-throughput silicon nanophotonic wavelength-insensitive switch for on-chip optical networks. *Nat Photonics*, **2**, 242–246.

Xu, Q., Schmidt, B., Pradhan, S., and Lipson, M. (2005). Micrometre-scale silicon electro-optic modulator. *Nature*, **435**, 325–327.

11 Photovoltaics

Photovoltaics, to become comparable in scale with other electric energy sources, should offer either cheaper or more efficient solar cells compared to their current level. Physical properties of nanostructures offer certain means to pursue both of these options. First, nanostructures offer flexibility in band gap engineering, which is very important in the context of fitting the solar spectrum in the optimal way. Second, colloidal quantum dots allow for a cheap production process, contrary to their monocrystalline counterparts. Third, quantum dots feature carrier multiplication under certain conditions, thus allowing overtaking the fundamental limit set by the assumption of a maximum of a single electron–hole pair created by every photon. This allows for internal quantum efficiency to exceed 100%. Additionally, plasmonics suggests enhancement in solar radiation absorption, which is crucial for cheap thin-film solar cells. Finally, bioinspired antireflection surface structure using the photonic crystal approach may additionally increase solar cell power conversion efficiency. The above phenomena and the first results of their implementation are discussed in this chapter.

11.1 WHAT CAN NANOPHOTONICS SUGGEST FOR PHOTOVOLTAICS?

11.1.1 Photovoltaics versus traditional electric power sources

The sun delivers to the earth surface approximately 1000 W of radiation energy per square meter. In the upper atmosphere layers the original sunlight spectrum can be reasonably fitted by the black body radiation spectrum described by the Planck formula for temperature about 5250 K. At the earth surface it changes and has the distribution shown in Figure 5.3. To become widespread as traditional electric energy sources, photovoltaics must offer the energy at a price comparable to the current prices for electric power (less than US$0.1 per kWh) or, better, much cheaper, since photovoltaics has to compete with well-developed power generation and delivery systems with well-organized service and maintenance infrastructures. Therefore it is reasonable to consider the current status of photovoltaics in the context of cost competitiveness and appeal for customers. The average photovoltaic module cost per square meter, the sun radiation intensity, and the power conversion efficiency define the cost per 1 W of peak power. Taking into account daily and seasonal radiance flux changes, service costs, etc., the cost of 1 kWh is usually estimated as 10% of the module cost per 1 W of peak

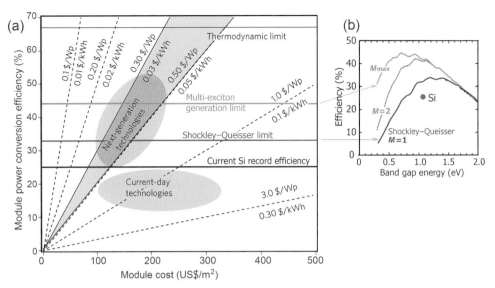

Figure 11.1 Photovoltaic conversion efficiency and cost issues. (a) Relationship between power conversion efficiency, module area costs, and cost per peak watt, Wp (in $/Wp) for 1000 W/m² solar radiation intensity. Adapted with permission from Beard et al. (2014), copyright Macmillan Publishers Ltd. (b) The fundamental Shockley–Queisser limit versus semiconductor band gap energy (red) under the assumption that every absorbed photon gives one ($M = 1$) electron–hole pair and its modification for two electron–hole pairs per photon ($M = 2$) and for the ultimate limit of maximal possible multiplication for the solar spectrum (M_{max}). Adapted from Nozik (2008), with permission from Elsevier.

power measured for the standard radiation spectrum (known as Reference AM 1.5 Spectra or ΛM1.5 Solar Spectrum). Figure 11.1(a) shows the current status and the desire for photovoltaics in the context of cost. Current technology offers the cost at about US$0.1 per kWh; this situation is referred to as "grid parity," but, as was mentioned above, to compete with or substitute for the traditional electric energy sources the price for solar energy conversion has to be reduced. This is marked as "next-generation technologies" in the figure and is shaded light blue. This need can be met either by making photovoltaic cells cheaper or making them more efficient. Using nanostructures in photovoltaics is therefore an area to be examined in the context of these two parameters.

11.1.2 The ultimate efficiency of solar energy conversion

The sun spectrum defines ultimate efficiency limits for radiation energy conversion into electric power. When using semiconductor material, only photons with energy $h\nu$ exceeding the band gap E_g will be absorbed. Thus it is reasonable to use narrow-gap semiconductors to consume the greater portion of the solar spectrum. In an ideal case every absorbed photon will give a single charge into a circuit and the voltage will be strictly defined by the band gap energy. One can easily see that all photons with $hv < E_g$ will be lost for power conversion and all photons with energy $hv > E_g$ will give the same contribution to the output power. The excessive energy $hv - E_g$ will be irreversibly lost, being converted into heat in the course of electron and hole relaxation to the bottom of the conduction band and to the top of the valence band, respectively. Thus the temptation to pick up more energy from the sun by taking a narrow-band material should be tempered by the issue of high efficiency loss for heating. Taking a wide-gap material will result in missing the low-energy portion of the solar spectrum. As a result of this tradeoff, a universal ultimate efficiency level versus semiconductor band gap can be derived under assumptions of (1) the real solar spectrum, and (2) every photon with energy higher than E_g is supposed to give one elementary charge e to the circuit at voltage $V_g = E_g/e$. This was reported for the first time by W. Shockley and H. J. Queisser in 1961 and is referred to as the *Shockley–Queisser limit*. Its absolute value is 33.7% and can be reached for $E_g = 1.34$ eV (corresponds to gap wavelength 925 nm). The power conversion efficiency versus band gap is shown as red lines in Figure 11.1. The records of efficiency achieved in the research laboratories with bulk monocrystalline semiconductors in 2017 are: 29% for GaAs ($E_g = 1.42$ eV) by Alta Devices; 26% for Si ($E_g = 1.12$ eV) by Kaneka; 22% for CdTe ($E_g = 1.47$ eV) by First Solar; and 22% for perovskites by KRIST (Green et al. 2017; NREL 2017). This value for silicon is shown by the blue line and dot in Figure 11.1.

Efficiency can also be raised by using a sandwich structure in which the solar spectrum is distributed over different layers, so that the upper layer consumes high-energy photons and the bottom layer picks up the lower-energy portion of the solar spectrum. The thermodynamic limit at "one sun" (without a solar concentrator) illumination is shown in Figure 11.1(a) as the brown line at 67%, which can be reached by an infinite stack of p–n junctions. Regretfully, these structures are expensive. The current record for a four-junction system with no solar concentrator was 38% in 2015 (Boeing Spectrolab). A solar concentrator allows this value to reach 45%; however, solar concentrators add cost and weight to the product.

Efficiency can be raised if a possibility exists that a high-energy photon can give more than one electron–hole pair, provided the condition $hv > 2E_g$ is met. However, the voltage will still be defined by the band gap rather than by the photon energy, and for higher photon energy smaller numbers of photons are available because of the solar spectrum cutoff at the high-energy side. Calculations show that multiple electron–hole pair generation can add approximately 10% to the classical Shockley–Queisser limit. In quantum dots, *multiple exciton generation* becomes possible and can be used for power conversion efficiency enhancement (Figure 11.2(a)). Owing to outstanding photoluminescent properties, quantum dots can also be used as luminescent concentrators (to substitute costly optical concentrators) enabling efficiency enhancement by converting high-energy photons (2–2.5 eV) into lower-energy ones (1.2–1.5 eV) if the appropriate Stokes shift occurs (Meinardi et al. 2014).

(a) Quantum dot
solar cell

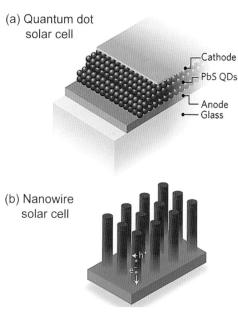

— Cathode
— PbS QDs
— Anode
— Glass

Benefits
Multiple exciton generation
Tunable band gap
Low-temperature fabrication

Challenges
Charge/energy transport through QD array
Monodispersity of QDs
QD long-term stability

(b) Nanowire
solar cell

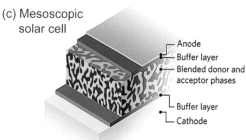

Benefits
Reduction in minority-carrier lifetime
Reduced material usage
Reduced reflectivity

Challenges
Positional stability of dopants
Achieving high areal density
Top contact

(c) Mesoscopic
solar cell

— Anode
— Buffer layer
— Blended donor and
acceptor phases
— Buffer layer
— Cathode

Benefits
Use of low dielectric materials,
use of metal oxide electrodes
Lightweight absorber layers
Cheap

Challenges
Stability of morphology
Photostability of polymer/dyes
Charge transport in polymer phase

Figure 11.2 Three examples of nanostructured solar cells. (a) A quantum dot solar cell allows multiple exciton generation, band gap tuning, and inexpensive fabrication. (b) A nanowire solar cell enable holes (h^+) to be extracted from the outer layer (red) and electrons (e^-) to flow through the core of the nanowire (blue); reflectivity can be reduced. (c) A mesoscopic solar cell suggests an extensive list of materials and low fabrication cost. Reprinted by permission from Macmillan Publishers Ltd.; Beard et al. (2014).

Nanostructures allow gains in reducing the cost of photovoltaic modules owing to the smaller amounts of material used. This is especially important when silicon counterparts are considered. Because of indirect-gap transitions in silicon, silicon cells must be thick to acquire the necessary absorption. Owing to quantum size effects in nanostructures, the list of solar cell materials can be extended by tuning the absorption spectrum of a narrow-band semiconductor to the desirable range of 1.2–1.4 eV according to the Shockley–Queisser prediction (Figure 11.2(a)). For example, PbS quantum dot solar cells are extensively examined in many research centers.

To reduce the amount of material in a solar cell, nanowire structures with a core–annular p–n junction were suggested (Figure 11.2(b)). The p–n junction is made along the length of the wire. When the incident radiation generates electrons and holes, minority carriers only need to traverse the nanowire diameter in order to be collected. Therefore, the rate of the minority-carrier recombination is not as critical as in the bulk cells, and lower-grade silicon can be used.

Nanostructures allow departing from traditional solar cell design when electrons and holes are created in the same semiconductor crystal and then separate and migrate to different contacts. With nanostructures, two different materials, one featuring n-type conductivity and another being of p-type conductivity can be mixed together (Figure 11.2(c)). Then, electron and holes are created in different phases of mixture, their interaction is inhibited, and their separation becomes more probable. This approach is used nowadays in organic and dye-sensitized solar cells. Colloidal quantum dots can be successfully used in this design as sensitizers instead of dyes.

All the above examples were related to electron confinement phenomena. Lightwave confinement can also be used in photovoltaics. One example is antireflective coatings. Because of the high refractive index of semiconductor materials, the monocrystal–air interface features more than 30% reflectance. The photonic crystal (PC) concept may become useful to develop antireflective coatings on top of a solar cell. Confinement (concentration) of incident radiation using metal nanostructures or nanoparticles on top promises enhancement of absorption of radiation, thus enabling picking up the longwave portion of sunlight and reducing cell thickness, which is again a cost-saving solution.

Finally, many types of nanostructures suitable for solar cell fabrication can be commercially fabricated, avoiding costly epitaxial growth techniques without using high vacuum and expensive equipment (a single reactor for epitaxial growth by MOCVD [metal–organic chemical vapor deposition] or MBE [molecular beam epitaxy] costs a few million US dollars). In this context, emergence of colloidal photovoltaics can be foreseen.

In the forthcoming sections the basic principles and research results of nanostructures' application to solar cells are summarized. At the time of writing (2017) all developments in this field are at the laboratory stage.

11.2 COLLOIDAL QUANTUM DOTS IN SOLAR CELLS

There are three major trends to utilize semiconductor quantum dots in solar cells. The first is the *metal–semiconductor junction* photovoltaic cell, often referred to as the *Schottky cell* by analogy with Schottky diodes formed by metal–semiconductor transition rather than a p–n junction. The second is the polymer–semiconductor *hybrid solar cell*. The third is the *quantum dot sensitized* solar cell.

Though colloidal quantum dot solar cells are at the research stage, this type of cell, along with perovskites, represents the most rapidly improving designs that have enabled efficiency growth from 1% (first report by McDonald et al. [2005] at the University of Toronto) to a certified 13% (NREL 2017).

A colloidal quantum dot film forms the core of quantum dot solar cells. Such a film is viewed as a semiconductor material with a dielectric constant, free carrier concentration, and free carrier mobilities. The function of a quantum dot film in a solar cell is to generate electron–hole pairs by absorbing solar radiation. Because of strong Coulomb interaction, electron–hole pairs are often referred to as excitons, though hydrogen-like states inherent in bulk semiconductor crystals do not occur in dots whose size is typically less than the exciton Bohr radius. Here, the principal advantage is band gap tuning by pronounced quantum size effect along with carrier mobility in dense quantum dot ensembles (quantum dot solids) owing to tunneling. Another important phenomenon is the enhanced probability of *multiple exciton generation* (MEG) with a single photon absorbed. MEG occurs by means of an internal impact ionization process promoted by vanishing momentum conservation in a dot. When exciton generation is managed, the design of a cell has to ensure charge separation by bringing electrons and holes to different electrodes. The overall process of electron–hole pair generation and charge separation is characterized by *internal quantum efficiency* (IQE), which measures the portion of photogenerated charges delivered to the device electrodes. IQE reaches 80% in good colloidal quantum dot solar cells. It is deteriorated by fast recombination, low mobility, surface trapping, and other side effects. Table 11.1 gives the best implementation of quantum dot solar cells of various designs reported as of 2016. PbS dominates in the design. Recalling Table 2.3, one can see this material features a narrow intrinsic band gap ($E_g = 0.41$ eV, $\lambda_g = 3020$ nm) and small electron and hole effective masses ($m_e = 0.04\,m_0$; $m_h = 0.034\,m_0$), which allows for absorption tuning over the wide range from a few micrometers to the visible, thus ensuring that the optimal band gap prescribed by the Shockley–Queisser condition is met. Figures 9.25 and 10.6(d) provide an example of the absorption tuning range for PbS

Table 11.1 The best colloidal quantum dot solar cells data for various designs

Design	Materials	Ligand	First absorption peak (nm)	Power conversion efficiency (%)	Reference
Schottky	ITO/PbS/LiF/Al	1,4-benzendithiol	1100	5.2	Piliego et al. 2013
Heterojunction	TiO$_2$/PbS	MPA and CdCl$_2$	950	7.0	Ip et al. 2012
p–i–n	ZnO/PbS/PbS	TBAI/EDT	935	8.6	Chuang et al. 2014
Fused heterojunction	TiO$_2$/PbS	MPA and CdCl$_2$	1000	9.2	Carey et al. 2015

Notes: MPA = mercaptopropionic acid; TBAI = tetrabutylammonium iodide; EDT = ethanedithiol.

quantum dots. However, Pb emergence in commercial photovoltaics may not be welcome, considering its toxicity.

11.2.1 Solar cells based on Schottky structures

In a Schottky device, a transparent conducting oxide with a relatively large work function (indium tin oxide, ITO) interfaces a p-type colloidal quantum dot film (typically PbS). On the bottom side, a low work function metal (e.g., Al) is used to form a charge-separating junction to extract electrons and repel holes. Charge separation is promoted by the built-in field. A sketch and characteristics of the most efficient device are shown in Figure 11.3.

Schottky devices, though being attractive owing to their simplicity, cannot offer active layer structures thicker than 200 nm. In thicker layers, absorption rises up accordingly, but since charge separation occurs far from the junction carrier, extraction becomes inefficient and internal quantum efficiency drops.

11.2.2 Heterojunction devices

A p–n heterojunction based on quantum dot–quantum dot or quantum dot–titania junction allows for higher efficiency compared to the Schottky design. In a heterojunction colloidal quantum dot solar cell, charge separation occurs at the front side of the cell via a junction between the active layer and a large band gap, shallow work function electron acceptor. The electron affinity of the electron acceptor promotes electron extraction with

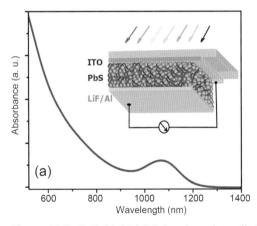
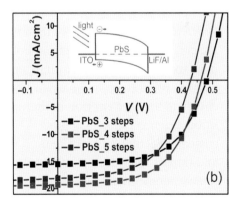

Figure 11.3 Colloidal PbS Schottky solar cell. (a) Absorption spectrum of oleic acid-capped PbS nanocrystals dispersed in chloroform. Inset: the schematic of the photovoltaic device structure with the nanocrystal layer inserted between the ITO and the LiF/Al electrode. (b) Current density–voltage characteristics (J–V) of devices fabricated using the PbS solution washed three, four, and five times. Inset: the diagram of the energy levels of the junction under illumination. Reproduced from Piliego et al. (2013) with permission of the Royal Society of Chemistry.

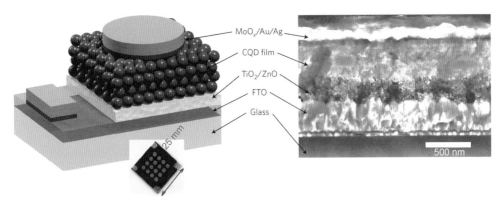

Figure 11.4 A sketch of colloidal heterojunction TiO_2/PbS quantum dot-based solar cell (FTO = fluorine-doped tin oxide; CQD = colloidal quantum dot). Power conversion efficiency is 7%. Reprinted by permission from Macmillan Publishers Ltd.; Ip et al. (2012).

no effect on the open circuit voltage. The ohmic contact design on the back side is also important and typically needs a large work function electrode such as MoO_3. For a heterojunction of TiO_2/PbS with mercaptopropionic acid single-junction power conversion efficiency of 7% has been reported by E. Sargent and co-workers from the University of Toronto (Figure 11.4). Better results can be obtained with p–i–n junctions like (p)quantum-dot/(i)quantum dot/(n)zinc oxide structures (8.6% power conversion efficiency found by Chuang et al. [2014]).

11.2.3 Multiple exciton generation

When photon energy is more than two times higher than the minimal energy necessary for the creation of a single electron–hole pair, the energy conservation rule allows for the two pairs to be created. This may happen owing to *impact ionization*. In this case, a high-energy electron of an electron–hole pair born by absorption of a high-energy photon passes extra energy to another electron in the valence band, thus initiating an interband upward transition (Figure 11.5). This process is the reverse of Auger recombination, which was discussed in Section 8.2 (see Figure 8.8). In bulk semiconductors, both Auger recombination and impact ionization processes need simultaneous conservation of the total energy and (quasi-)momentum. In a quantum dot, the quasi-momentum conservation rule applies because of the lack of translational symmetry and only energy conservation should be met. Thus, both Auger recombination and impact ionization become much more probable in dots versus parent bulk crystals. MEG (also referred to as carrier multiplication) allows for the classical *Shockley–Queisser limit* to be surmounted since this limit is derived with the assumption of a single electron–hole pair per photon at best.

Shaller and Klimov (2004) at Los Alamos National Laboratory were the first to discover MEG and immediately suggested its importance for quantum dot solar cells. The

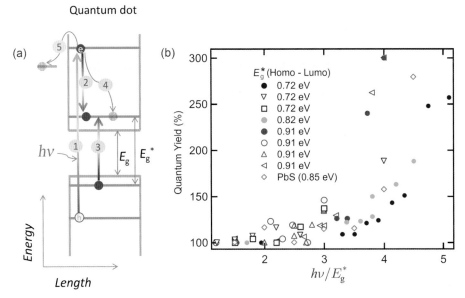

Figure 11.5 Multiple exciton generation by means of impact ionization in narrow-band semiconductor quantum dots. (a) Energy diagram. (b) Experimental data on internal quantum yield for various photon energies versus band gap. Reprinted with permission from Ellingson et al. 2005. Copyright 2005 American Chemical Society. See text for detail.

phenomenon is explained in Figure 11.5. A photon whose energy hv many times exceeds the modified band gap energy E_g^* of a quantum dot creates an electron–hole pair (process "1") with electron kinetic energy higher than E_g^*. Then, if energy relations are perfectly met, the energy conservation law allows that excess energy to be passed to another electron in the valence band to create the second electron–hole pair (processes "2" and "3"). This process, though possible, needs a number of conditions to be met. First, the energy difference between the second and the first electron states in a dot should be equal to the modified band gap energy. Second, the impact ionization should occur much faster than the competitive bypassing processes, namely electron cooling (process "4"), with energy passed to the crystal vibrations, and electron trapping (process "5") by surface states or defects.

Multiple exciton generation has been successfully observed by several groups in narrow-band quantum dots (PbSe, PbS) though its implementation with solar radiation for wider-gap dots (CdSe) was questionable, probably because of the lack of radiation well exceeding the CdSe quantum dot band gap energy E_g^*. Figure 11.5(b) shows an example of experimental results reported for PbSe and PbS dots. Here, quantum yield for exciton formation from a single photon is shown versus photon energy expressed as the ratio of the photon energy to the quantum dot modified band gap (HOMO-LUMO energy) for three PbSe and one PbS quantum dot sizes (diameters 3.9, 4.7, 5.4, and 5.5 nm, respectively, and $E_g^* = 0.91$, 0.82, 0.72, and 0.85 eV, respectively). One can see that two and even three electron–hole pairs can be created by a single absorbed photon. However, calculations and experiments for application of this phenomenon in real solar cells are not so impressive. Several groups reported on evaluated contribution to photocurrent from MEG, but it measures just a few percent from the value defined by single-exciton generation. The main reason is the low content of high-energy photons in solar radiation, the need for a perfect energy match for the processes involved, and presence of competitive processes like electron cooling.

11.2.4 Quantum dots as sensitizers in mesoscopic solar cells

A typical photovoltaic cell converts light energy into electricity by means of light harvesting and charge separation through electron and hole transport in a single material. In 1985, Michael Grätzel and co-workers at École Polytechnique Fédérale at Lausanne (EPFL, Switzerland) introduced another type of photovoltaic cell, a dye-sensitized cell in which the light-harvesting function and charge carrier transport are separated. In this cell, a sensitizer is placed between electron conducting (n-type) and hole conducting (h-type) materials. Upon photoexcitation by sunlight, sensitizing material injects an electron to the conduction band of the n-type adjacent component (a wide-gap semiconducting oxide, e.g., TiO_2) whereas a hole is injected into the sensitizer from a p-type conductor. The charges then diffuse to the front and back contacts to generate electric current in the circuit. The open circuit voltage equals the difference in the Fermi levels of n- and p-type conductors upon illumination. A *redox* electrolyte is used to regenerate sensitizer by hole injection.[1] This type of solar cell is referred to as a mesoscopic cell or *Grätzel cell*. In 1997 dye-sensitized solar cells reached 10% power conversion efficiency and since then only incremental progress has occurred, up to 12% in 2013 (NREL 2017). All successful experimental implementations use TiO_2 powder consisting of 10–100 nm crystallites (Grätzel 2009).

Colloidal quantum dots can be successfully integrated in this type of solar cell to substitute organic dyes as sensitizers (Figure 11.6). By means of quantum size effect their light-harvesting function can be tuned over the visible and near-infrared (IR), which is not possible with typically used dye sensitizers. Most of the quantum dot-sensitized solar cells reported to date use CdSe and CdS nanocrystals deposited on mesoscopic oxide films; the sulfide/polysulfide electrolyte acts as a hole transporter (Kamat 2013). Power conversion of 5% efficiency has been reached in the standard AM 5.1 tests. This means that a lot of research needs to be done toward practical applications. In this context, ternary compounds (for example, $CdSe_xS_{1-x}$ and $CuInS_2$) are seen as promising materials to enhance the efficiency. The critical issue is unwanted loss of electrons through interfacial recombination processes and discharge of electrons at the counter electrode.

---

[1] *Redox* is the short term for *red*uction–*ox*idation reaction, a general notation for reactions involving transfer of electrons from chemical species. The word *oxidation* originally was coined for reaction with oxygen to form an oxide, but later it was extended to all reactions involving loss of electrons. The meaning of *reduction* was also generalized to all processes involving gain of electrons.

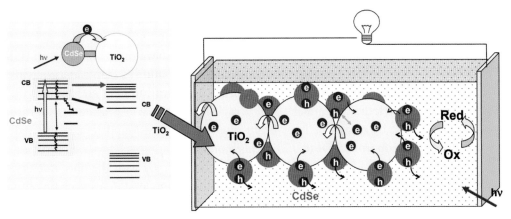

Figure 11.6 Principle of operation of quantum dot-sensitized solar cell. Charge injection from excited CdSe quantum dots into TiO$_2$ nanoparticles is followed by collection of charges at the electrode surface. The redox electrolyte (e.g., sulfide/polysulfide) scavenges the holes and thus ensures regeneration of the CdSe. Reprinted with permission from Kamat (2008). Copyright 2008 American Chemical Society.

11.2.5 Quantum dots as spectral converters

Efficiency of solar cells in many cases can be enhanced by more efficient conversion of shortwave radiation. Sensitivity of typical solar cells features a maximum in the near-IR to meet the Shockley–Quessier condition for highest overall efficiency. Efficiency of typical monocrystalline cells drops for the blue–violet portion of the solar spectrum because of very high absorption coefficient, which leads to absorption of light in a very thin surface layer. Then, charge separation from this layer becomes less efficient because of fast recombination promoted by surface defects. Using spectral converters to convert shortwave radiation into the red and near-IR range enables more efficient "consumption" of the blue–violet portion of the solar spectrum. Semiconductor quantum dots are considered as promising luminescent materials featuring high quantum yield, size-controllable spectral tuning, a wide excitation spectrum, and high photostability. A wide absorption spectrum is the definite advantage of quantum dots versus rare earth-based phosphors. However, the typical emission spectrum of quantum dots represents a relatively narrow band close to the first optical transition with relatively small shift to the red (Stokes shift, see Figure 8.6(a)). Stokes shift can be made much bigger if doping of quantum dots is used by rare earth or transition elements (Eu, Mn). However, quantum yield of doped quantum dots is low. The group from Los Alamos National Laboratory suggested a way to increase Stokes shift in quantum dots (Meinardi et al. 2014). The key issue is using core–shell quantum dots with very thick shells (CdSe/CdS). This allows efficient light conversion from the blue–green to the red (600–700 nm).

11.3 REMOVING REFLECTIONS WITH PERIODIC STRUCTURES

In 1973 P. Clapham and and M. Hutley at Physical Laboratory in Teddington (UK) ingeniously suggested that the regular submicron surface structure of many nocturnal insects' eyes may be evolutionarily designed to minimize reflection (Clapham and Hutley 1973). They fabricated a periodic two-dimensional array on a glass surface using a pattern developed on photoresist by means of laser beam interference, and found remarkable reduction of Fresnel reflection from 5.5% to 0.2%. This biomimetic approach has been further elaborated in detail (Wilson and Hutley 1982) and lately has merged with the photonic crystal paradigm. It is widely used as an efficient antireflection design, first in research on photovoltaics.

Semiconductor crystals used in photovoltaics feature reflection coefficients exceeding 20% and growing to over 30% in the blue because of the high refractive index (see Table 4.2). The traditional approach to antireflective coating is a single layer of a material with refractive index

$$n = \sqrt{n_1 n_2},\qquad(11.1)$$

where n_1 and n_2 are refractive indices of interfacing materials (media). This approach works only for the band of the order of 100 nm centered at $\lambda = n/(4d)$, where d is the layer thickness. It is not possible to compensate for reflection in a wide range even for the glass–air interface with a relative small refractive index step. For example, taking the central wavelength in the green for the glass–air interface in a photocamera results in a pink coloring of a lens since reflection in the red and the blue remains uncompensated. The situation is much more dramatic with a semiconductor–air interface. Using traditional single-layer antireflective coating with a material having refractive index between a semiconductor and air values does not provide an antireflective function in the whole operating range. Usually an antireflective layer is designed to remove reflectance in the range of maximal cell response, around 800–900 nm, whereas growing reflectance in the green and blue remains uncompensated. For example, using silicon nitride coating on silicon enables compensating for reflection in the range 600–900 nm, whereas reflection in the green–blue range progressively grows from 5% to 50% with descending wavelength.

Here, the "moth eye" solution was shown to be efficient. An example of its efficiency for the Si–air interface is shown in Figure 11.7. The pillar periodic structure drops reflection by a factor of ten, and even more over the range of more than 1000 nm! Regular pillars can be developed by etching with the pattern formed by various approaches: (1) using interference of beams on a photoresist; (2) using colloidal self-organized globules as a template for etching, so-called *colloidal lithography*; (3) using nanoporous alumina plate, which in turn is developed by template-free self-organized electrochemical processing. Pillars of ZnO can also be grown directly on a semiconductor surface and have been successfully demonstrated for InGaN (Lin et al. 2011). There are a number of reports of successful applications of the moth-eye design for surfaces of Si (Sun et al. 2008), GaAs (Yu et al. 2009), and GaSb (Min et al. 2008b).

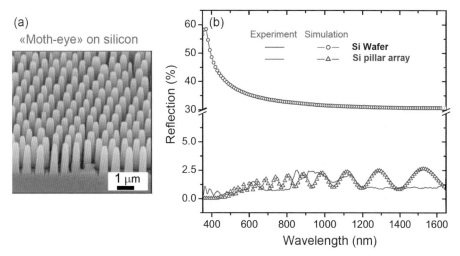

Figure 11.7 Bioinspired moth-eye design of antireflective surface treatment of a silicon wafer. (a) Electron microscopy image of the pillar silicon surface developed by templated etching. (b) Calculated and measured reflectance from the plane silicon wafer and etched pillar structure. Note change in the vertical scale for wafer and pillars in panel (b). Reprinted from Min et al. (2008a), with permission from John Wiley and Sons. Copyright © 2008 WILEY-VCH Verlag GmbH & Co. KGaA, Weinheim.

11.4 INCREASING ABSORPTION WITH METAL NANOSTRUCTURES

The remarkable properties of metal nanostructures to enhance local amplitude of incident electromagnetic radiation is believed to be useful for enhancement of optical absorption for solar radiation, especially in the range where, for example, silicon cell features low absorption because of the indirect character of transitions (wavelength about 1000 nm). The phenomenon occurs in close vicinity to the metal nanotextured surface (50 nm or less) in the spectral range where metal nanostructure features high extinction (sum of absorption and scattering) as was discussed in detail in Section 4.7 (Figure 4.34). It is considered to be especially efficient in thin-film cells. For example, a 2 μm-thick silicon film is missing a valuable portion of the longwave solar spectrum (Figure 11.8).

However, it should be noted that implementation in solar cells is not so easy. First, incident field enhancement occurs in a very small portion of space near a metal particle; second, it is wavelength-dependent; and third, metal proximity brings inevitable losses from intrinsic metal absorption as well as because of metal-induced recombination rate enhancement negatively affecting charge separation efficiency. Nevertheless, there are a number of early promising reports on plasmonic enhancement of solar cell performance. Pryce et al. (2010) achieved three-fold enhancement of a 200 nm-thick InGaN cell in the range 350–400 nm with silver nanoparticles that has resulted in 6% overall enhancement of photocurrent for the solar spectrum. Kholmicheva et al. (2014) reported on 12.5% power conversion efficiency improvement for PbS colloidal solar cells with 5 nm gold nanoparticles. There are many other reports which, however, are related to cheap solution-processed

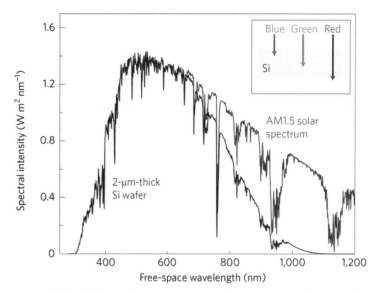

Figure 11.8 AM1.5 solar spectrum, together with a graph that indicates the solar energy absorbed in a 2 μm-thick crystalline Si film (assuming single-pass absorption and no reflection). Clearly, a large fraction of the incident light in the spectral range 600–1100 nm is not absorbed in a thin crystalline Si solar cell. Reprinted with permission from Macmillan Publishers Ltd.; Atwater and Polman (2010).

cells with low intrinsic efficiency below 10% (Arinze et al. 2016). There seems to be a lack of data on the positive effect of metal nanostructures on highly efficient semiconductor crystalline solar cells, probably because of unwanted effects on charge transport processes.

11.5 CHALLENGES AND OUTLOOK

According to the National Renewable Energy Laboratory data on progress in power conversion efficiency of solar cells, quantum dot cells represent a rapidly growing trend with the highest rate of improvement, namely 13 times in 12 years from 1% in 2005 to 13% in 2017 (Figure 11.9). This rapid efficiency improvement competes with that of solar cells based on perovskites and organic ones, whereas more traditional Si- and GaAs-based cells demonstrated only incremental progress in efficiency. In the two possible paths toward higher competiveness of solar cells – higher efficiency and lower price – colloidal quantum dot solar cells should be explored in the context of the lower-price approach rather than the high-efficiency one. They have well-defined favorable physical characteristics (band gap tuning, multiple exciton generation, antireflective solutions, plasmonic enhancement of absorption) that suggest successful progress toward cheap nanostructured solar cells, but a lot of research is still required to push nanostructured solar cells to the market.

However, as often happens, favorable physical processes of nanostructures exist alongside less favorable ones. For example, MEG needs lower band gap materials and

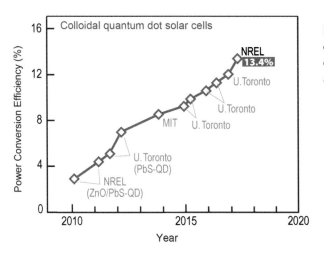

Figure 11.9 Progress in power conversion efficiency of research-grade colloidal quantum dot solar cells according to NREL (2017) data.

strict energy conservation conditions; development of an antireflective surface structure can deteriorate IQE through undesirable surface defects; and metal nanoparticles used to enhance absorption by means of local incident field enhancement bring additional losses from their intrinsic absorption of solar radiation. The currently dominating lead salts used as colloidal solar cell materials will probably be replaced by ternary compounds to exclude the toxic Pb content in mass production. Therefore, nanophotonics for photovoltaics will be an active field of research in the future decade and probably will bring cheap colloidal solar cells to the market.

Conclusion

Nanostructures have not had a strong impact on photovoltaics to date, but nanophotonics is seen as a field that can contribute to commercial solar cell improvement in the near future. There are several ways nanostructures can be involved in photovoltaics. First, colloidal thin-film solar cells with tunable absorption spectrum and MEG can be used to develop cheap cells with efficiency of the order of 10% or more. Colloidal quantum dot solar cells offer the solar cell research sector the most improvement in efficiency compared to other cells during last decade. Second, metal nanostructures can be used to increase thin-film solar cells efficiency by increasing the portion of solar energy absorbed by a cell near the absorption onset. This is a field of very active research. Third, periodic nanostructures on top of a solar cell are efficient in their antireflection effect across a wide spectral range, which is not possible with a multilayer design. The challenges are to bring nanostructure-based solar cells from the research to the industrial level. Most probably, nanostructures will mainly contribute to progress in cheap cells with moderate efficiency for massive applications in daily life.

Problems

11.1 Explain the origin of the ultimate limit for solar energy conversion.

11.2 Recall the ways to increase power conversion efficiency for solar cells.

11.3 Explain what multiple exciton generation is and why it cannot raise power conversion efficiency considerably.

11.4 Recalling the physical reason for quantum size effects, explain why small electron and hole masses mitigate wide-band gap tuning for PbS quantum dots.

11.5 Outline the case in which colloidal nanostructures can be efficient for photovoltaics.

11.6 Explain options and obstacles for plasmonics application to photovoltaics.

Further reading

Beard, M. C., Luther, J. M., Semonin, O. E., and Nozik, A. J. (2012). Third generation photovoltaics based on multiple exciton generation in quantum confined semiconductors. *Acc Chem Res*, **46**, 1252–1260.

Borchert, H. (2014). *Solar Cells Based on Colloidal Nanocrystals*. Springer.

Chattopadhyay, S., Huang, Y. F., Jen, Y. J., et al. (2010). Anti-reflecting and photonic nanostructures. *Mater Sci Eng, R*, **69**, 1–35.

Kim, J. Y., Voznyy, O., Zhitomirsky, D., and Sargent, E. H. (2013). 25th anniversary article: colloidal quantum dot materials and devices – a quarter-century of advances. *Adv Mater*, **25**, 4986−5010.

Kramer, I. J., and Sargent, E. H. (2014). The architecture of colloidal quantum dot solar cells: materials to devices. *Chem Rev*, **114**, 863–882.

Otnes, G., and Borgström, M. T. (2016). Towards high efficiency nanowire solar cells. *Nano Today*, **12**, 31–45.

Polman, A., Knight, A., Garnett, E. K., Ehrler, B., and Sinke, W. C. (2016). Photovoltaic materials: present efficiencies and future challenges. *Science*, **352**, 307–318.

References

Arinze, E., Qiu, B., Nyirjesy, G., and Thon, S. M. (2016). Plasmonic nanoparticle enhancement of solution-processed solar cells: practical limits and opportunities. *ACS Photonics*, **3**, 158–173.

Atwater, H. A., and Polman, A. (2010). Plasmonics for improved photovoltaic devices. *Nat Mater*, **9**, 205–213.

Beard, M. C., Luther, J. M., and Nozik, A. J. (2014). The promise and challenge of nanostructured solar cells. *Nat Nanotechnol*, **9**, 951–954.

Carey, G. H., Levina, L., Comin, R., Voznyy, O., and Sargent, E. H. (2015). Record charge carrier diffusion length in colloidal quantum dot solids via mutual dot-to-dot surface passivation. *Adv Mater*, **27**, 3325−3330.

Chuang, C.-H. M., Brown, P. R., Bulović, V., and Bawendi, M. G. (2014). Improved performance and stability in quantum dot solar cells through band alignment engineering. *Nat Mater*, **13**, 796–801.

Clapham, P. B., and Hutley, M. C. (1973). Reduction of lens reflexion by the "moth eye" principle. *Nature*, **244**, 281–282.

Ellingson, R. J., Beard, M. C., Johnson, J. C., et al. (2005). Highly efficient multiple exciton generation in colloidal PbSe and PbS quantum dots. *Nano Lett*, **5**, 865–871.

Grätzel, M. (2009). Recent advances in sensitized mesoscopic solar cells. *Acc Chem Res*, **42**, 1788–1798.

Green, M. A., Emery, K., Hishikawa, Y., et al. (2017). Solar cell efficiency tables (version 49). *Prog Photovoltaics: Res Appl*, **25**, 3–13.

Ip, A. H., Thon, S. M., Hoogland, S., et al. (2012). Hybrid passivated colloidal quantum dot solids. *Nature Nanotechn*, **7**, 577–582.

Kamat, P. V. (2008). Quantum dot solar cells: semiconductor nanocrystals as light harvesters. *J Phys Chem C*, **112**, 18737–18753.

Kamat, P. V. (2013). Quantum dot solar cells: the next big thing in photovoltaics. *J Phys Chem Lett*, **4**, 908–918.

Kholmicheva, N., Moroz, P., Rijal, U., et al. (2014). Plasmonic nanocrystal solar cells utilizing strongly confined radiation. *ACS Nano*, **8**, 12549–12559.

Lin, G. J., Lai, K. Y., Lin, C. A., Lai, Y.-L., and He, J. H. (2011). Efficiency enhancement of InGaN-based multiple quantum well solar cells employing antireflective ZnO nanorod arrays. *IEEE Electron Device Lett*, **32**, 1104–1106.

McDonald, S. A., Konstantatos, G., Zhang, S., et al. (2005). Solution-processed PbS quantum dot infrared photodetectors and photovoltaics. *Nat Mater*, **4**, 138–142.

Meinardi, F., Colombo, A., Velizhanin, K. A., et al. (2014). Large-area luminescent solar concentrators based on "Stokes-shift-engineered" nanocrystals in a mass-polymerized PMMA matrix. *Nat Photonics*, **8**, 392–399.

Min, W.-L., Jiang, B., and Jiang, P. (2008a). Bioinspired self-cleaning antireflection coatings. *Adv Mater*, **20**, 1–5.

Min, W.-L., Betancourt, A. P., Jiang, P., and Jiang, B. (2008b). Bioinspired broadband antireflection coatings on GaSb. *Appl Phys Lett*, **92**, 141109.

Nozik, A. J. (2008). Multiple exciton generation in semiconductor quantum dots. *Chem Phys Lett*, **457**, 3–11.

NREL (2017). Best research-cell efficiency chart. Available at https://www.nrel.gov/pv/assets/images/efficiency-chart.png (accessed June 22, 2017).

Piliego, C., Protesescu, L., Bisri, S. Z., Kovalenko, M. V., and Loi, M. A. (2013). 5.2% efficient PbS nanocrystal Schottky solar cells. *Energy Environ Sci*, **6**, 3054–3059.

Pryce, I. M., Koleske, D. D., Fischer, A. J., and Atwater, H. A. (2010). Plasmonic nanoparticle enhanced photocurrent in GaN/InGaN/GaN quantum well solar cells. *Appl Phys Lett*, **96**, 153501.

Schaller, R. D., and Klimov, V. I. (2004). High efficiency carrier multiplication in PbSe nanocrystals: implications for solar energy conversion. *Phys Rev Lett*, **92**, 186601.

Sun, C.-H., Jiang, P., and Jiang, B. (2008). Broadband moth-eye antireflection coating on silicon. *Appl Phys Lett*, **92**, 061112.

Wilson, S. J., and Hutley, M. C. (1982). The optical properties of "moth eye" antireflection surfaces. *Optica Acta*, **29**, 993–1009.

Yu, P., Chang, C. H., Chiu, C. H., et al. (2009). Efficiency enhancement of GaAs photovoltaics employing antireflective indium tin oxide nanocolumns. *Adv Mater*, **21**, 1618–1621.

Emerging nanophotonics

In this chapter a number of challenging trends in modern nanophotonics are highlighted that can be traced based on extensive research during the last decades. These are the colloidal technological platform, nanoplasmonics to enhance light–matter interaction, novel optical sensors based on nanostructures, advances toward the silicon photonic platform, negative refractive index materials, and single photon emitters. These novel trends and ideas, when seen together, provide a definite forecast for new exciting devices and systems to appear in the next decades and present nanophotonics as an extremely active and promising field of research and development.

12.1 COLLOIDAL TECHNOLOGICAL PLATFORM

The core of the colloidal nanophotonics platform is formed by the quantum confinement phenomena in solution-processed nanocrystals of semiconductor compounds (see Section 3.5). For many decades, semiconductor nanocrystals have been used unintentionally in color cutoff filters and have been the subject of academic studies in solutions, including photophysical and photochemical processes. This period is labeled in Figure 12.1 as the "pre-quantum period." It was followed by a burst of activity in the 1980s when two groups identified systematic size-dependent optical properties of nanocrystals in glasses (Ekimov and Onushchenko 1981, 1984) and in solutions (Brus 1983). At the same time, Efros and Efros (1982) suggested the simple particle-in-a-box approach model that has been improved by accounting for electron–hole Coulomb interaction (Brus 1984; also see Box 3.2). These ideas triggered the new field of photonics, which has resulted in the emergence of new technological platforms. The role of quantum confinement effects has been identified in commercial color glass cutoff filters (Borelli et al. 1987; Zimin et al. 1990).

Since this time, colloidal nanophotonics have become a mature field of research and development. In the context of materials used, CdSe, CdTe, CdS, InP, and PbS, PbSe, dominated, with II–VI and III–V ones considered as promising compounds for visibly emitting devices and lead salts looking important for infrared (IR) applications. Hg salts have also been involved as IR materials. Recently, this list has been complemented by

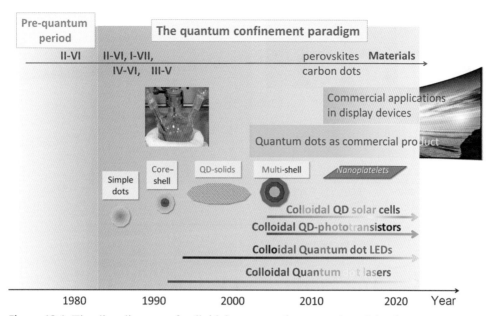

Figure 12.1 Timeline diagram of colloidal quantum dot research and development.

unexpected but already extensive studies of carbon dot (Reckmeier et al. 2016) and perovskite (Docampo and Bein 2016; Gonzalez-Carrero et al. 2016) nanocrystals. This is shown in Figure 12.1 in the "Materials" line. Perovskites nanocrystals feature well-defined size-dependent absorption and emission spectra, high quantum yield, and lasing under optical excitation (Figure 12.2).

In the context of nanocrystalline structures, there are a few important steps to be outlined. The first experiments with as-grown or capped nanocrystals in solutions were followed by core–shell structures with high quantum yield and multiple or gradient shells for higher photostability. In 1995 the important notion of *quantum dot solids* was coined by M. Bawendi and co-workers (Murray et al. 1995), and evolution from individual to collective electron states was demonstrated (Artemyev et al. 1999). These ideas served as important prerequisites for research toward quantum dot-based photodetectors. Lately, colloidal quantum dots have been complemented by nanoplatelets representing a colloidal counterpart to epitaxial quantum wells in double heterostructures, with an important advantage of stronger electron–hole coupling owing to interaction via ambient polymer material with low dielectric permittivity (Achtstein et al. 2012). Within a few years, nanoplatelets were found to feature strong efficient luminescence, strong electroabsorptive behavior, and optical gain (Achtstein et al. 2014; Guzelturk et al. 2014b).

Soon after 2000, colloidal quantum dots in solutions became available as a commercial product (Quantum Dot Corporation, QD Vision, Evidot, Sigma-Aldrich) and a decade afterwards several companies announced their application as color-converting phosphors in handheld tablets, cell phones, and TV sets (e.g., Sony, Samsung). These decisive events

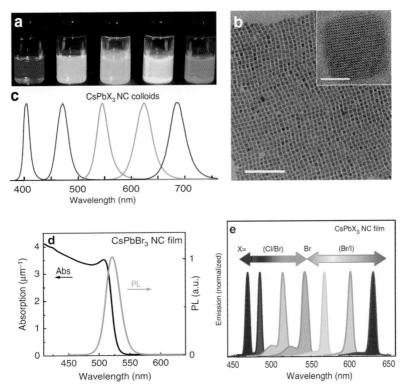

Figure 12.2 Perovskite colloidal quantum dots: novel light-emitting nanostructures. Data for cesium lead halide perovskite nanocrystals are presented: (a) stable dispersions in toluene under excitation by an ultraviolet lamp (365 nm); (b) low- and high-resolution transmission electron microscopy images of $CsPbBr_3$ NCs, corresponding scale bars are 100 and 5 nm; (c) photoluminescence emission spectra of the solutions shown in (a); (d) optical absorption and photoluminescence spectra of a $CsPbBr_3$ nanocrystalline film; (e) spectral tunability of amplified spontaneous emission via compositional modulation. Reprinted from Yakunin et al. (2015) under CCA 4.0 International License.

are indicative of the emerging technology platform and are shown in Figure 12.1 by the two color boxes.

Many groups over the world continue to work on elaboration of various photonic devices based on colloidal quantum dots. For colloidal photovoltaics the idea of multi-exciton generation (MEG) was of crucial importance (Schaller and Klimov 2004; Ellingson et al. 2005), as has been the approach to use quantum dot sensitizers for light harvesting (Kamat 2008).

BOX 12.1 SCIENTISTS SHAPING THE FUTURE OF COLLOIDAL PHOTONICS

Initiated by A. Ekimov, A. Efros, and L. Brus in the 1980s, photophysics of colloidal nanostructures exhibiting quantum size effects evolved into a mature field of research and development, resulting in existing and emerging applications in TV sets, tablets, LEDs, solar cells, lasers, and photodetectors. A number of teams headed by outstanding leaders are today shaping the future of the colloidal technological platform in photonics. Here, a few of them are presented and many are working with and around them in different countries.

A. P. Alivisatos   M. Bawendi   V. Klimov   P. Kamat

A. Nozik   A. Rogach   D. Talapin   N. Halas

Optical gain for nanocrystals in glasses was first observed by V. Klimov and co-workers in 1991 at Moscow State University and confirmed by several groups, but the undesirable Auger recombination prevented development of reliable laser sources. A decade ago, the ingenious idea of single-exciton optical gain in type II quantum dots advanced by Klimov et al. (2007) made optically pumped multicolor lasing with colloidal quantum dots a reality.

The history of colloidal quantum dot LEDs dates back to 1994 (Colvin et al. 1994) and is marked by remarkable progress owing to multi-shell and gradient shell structures improving photostability (Su et al. 2016 and references therein).

Recently discovered perovskite nanocrystals and II–VI nanoplatelets have been immediately identified as laser materials. Optical gain and luminescence band narrowing as a lasing prerequisite have been observed (Wang et al. 2015; Yakunin et al. 2015).

Table 12.1 **Colloidal nanostructures in photonics**

Photonic component/technology	Current status	Type of colloids used			Sample reference
		Semiconductor	Dielectric	Metal	
Filters	Commercial	+		+	Borelli et al. 1987; Woggon 1997; Gaponenko 1998
Luminophores for display devices	First commercial applications	+	+		Shirasaki et al. 2013; Guzelturk et al. 2014a
Luminophores for lighting	Active research	+	+		Erdem and Demir 2013; Guzelturk et al. 2014a; Su et al. 2016
LEDs	Active research	+		+	Wood and Bulović 2010; Demir et al. 2011; Bae et al. 2013; Su et al. 2016
Lasers	Active research	+			Klimov et al. 2007; Dang et al. 2012; Kovalenko et al. 2016
Solar cells	Active research	+	+	+	Kamat 2008; Sargent 2012
Electro-optical devices	Research	+			Achtstein et al. 2014
Photodetectors	Research	+		+	Konstantatos and Sargent 2009; Talapin et al. 2009; Chen et al. 2014
Sensors and biocompatibility	Active research	+		+	Lesnyak 2013; Zenkevich et al. 2013; Palui 2015
Single-photon emitters	Research	+			Lodahl et al. 2015
Nanolithography	Research		+		Vogel et al. 2012
Hyperbolic metamaterials	Research	+		+	Zhukovsky et al. 2014

In many cases the device performance with colloidal quantum dots as the core components can be further improved using metal colloidal nanoparticles, which bring a realm of plasmonic phenomena into action. This is shown in a separate column in Table 12.1 and will be the subject of Section 12.2. In certain cases, colloidal dielectric particles can be used. These are important for solar cells (e.g., TiO_2, ZnO nanoparticles) and are also shown in a column in Table 12.1.

Colloidal dielectric nanostructures of a few hundred nanometers in size can be further involved in devices as photonic crystals (PCs) (see Section 4.3). Various chemical techniques can readily form luminescent molecules (Petrov et al. 1998), quantum dots (Gaponenko et al. 1998), and rare earth ions (Gaponenko 2001). Dielectric microcavities with embedded emitters, templates for plasmonics nanostructures (Figure 12.3), and masks for lithography to develop subwavelength periodic patterns can be implemented with dielectric colloidal structures as well.

Various applications of colloidal nanocrystals to LEDs, lasers, and solar cells can be seen in Chapters 8, 9, and 11. Therefore, in what follows we briefly discuss the *photodetector* issue, which has not been covered hitherto.

Colloidal quantum dot photodetectors have become a field of extensive research as a promising novel sensing platform for high-sensitivity, low-cost photodetectors from the ultraviolet (UV) to the shortwave and mid-IR spectrum. Its concept is essentially based on quantum dot solids with controllable spacing between dots provided by inorganic ligands. Inorganic ligands typically provide colloidal stability in polar solvents, which is needed for solution-based fabrication of electronic and optoelectronic devices. At the same time, inorganic ligands do not block electron transport, enabling efficient solution-processed photodetectors. Ligands could form conductive "bridges" between NCs to facilitate charge transport. Sometimes, ligand change from organic to inorganic is performed during synthesis.

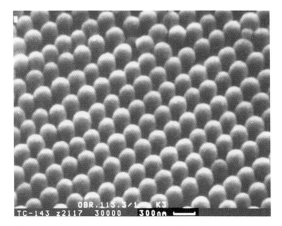

Figure 12.3 Gold-coated periodic arrangement of 250 nm silica spheres (a colloidal crystal layer), a template for surface enhanced spectroscopy (fluorescence and Raman scattering) and colloidal nanolithography.

Figure 12.4 shows the structure and the performance of a photodetector based on a CdSe/CdS colloidal quantum dot solid with nanocrystals capped with $In_2Se_4^{2-}$. The energy-level diagram (Figure 12.4(b)) shows easy transport of photogenerated electrons between the nanocrystals, whereas holes are confined within CdSe cores. The combination of highly mobile electrons and trapped holes provides high internal photoconductive gain. The detector responsivity (signal current per input radiation power, A/W) depends on applied voltage

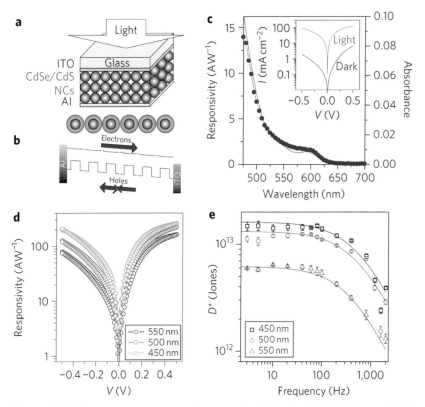

Figure 12.4 Charge transport and photoconductivity in CdSe/CdS core–shell nanocrystal films. (a) Device structure. (b) Energy-level offsets in CdSe cores and CdS shells. (c) Responsivity measured at 200 mV bias (red circles) compared to the absorption spectrum (blue). Inset: $I–V$ characteristics measured in the dark and under illumination with 450 nm light, 0.75 mW/cm². (d) Responsivity of a 60 nm-thick layer of $In_2Se_4^{2-}$-capped CdSe/CdS core–shell nanocrystals with 2.9 nm CdSe cores and 2.6 nm-thick CdS shells at different wavelengths measured at 5 Hz light modulation frequency. (e) Frequency dependence of the normalized detectivity D^*, measured under 21 V for a device using $In_2Se_4^{2-}$-capped CdSe/CdS core–shell nanocrystals. The 3 dB bandwidth is 0.4 kHz. Reprinted by permission from Macmillan Publishers Ltd.; Lee et al. (2011).

and correlates remarkably with the absorption spectrum. Photodetectors feature specific detectivity $D* > 10^{13}$ Jones (see Box 12.2 for detail, and Nudelman [1962] for an introduction to photodetector characterization). The approach does not require high processing temperatures and can be extended to different nanocrystals and inorganic surface ligands.

For various applications, near-IR and mid-IR detectors are used, which are made of cooled monocrystalline semiconductors and are very expensive. A highly sensitive night-vision camera can cost US\$50,000. Colloidal nanocrystals look promising for cheap IR detectors. For nanocrystal-based detectors, either gapless or narrow-band semiconductor compounds (like HgS, HgTe, PbS, or PbSe) should be used. Konstantatos and Sargent (2009) reported on a PbSe-based detector with a spectral response in the range 800–1500 nm and peak normalized detectivity $D* = 2 \cdot 10^{13}$ Jones. To go deeper in the IR, Deng et al. (2014) suggested intraband HgSe detectors using photoinduced current within discrete electron states in the c-band. A spectral response in the range 3.3–5 μm has been obtained, though the absolute sensitivity data are not provided.

In Section 4.7 we saw that metal nanoparticles feature a strong incident local field enhancement, the spectral range of enhancement correlating with the extinction spectrum. The latter in turn can be tuned to longer wavelengths by means of nanoparticle size and shape. An elongated shape shifts the extinction spectrum to the longer wavelengths. Chen

BOX 12.2 PHOTODETECTOR SENSITIVITY

At first glance, a photodetector responsivity R (ampere/watt) measured as current, I, per unit input radiation power, W, can serve as the principal characteristic of any detector. However, detection of impinging external radiation is possible only if radiation-induced current I exceeds the noise current, I_{noise}, i.e., if *signal-to-noise ratio* (SNR),

$$SNR = R\frac{W}{I_{noise}},$$

is more than 1. The related power figure of merit is noise equivalent power, $\text{NEP} = \dfrac{I_{noise}}{R}$. When impinging radiation power equals NEP, the SNR equals unity. A detector is described by *detectivity*, $D = 1/(\text{NEP})$ measured in W^{-1}. NEP describes a device but not the material it is made of. NEP is proportional to $(S\Delta f)^{1/2}$, where S is a detector surface and Δf is its frequency bandwidth. To characterize a detector *material* property, *specific detectivity* $D*$ (also referred to as *normalized detectivity*) is the figure of merit. It reads:

$$D^* = D\sqrt{S\Delta f} = \sqrt{S\Delta f} \, / \, (\text{NEP})$$

and the relevant unit is termed *Jones* after R. C. Jones, who introduced this parameter in 1953. 1 Jones = cm $\cdot$ Hz$^{1/2}$ $\cdot$ W^{-1}.

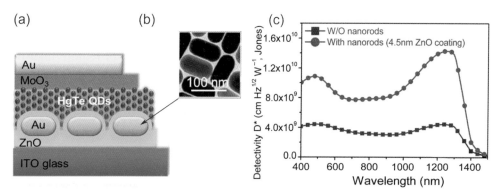

Figure 12.5 Infrared HgTe quantum dot-based photodetector with Au nanorods to enhance detectivity owing to the plasmonic effect of incident field enhancement. (a) The design; (b) Au nanorods SEM image; (c) detectivity with and without nanorods. Reprinted with permission from Chen et al. (2014). Copyright 2014 American Chemical Society.

et al. (2014) suggested that elongated Au nanoparticle (nanorods) would enhance incident local radiation intensity for near-IR and proposed using this effect to achieve higher photodetector sensitivity owing to stronger radiation absorption. These authors successfully implemented the approach in a HgTe quantum dot-based photodetector (Figure 12.5). The metal-induced enhancement factor was more than two-fold.

It is important that the colloidal paradigm allows for the problem of lattice matching in epitaxial growth to be completely eliminated. Thus, colloidal light emitters and detectors can be integrated with other technological platforms; primarily, this is important for Si-based circuitry.

12.2 NANOPLASMONICS

Plasmonic phenomena do not constitute a technological platform since these cannot suggest novel devices. However, plasmonic phenomena essentially alter the properties of photonic devices resulting from enhanced light–matter interaction promoted by proximity of a metal. Additionally, plasmonics offer a way to possible miniaturization of waveguiding components and cavities owing to the shorter wavelength inherent in surface plasmon polaritons. Therefore, step by step, plasmonics can contribute to advanced lasers, LEDs, sensors, and photodetectors, as well as to optical circuitry in data transfer/routing/switching. In this section we highlight the impact of plasmonics on nanophotonic devices development.

12.2.1 Enhancement of light–matter interaction

In Section 4.7 we saw that metal nanostructure enhances drastically incident electromagnetic radiation in the spectral range corresponding to noticeable extinction (absorption plus scattering). At the same time, proximity of metal nanotextured surfaces promotes higher density of photon states (DOS) and faster nonradiative and radiative dissipation of energy stored in an excited quantum system (atom, molecule, quantum dot, and even micrometer-size crystallites). These effects essentially modify light–matter interaction, including absorption, emission, and scattering of light, as well as a photonic device response time. Intensity-dependent nonlinear processes (lasing, absorption saturation, second and higher harmonics generation, photochemical and photothermal processes) will be affected as well. This is shown in Table 12.2, where positive and negative impacts of metal nanostructures are summarized for different processes. Consider these effects in more detail.

Table 12.2 **Plasmonic enhancement of light–matter interaction and its possible effects on device performance**

Photoprocess	Plasmonic effect		
	Local field enhancement	Density of states (radiative decay rate) enhancement	Nonradiative rate enhancement
1. Scattering (elastic, inelastic, i.e., Raman and Mandelstam–Brillouin)	+	+	0
2. Photoluminescence	+	+	–
3. Electroluminescence (intensity)	0	+	–
4. Electroluminescence (modulation rate)	0	+	+
5. Photovoltaics	+	–	–
6. Photoinduced processes (e.g., photoionization)	+	–	–
7. Photostability	–	+	+
8. Photodetector (sensitivity)	+	–	–
9. Photodetector (response rate)	0	+	+
10. Photothermal action	+	–	+
11. Nonlinear optics: second harmonic generation	+	0	0
12. Nonlinear: absorption saturation	+	–	–
13. Nonlinear: lasing threshold decrease	+	–	–

Note: "+" positive influence; "–" negative influence; "0" no effect.

1. **Scattering** All scattering phenomena in optics will be necessarily enhanced by plasmonic structures. First, scattering is proportional to the incident radiation intensity that is enhanced, and second, scattering in quantum optics is viewed as the result of virtual excitation of a quantum system and subsequent prompt (without time delay) emission of a new photon with the same (elastic scattering) or shifted (inelastic scattering) frequency. Photon emission into a mode with frequency ω is directly proportional to the density of photon states $D(\omega)$ similar to spontaneous emission of photons by an excited quantum system (see Gaponenko (2010) and Gaponenko and Guzatov (2009) for more detail). Since the scattering process occurs via virtual excitation, it goes instantaneously and the known nonradiative decay enhancement by metal nanostructures has no effect.

2. **Photoluminescence** has two promoting factors and one quenching factor. It can be enhanced under certain conditions when incident field enhancement promoting higher excitation rate is not overcome by nonradiative decay enhancement, resulting in quenching. We showed in Section 5.9 that under certain conditions considerable gain in photoluminescence intensity can be obtained up to 10^3 times for emitters with low intrinsic quantum yield. A representative example of 30-fold enhancement is presented in Figure 12.6.

3. **Electroluminescence intensity** is favored by radiative decay enhancement, which, however, competes with luminescence quenching, the incident field enhancement having no effect. We have discussed the possibility of positive radiative/nonradiative processes tradeoff to get the net gain in electroluminescence intensity for materials with intrinsic quantum efficiency less than 1 (see Section 8.6).

4. **Electroluminescence modulation rate** will always be faster, i.e., modulation performance will become better, but special care is to be taken to remain at intensities that are possible only for materials with intrinsic quantum efficiency less than 1 (see Section 8.6). An enhanced modulation rate is important for data transfer using lighting sources (Li-Fi).

5. **Photovoltaics** has only one promoting plasmonic factor, i.e., incident intensity enhancement resulting in stronger radiation absorption. However, the two factors of radiative and nonradiative decay enhancement can diminish this positive influence because faster electron–hole recombination will not allow electrons and holes to be efficiently separated in a photovoltaic cell. That is why the reports on positive influence of metal nanoparticles on solar cell sensitivity are very rare (see Section 11.4 for detail).

6. **Photoinduced processes** (e.g., photoionization) experience the same effects as photovoltaic cells, i.e., promotion from enhanced incident radiation intensity deteriorated by the two factors resulting in enhanced excited state decay rate, e.g., electron–hole recombination rate. The latter forms a bypass to photoinduced processes like photoionization.

7. **Photostability** of quantum dot LEDs and luminophores, as well as that of quantum well LEDs, is believed to be deteriorated by high population-induced Auger processes resulting

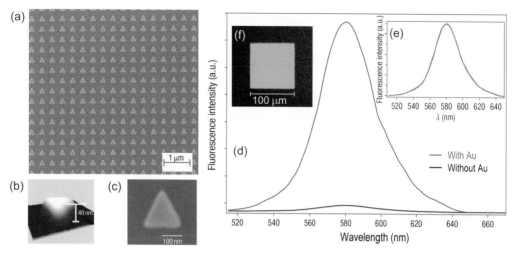

Figure 12.6 Enhanced photoluminescence of CdSe/ZnS core–shell quantum dots dispersed in a polymer blend on top of a regular Au nanopattern. (a) SEM image of a portion of periodic Au pattern consisting of triangular pyramids; (b,c) AFM and SEM images of a single triangle; (d) photoluminescence intensity spectra of quantum dots with (red) and without (black) Au pyramids; (e) normalized emission spectra with (red) and without (black) metal pattern; (f) video-image of luminescent CdSe/ZnS quantum dots. The bright central square 100×100 μm^2 contains a metal triangle pattern, whereas the darker area around it does not. Reprinted by permission from Macmillan Publishers Ltd.; Pompa et al. (2006).

either in nonradiative decay or even in photoionization. All Auger processes are bypassed by enhanced decay (recombination) rate, both radiative and nonradiative. Therefore, plasmonics can enhance photostability of light-emitting devices provided that a reasonable balance between radiative and nonradiative decay enhancements can be retained. This issue has not been thoroughly examined to date and needs systematic studies since it looks promising for improvement of durability of quantum dot-based emitters and prevention of efficiency droop at high current for quantum well LEDs.

8. **Photodetector sensitivity** can be enhanced owing to enhanced absorbance, but is deteriorated by the two competing factors resulting in enhanced recombination rate preventing efficient charge separation. An example of successful tradeoff of these factors was presented in Figure 12.5 for a HgTe quantum dot-based photodetector. Another representative example is using an array of metal nanoparticles (nanoantenna) to generate carriers in Schottky photodiodes (Knight et al. 2011).

9. **Photodetector response rate** (bandwidth) will always be improved by plasmonic effects, but often at the expense of lower sensitivity (see above).

10. **Photothermal action** (e.g., in laser medicine or material processing) will be typically enhanced by plasmonic effects (two promoting factors versus one deteriorating factor).

11–13. Nonlinear phenomena. All nonlinear processes featuring nonlinear dependence on incident radiation intensity will be enhanced though absorption saturation and lasing may be deteriorated by faster decay times. However, in the case of absorption saturation, the faster Q-switching and more efficient mode-locking will become possible owing to faster decay (recombination) rates. Enhanced second harmonic generation was successfully observed in numerous experiments in the 1980s. Absorption saturation and lasing influenced by plasmonic phenomena have not been systematically studied.

12.2.2 Lightwave confinement beyond the wavelength scale

In addition to the above metal-induced modification of optical and photophysical processes, plasmonics promises ultra-dense optical data recording and read-out owing to deep subwavelength focusing as well as guiding electromagnetic waves within tapered metal structures, well beyond the diffraction limit. Here, the challenging issue is minimizing or avoiding undesirable losses. Subwavelength optics with metal nanostructures forms an active field of research (Gramotnev and Bozhevolnyi 2010). A transition from classical or semi-classical consideration toward quantum plasmonics suggests further interesting phenomena (Törmä and Barnes 2015). The subwavelength scale of light focusing and guiding is important for better integration of electronic and optical counterparts in optoelectronic chips, which is still rather questionable because of the different wavelength scales for electrons in semiconductors and electromagnetic radiation.

12.3 SENSORS

Optical properties of nanostructures combined with fine surface chemistry allow for multiple applications as various sensors based on optical signal detection. These sensors can be broken down into at least two big groups. One group is based on recognition of target molecules, e.g., antigens in human blood. Another group is based on identification of environmental parameters, e.g., the presence of heavy metal atoms, pH level, or refractive index of liquids.

12.3.1 Bioconjugates

The notion of bioconjugate is of principal importance in understanding the prospects for colloidal quantum dots as biosensors. Bioconjugate is a colloidal nanocrystal covalently linked to a biomolecule (Figure 12.7(a)). It becomes possible by means of additional chemical groups attached to the nanocrystal surface using various strategies, provided that the mandatory condition of solubility in water is met. These linking groups also should

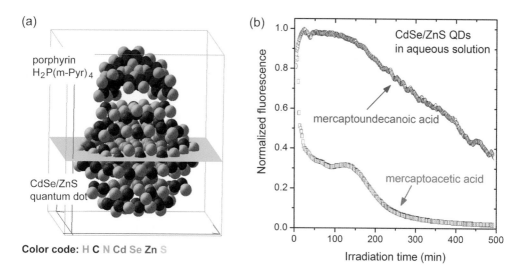

(a)

porphyrin
H$_2$P(m-Pyr)$_4$

CdSe/ZnS
quantum dot

Color code: H C N Cd Se Zn S

(b)

CdSe/ZnS QDs
in aqueous solution

mercaptoundecanoic acid

mercaptoacetic acid

Normalized fluorescence

Irradiation time (min)

Figure 12.7 Quantum dot bioconjugates. (a) Quantum chemical simulation of a CdSe/ZnS core–shell quantum dot bound to a porphyrin H$_2$P(m-Pyr)$_4$ molecule via an n-trioctylphosphine oxide (TOPO) molecule. Courtesy of D. S. Kilin. (b) Photoluminescence intensity of semiconductor colloidal nanocrystals in water with different organic compounds under prolonged irradiation by a continuous wave laser (5 mW/cm^2 at 532 nm).

not interfere with capping groups or shells, which are necessary for nanocrystal stability, durability, and high photoluminescence yield, since it is the photoluminescence signal that is typically used in detection schemes. Chemical aspects of bioconjugate synthesis are beyond the scope of this book and can be retrieved from topical reviews (Medintz et al. 2005; Resch-Genger et al. 2008).

12.3.2 Fluorescent labels

Fluorescent labeling of biomolecules is widely used in immunoassay analyses. The *immunoassay* approach is based on detection of specific biomolecules called *antigens* in blood serum. Antigens are produced in correlation with development of a malignant tumor in a human body and therefore are termed *tumor markers*. A number of markers have been identified for various cancer types and the search for possible new ones is an active field of biomedical studies. Antigens are present in healthy organisms but their concentration grows significantly in the case of tumors. High antigen concentration cannot offer a precise diagnosis, but gives a hint to thorough patient examination including biopsy probing. Thus, systematic large-scale monitoring of antigens in human blood is a good practice to reveal early stages of cancer. Antigens can be recognized by specially synthesized or identified molecules called *antibodies*. In turn, an antibody can be labeled with a small but strongly luminescing dye molecule(s) such as fluorescein. Then, antigen concentration can be traced *in vitro* by fluorescence intensity measurements from labeled antibodies linked to the antigen.

Quantum dots are considered to be promising fluorescent labels to substitute dyes. This idea was advanced in 1998 based on the chemical feasibility and a number of advantageous features of colloidal quantum dots versus dyes (Bruchez et al. 1998; Chan and Nie 1998). These advantages were discussed in Section 8.2, namely higher photostability, narrow emission spectrum, and wide excitation spectrum. High photostability is illustrated in Figure 12.7(b). Colloidal nanocrystals exhibit bright emissions for many hours under continuous wave (CW) laser irradiation, whereas the organic fluorescent dyes degrade considerably in a few minutes under the same conditions. It allows for lower levels and precise concentration detection of emitting nanocrystals in biosolutions. A narrow emission spectrum favors higher signal-to-noise values. A wide excitation spectrum allows for a number of probes to be excited by the same light source which, along with a narrow emission spectrum, allows for multiplexing, i.e., probing of a number of tumor markers in a single run using different colors for different antibodies.

Colloidal quantum dot fluorescent labels can also be used in various bioimaging applications, including *in vivo* imaging during surgery, where again their luminescent properties are very useful.

12.3.3 Raman labels

Raman scattering is inelastic photon scattering by molecules and solids viewed as virtual excitation of a quantum system with immediate emission of another photon whose frequency is either lower or higher than the incident light frequency by the frequency of atomic vibrations forming chemical bonds. Atoms in molecules and solids experience continuous vibrations, with frequency depending on atomic mass and bond strength, similar to mass–spring pendulums. Heavier atoms feature lower frequencies. Every molecule and solid exhibits a characteristic set of intrinsic vibration frequencies resulting in a characteristic Raman spectrum, i.e., intensity of scattered light versus detuning from incident light frequency, typically presented on a $1/\lambda$ [cm^{-1}] scale. The Raman spectrum is considered as a "fingerprint" of substances and forms a solid field of analytical spectroscopy. Organic molecules and especially biological ones have tens of lines which sometimes are difficult to detect and resolve. Semiconductors possess just a few characteristic lines that can be readily identified. Therefore, semiconductor nanocrystals linked to antibodies can serve as Raman labels in immunoassays. There have been a few attempts to commercialize small organic molecules as Raman labels for immunoassay, but without commercial success, mainly because of extremely low Raman scattering probability (several orders of magnitude lower than luminescence), which results in expensive instrumentation (for Raman spectrometers, the price scale is of the order of US$100,000) and qualified personnel. Plasmonics may turn the idea of Raman markers into real practice. Molecules and quantum dots adsorbed on a nanotextured metal surface or mixed with metal nanoparticles in a solution experience

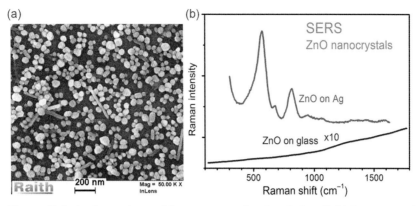

(a)

(b)

Figure 12.8 Surface enhanced Raman scattering for ZnO colloidal quantum dots on a substrate covered with dense Ag nanoparticles with average diameter about 50 nm. (a) substrate image; (b) Raman intensity spectrum at excitation by a He–Ne 1 mW laser (632 nm).

giant enhancement of Raman scattering intensity owing to incident field enhancement and local density of states enhancement (see Table 12.2). The phenomenon gained the notation SERS (surface enhanced Raman scattering). Experiments demonstrated the possibility of single-molecule detection by SERS (Kneipp et al. 2006). Figure 12.8 shows a representative example of strong enhancement of Raman scattering from colloidal quantum dots adsorbed on an Ag-coated substrate (Rumyantseva et al. 2013).

12.3.4 Plasmonics for fluorescent sensing

In Section 5.9 we saw that metal nanoparticles and nanotextured surface can enhance photoluminescence of molecules; the example of nine-fold enhancement of fluorescein labeled protein was presented (Figure 5.20). Much higher enhancement can be performed with regular nanotextured surfaces (Figure 12.6). In a similar way, the fluorescent signal enhancement from antibodies labeled either with organic dyes or with quantum dots can be implemented. Therefore, immunofluorescent testers can be made cheaper and more compact for routine use in medical practice.

Another example of plasmonics application in fluorescence-based sensing is possibly glucose detection (important for people suffering from diabetes) using competitive binding of aminophenylboronic acid-functionalized CdSe colloidal quantum dots, either with glucose or with mercaptoglycerol-modified Ag nanoparticles in a solution (Tang et al. 2014). Binding with Ag nanoparticles enhances quantum dot fluorescence, whereas binding to glucose does not. Therefore, fluorescence intensity falls with increasing glucose concentration. This approach is a clear example of efficient use of both physical and chemical techniques in sensor development.

12.3.5 Plasmonics for Raman sensing

Plasmonic enhancement of Raman scattering can reach 10^{14} times in certain "hot spots" between metal nanoparticles, enabling single-molecule detection in model samples in which an extremely low concentration of molecules presents in extremely pure solution, which evaporates after molecule adsorption. However, in routine practice the target molecules (say, toxic ones in food or water probes) should be detected in the presence of other substances that are often unknown. Notably, it is very difficult to enhance Raman scattering selectively and typically many species presented in a probe show enhanced scattering. Moreover, fluorescent background may also be enhanced. These circumstances create the problem of high signal-to-noise ratio along with signal enhancement to ensure enhanced detectivity (lower detection level) for a substance in question. The overall signal enhancement allows using cheaper detectors but deteriorated signal-to-noise ratio may lead to worse detectivity. It is for this reason that SERS techniques have not found commercial applications to date, despite being known for several decades (since 1974). Nowadays an approach is followed by many groups searching for specific reactions for each analytical problem to make use of enhanced Raman signals in plasmonic structures. In this approach, the formation of new chemical bond(s) is detected, which for purposeful detection needs just one or a few lines in correlation with the specific chemical agent added to the probe. In a sense, this approach is similar to the antibody-based technique in biomedicine. The representative examples are as follows. A new substrate material comprising a poly(styrene-*co*-acrylic acid) core covered with a highly packed Fe_3O_4/Au nanoparticle shell has been proposed as an efficient SERS substrate for the detection of ultra-trace amounts of TNT (Mahmoud and Zourob 2013). Lignin was introduced to obtain high selectivity of detection due to the proven affinity of lignin toward trinitrotoluene (TNT) detection. Antimony detection has been suggested by SERS based on binding to phenylfluorone (Panarin et al. 2014). Kulakovich et al. (2016) have proposed a SERS approach for detecting bromate (a potentially carcinogenic agent) concentration in desalinated drinking water based on catalytic oxidation of Rhodamine 6 G (R6 G) dye. In this technique, R6 G is added to a water probe and the R6 G Raman signature is then detected by SERS using colloidal silver particles. A detection limit well below the World Health Organization safety level has been reported.

Plasmonic enhancement of Raman scattering has been used for decades in experiments with organic molecules exclusively, whereas inorganic species have not been examined. However, it has been shown that inorganic nano- and microcrystallites also experience Raman scattering enhancement when mixed with or adsorbed on metal nanoparticles. High Raman scattering enhancement for nanocrystals (colloidal quantum dots) has even been suggested as a base for Raman markers using bioconjugates (Rumyantseva et al. 2013). Surprisingly, considerable Raman enhancement with silver or gold nanostructures has been observed for micrometer-sized inorganic crystallites in powders in spite of the fact that plasmonic effects rapidly fall with distance and vanish at distances greater than 50 nm (Klyachkovskaya et al. 2011; Shabunya-Klyachkovskaya et al. 2016). These experiments have been successfully extended toward examination of cultural heritage items since pigments in art paintings are typically inorganic crystallites (CdSSe as orange–yellow, HgS as red, ZnO as white, cobalt green, malachite green, ultramarine blue, and many others). These authors have also demonstrated successful application of Au- and Ag-coated self-organized SiGe dots on Si as efficient SERS substrates, thus paving the way for a cheap Si-based approach to SERS sensors.

12.3.6 FRET-based fluorescent quantum dot sensors: colloidal quantum dots

Förster resonance energy transfer (FRET), discussed in detail in Chapter 7, forms the basis for development of various sensors using colloidal quantum dots. Since quantum dots feature a wide absorption spectrum, it is reasonable to use quantum dots as donors rather than acceptors. Figure 12.9 shows the principle of a so-called "switch-on" sensor. A colloidal quantum dot is linked to a dye serving as a photoluminescence quencher owing to efficient FRET. An analyte competes with the quenching dye and substitutes it by chemical binding with a dot. Then photoluminescence recovers, its intensity being in proportion to the analyte concentration in a solution. This approach was used by Goldman et al. (2005) to develop a TNT detector based on CdSe/ZnS quantum dots in water with an oligohistidine tag serving as the TNT antibody.

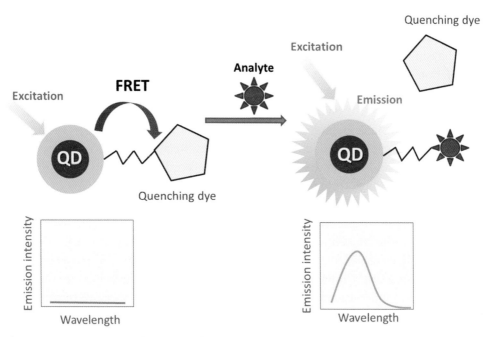

Figure 12.9 A "switch-on" concept of a FRET quantum dot sensor. Photoluminescence of a colloidal quantum dot is quenched by the purposefully linked dye molecule serving as an efficient acceptor. Analyte molecules in solution compete with quenching dye and link to quantum dots. Their photoluminescence then recovers.

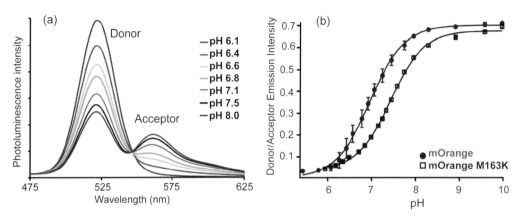

Figure 12.10 Detection of pH levels with functionalized colloidal quantum dots conjugated with pH-sensitive fluorescent protein (mOrange or mOrange M163 K). (a) Photoluminescence spectra of dots (donor) and protein (acceptor) depending on pH level. (b) Ratio of donor to acceptor emission intensities versus pH for two conjugated pH-sensitive proteins. Reprinted with permission from Dennis et al. (2012). Copyright American Chemical Society.

Another approach to quantum dot FRET sensors is control of spectral overlap between quantum dot emission and an acceptor absorption spectrum. The latter can be sensitive not only to the chemical compounds in solution but also to the properties of the microenvironment, including pH level. Since a few quantum dots can be detected using the microfluorescence technique, intracellular pH sensing becomes feasible. Cell pH level is indicative of physiological processes and pathophysiology of cells, including cancers. Thus routine detection of pH level in cells is topical. Figure 12.10 shows an approach implemented by Dennis et al. (2012). Using carbodiimide chemistry, carboxyl-functionalized colloidal quantum dots have been conjugated with pH-sensitive fluorescent protein mOrange and its more photostable homolog mOrange M163 K. Change in pH shifts the protein absorption spectrum, thus modulating FRET efficiency. Relative weights of donor/acceptor (dot/protein) photoluminescence intensities change accordingly. The approach shows functionality in the pH range from 6 to 8.

Heavy metal contamination is a serious issue for human health. Certain heavy metal ions such as Pb^{2+}, Hg^{2+}, and Cd^{2+} are not biologically essential and are harmful to humans, even at very low concentration levels. Some heavy metals (e.g., Fe, Cu, Mn, and Zn) are nutritionally essential for health. Therefore, tracing heavy metal ions is important and a number of organic dye-based metal ion probes have been developed. In recent years, colloidal quantum dot-based sensors have become a field of extensive research. Depending on specific surface chemistry-related processes, interaction of metal ions with quantum dots can either quench or enhance fluorescence. Generally, the direct interaction of metal ions with quantum dots is a very complicated process and often it is not element-specific. Therefore, sensors based on direct linking of heavy metal ions to quantum dots are not the main trend. Instead, dyes sensitive to specific heavy metal ions are used conjugated to

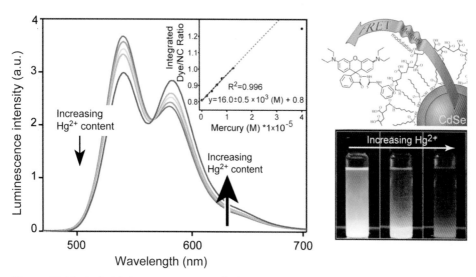

Figure 12.11 Colloidal quantum dot Hg^{2+} fluorescent sensor using Hg^{2+}-sensitive dye conjugated to CdSe/ZnS core–shell quantum dot. Reproduced from Page et al. (2011) with permission of the Royal Chemistry Society.

fluorescent colloidal quantum dots and, as in the above examples, presence of the target ions in solution modulates fluorescence intensity via FRET efficiency. An example is shown in Figure 12.11 for mercury ion detection. Ratiometric detection of toxic Hg^{2+} ions is demonstrated using a semiconductor nanocrystal (CdSe/ZnS core–shell structure) energy transfer donor coupled to a mercury-sensitive "turn-on" dye acceptor (thiosemicarbazide functionalized rhodamine B). The dye reacts with Hg^{2+} ions to form HgS, which is highly insoluble in water. This prevents quantum dot luminescence quenching by the metal. At low Hg^{2+} content the green emission band from quantum dots dominates because of small donor–acceptor emission–absorption spectral overlap. With growing mercury content the dye emission grows, whereas dot emission falls. This happens owing to the desulfurization reaction that creates HgS and also results in the restoration of the dye's absorption and emission properties, causing it to act as a good energy acceptor for quantum dots.

12.3.7 Plasmonics for refractive index sensing

Refractive index n of liquid or polymer samples can be monitored using the dependence of the extinction spectrum of metal nanoparticles dispersed therein on n. It is of practical importance since the refractive index value gives information on protein concentration in blood serum, its value for intracellular liquid is important in biomedical research, and data on refractive index and its change are important in many technological applications. Recalling Figure 4.32 and comparing it with Figure 5.17, one can see that the extinction spectrum of metal nanoparticles dispersed in a dielectric medium strongly depends on n.

Consider the simplest case of very small metal nanoparticles (much smaller than optical wavelengths) for which size- and shape-dependent scattering can be ignored. Suppose, then, that the metal dielectric function obeys the simple form inherent in the case of ideal electron gas (see Eq. (4.52)):

$$\varepsilon(\omega) = 1 - \frac{\omega_p^2}{\omega^2}. \tag{12.1}$$

Then, if such nanoparticles are dispersed in a medium with dielectric permittivity $\varepsilon_m = n^2$, using the resonance condition $\varepsilon = -2\varepsilon_m$ (see, e.g., Gaponenko 2010; see also Section 6.3) we arrive at the expression for the extinction maximum wavelength,

$$\lambda_{max} = \lambda_p \left(2n^2 + 1\right)^{1/2}, \tag{12.2}$$

where λ_p is wavelength corresponding to ω_p. The function (12.2) is plotted in Figure 12.12(a) for the reasonable value of $\lambda_p = 300$ nm. It is important that for the rather wide and practically important range of n value this function obeys a linear law. Therefore, a reasonable figure of merit for plasmonic refractive index sensors is *sensitivity S*, expressed as

$$S = \frac{d\lambda_{max}}{dn}(\text{nm/RIU}), \tag{12.3}$$

where RIU stands for refractive index units. One can see that for the selected λ_p sensitivity $S = 380$ nm/RIU, allowing for refraction index changes of the order of 0.01 measurements provided that extinction maximum can be evaluated from experimental spectra with the accuracy of order of 1 nm.

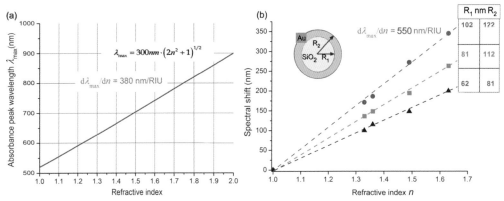

Figure 12.12 Principle of plasmonic refractive index sensors. (a) Peak wavelength λ_{max} versus refractive index n calculated in the approximation of very small metal particles (Eq. (12.2)) for $\lambda_p = 300$ nm. (b) Experimentally observed shift of the extinction maximum for silica–gold nanoshells with different core–shell sizes in different environments (Tam et al. 2004; air, $n = 1.003$, water, $n = 1.33$, ethanol, $n = 1.36$, toluene, $n = 1.49$, and carbon disulfide, $n = 1.63$). Dots are experimental data; lines are guides for the eye.

Relation (12.2) is valid for nanoparticles no bigger than 20 nm. For larger particles their size should be explicitly taken into account and extinction should necessarily involve analysis of scattering. Shape in this case will also matter. Nanorods, nanoshells, and other nanoparticles with complex shape (e.g., stars) with size 50–100 nm were found to feature higher sensitivity to ambient refractive index than the simple prediction by Eq. (12.2) (Sun and Xia 2002; Tam et al. 2004; Miller and Lazarides 2005; Liao et al. 2006; Mayer and Hafner 2011; Zhu et al. 2013). Figure 12.12(b) shows a representative example of high-sensitivity spectral shift of extinction peak wavelength to ambient refractive index obtained with core–shell dielectric–metal structures (nanoshells). One can see the slope $d\lambda_{max}/dn$ grows with size and the values of $d\lambda_{max}/dn > 500$ nm/RIU are realistic. For nanoshells, Jain and El-Sayed (2007) have shown that for the same dielectric core diameter, a thinner metal shell offers higher sensitivity to n variation.

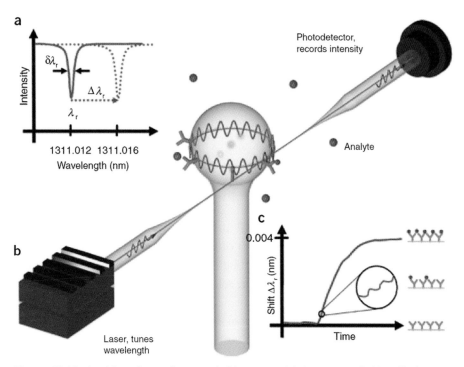

Figure 12.13 A whispering gallery mode biosensor. (a) Resonance is identified at a specific wavelength from a dip in the transmission spectrum acquired with a tunable laser. The resonance wavelength, λ_r, is measured by locating the minimum of the transmission dip within its linewidth (arrows). A resonance shift associated with molecular binding, $\Delta\lambda_r$ is indicated by the dashed arrow. (b) Whispering gallery mode in a dielectric sphere driven by evanescent coupling to a tapered optical fiber. The lightwave (red) circumnavigates the surface of the glass sphere (green) where binding of analyte molecules (purple) to immobilized antibodies (blue) is detected from a shift of the resonance wavelength. (c) Binding of analyte is identified from a shift $\Delta\lambda_r$ of resonance wavelength (also see (a)). Single-molecule binding is theorized to appear for the "reactive" effect in the form of steps in the wavelength shift signal (magnified inset). Reprinted by permission from Macmillan Publishers Ltd.; Vollmer and Arnold (2008).

12.3.8 Whispering gallery modes in sensing applications

Whispering gallery modes develop in a microsphere, microdisk, microring, microcylinder, and microtoroid. Whispering gallery modes feature typically very high Q-factor, i.e., very high photon local density of states (LDOS). High LDOS and Q-factor promotes efficient coupling of radiation propagating in a guide near a whispering gallery mode object, provided that its frequency meets the whispering gallery mode resonance value. Coupling occurs by means of evanescent wave (light tunneling) and can also be treated as enhanced photon scattering from a guide into a high-LDOS radiation mode (i.e., the whispering gallery mode). Coupling is very sensitive to the accurate resonance condition. Therefore, if whispering gallery mode frequency of a microsphere or another object supporting the whispering gallery mode regime is disturbed by an environmental change, coupling efficiency will be severely relaxed. This property has been proposed for sensing of refractive index changes in liquids as well as detecting adsorbance of certain molecules presenting therein. The concept of a whispering gallery mode sensor is explained in Figure 12.13.

A microdisk or a microsphere laser itself can be used as a sensor since a laser cavity Q-factor changes depending on its environment. He et al. (2011) implemented a whispering gallery mode microlaser for detection of single viruses and nanoparticles in solutions by using an ultra-narrow linewidth whispering gallery microlaser, whose lasing frequency spectrum split upon the binding of individual nano-objects. Influenza A virion detection has been demonstrated by monitoring changes in lasing modes.

12.4 SILICON PHOTONICS

Silicon is the dominant electronic material and the estimate of spending for Si electronics R&D during the last decades is of the order of trillions of US dollars. Silicon features a high refractive index ($n = 3.4$), enabling efficient lightwave confinement in microcavities and PCs, waveguiding, multiplexing/demultiplexing, selective filtering (as was shown in Chapters 4 and 10) using e-beam lithography, and advanced anisotropic etching. It is therefore a challenge to extend the Si platform to photonic devices as far as possible. Silicon technology offers efficient photovoltaic cells, fine detectors for the range from 1000 to 400 nm (regretfully no Si detector for 1.3–1.5 μm!), but no emitter because of the indirect-gap transitions inherent in Si. Extensive attempts to "squeeze" light from silicon nanostructures have not resulted in a commercial light-emitting device (Pavesi et al. 2012) and even under strong confinement regimes optical transitions in silicon have been shown to be indirect. Unexpectedly bright photoluminescence was attributed to statistical lack of defects (Gaponenko et al. 1994). The latter means that emission comes from a small portion of nanocrystals within a statistically big ensemble which are free of quenching centers (trap states) but since the portion of such nanocrystals is always small the overall ensemble-averaged quantum yield will always be low. Since silicon waveguides, microcavities, and other guiding and confining components are feasible, in this section we consider activity toward light emitters, detectors, and modulators based on or compatible with the Si platform.

12.4.1 Hybrid III–V on Si structures

In Chapters 6 and 9 we saw that the main materials for semiconductor lasers are III–V compounds (InP for 1–2 µm, GaAs for the 0.8 µm range, InAs quantum dots for 1.3 µm, and GaN-based for the near-UV and blue–violet range). Since lasers cannot be made of Si, it is reasonable to consider possible integration of III–V semiconductors on silicon wafers. There are several approaches, namely flip-chip integration, bonding approaches, and hetero-epitaxial growth. In the flip-chip approach, devices are grown on the native substrate and then need precise alignment on a silicon wafer in the assembly process. Hetero-epitaxial growth needs accurate lattice matching. For example, a 1.3 µm InAs quantum dot laser epitaxially grown on silicon with low thresholds (16 mA), high output power (176 mW), and high temperature lasing (up to 119 °C) has been reported

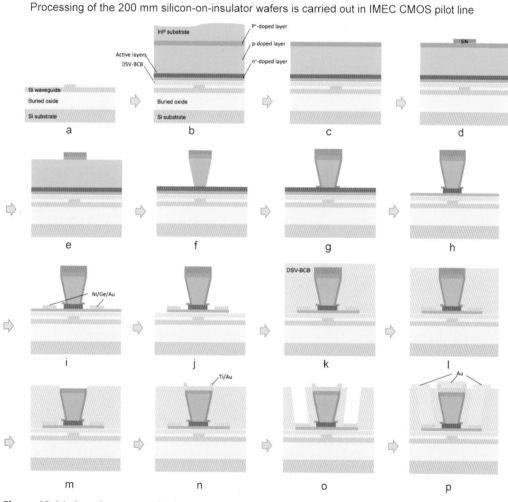

Figure 12.14 Step-by-step typical process flow for heterogeneously integrated III–V on Si devices (IMEC, Belgium). See text for detail. Reprinted from Roelkens et al. (2015) under CCA license.

(Liu et al. 2014, University of California at Santa Barbara). The bonding technique uses growth of the active component on the native substrate, but needs the appropriate bonding agent (molecular, metal, adhesive). An example of GaAs 840 nm VCSEL (vertical cavity surface-emitting laser) fabricated by adhesive bonding was shown in Figure 9.7. The representative example of the process flow based on adhesive bonding for 200 mm silicon-on-insulator (SOI) wafer processing is shown in Figure 12.14 (CMOS pilot line at IMEC, Belgium). The adhesive is DVS-BCB (divinylsiloxane-bis-benzocyclobutene) and the thickness is <50 nm. The process flow breaks down into the following steps:

1. Si wafer with waveguide structure is made;

2. InP structure is bonded and thermally cured;

3. the InP substrate is removed using HCl:H_2O until the InGaAs stop layer is reached;

4. SiN hard mask is deposited by PECVD (plasma-enhanced chemical vapor deposition) and the waveguide pattern is transferred on top;

5. the waveguide pattern is etched into the p+ contact layer by ICP RIE (inductively coupled plasma reactive ion etching);

6. wet chemical etching is used to develop an inverted-trapezoid-shaped waveguide;

7. SiN (~200 nm) is deposited and patterned by optical lithography and RIE;

8. etching of the active layers using H_2SO_4:H_2O_2:H_2O;

9. Ni/Ge/Au contacts are evaporated onto the InP n+ contact layer;

10. the n+ contact layer between neighboring devices is etched using HCl·H_2O;

11. DVS-BCB is spin-coated on top of the wafer to planarize the top surface;

12. thinning the DVS-BCB layer by RIE;

13. removal of the SiN layer by RIE;

14. the metal p-contact layer (Ti/Au) is formed on top of the p+ contact;

15. lithography and etching into the DVS-BCB layer down to the n-metal contact;

16. Au is deposited on top of the p- and n-metal contacts.

Optical coupling of a III–V gain section to a Si substrate is implemented by means of the tapered guides approach. Strong wave vector mismatch between the III–V gain section and silicon waveguide layer can be minimized by tapering either III–V or Si guiding sections, or both. If waveguide dimensions vary slowly along the propagation direction, no energy exchange occurs between the fundamental and the higher-order waveguide modes.

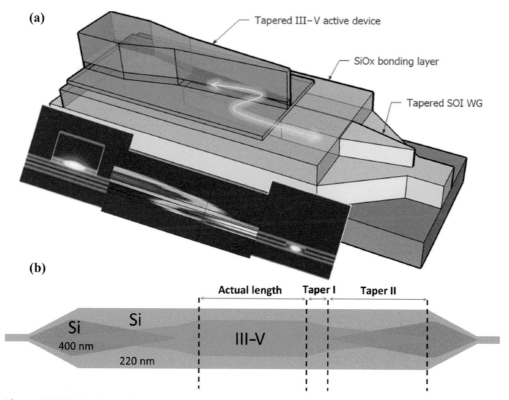

Figure 12.15 (a) Three-dimensional view of the coupling structure in the gain section with representative mode profiles in two cross-sections. (b) Detailed top view of the gain structure. Reprinted from Roelkens et al. (2015) under CCA license.

In this case, tapering is referred to as adiabatic. Group III–V–Si hybrid laser with double adiabatic coupling is shown in Figure 12.15.

12.4.2 Coupling of Si waveguiding structure to a III–V photodetector

Radiation confined in Si circuitry can be diffracted vertically to a photodetector (Figure 12.16). In this approach it is important to ensure efficient diffraction in the vertical direction. Adding an antireflective coating between the DVS-BCB layer and the III–V layer on top results in more efficient coupling (design B).

12.4.3 Epitaxially grown GeSn photodetectors

In spite of existing advanced Si photodetectors for the visible, including high-sensitivity high-resolution CCD cameras (Nobel Prize in 2009 to Willard S. Boyle and George E. Smith), the main communication bands at 1.3 and 1.5 μm are not covered because optical absorption of Si is negligible for wavelengths longer than 1.1 μm. Ge can be epitaxially grown on an Si wafer and Ge detectors can extend the operation range toward 1550 nm,

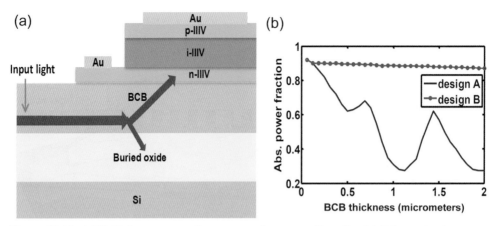

Figure 12.16 A III–V detector coupling to a grating on an Si wafer. (a) Schematic view of grating coupler interface; (b) simulation example of grating-based coupling to a III–V photodetector. Design B has an antireflective coating while design A does not. BCB stands for the benzocyclobutene-based adhesive. Reprinted from Roelkens et al. (2015) under CCA license.

though sensitivity falls because of the band gap threshold (recall Figure 2.13 for band structures and band gaps of Si and Ge). With a few percent content of Sn, $Ge_{1-x}Sn_x$ solid solution gains a direct gap slightly lower than the intrinsic Ge indirect gap and $Ge_{0.96}Sn_{0.04}$ thin-film detector epitaxially grown in Si (p–i–n photodiode) was shown to feature three times higher sensitivity at 1550 nm versus the pure Ge-on-Si analog (Werner et al. 2011). This approach is essentially based on heterostructures, though ideas of quantum size effects or lightwave confinement are not yet involved. The GeSn layers on an Si substrate feature tensile strain (up to +0.34%), which lowers the difference between direct and indirect band transition and makes this method even more promising for obtaining direct band gap Ge-based layers.

12.4.4 Ge photoluminescence promoted by predefined tensile strain

Jain et al. (2012) implemented externally induced tensile strain for Ge on SiO_2/Si structures to lower direct-gap transition energy and thus to make it dominant over indirect-gap ones. These authors reported on the 260-fold overall photoluminescence intensity enhancement and 130-fold enhancement of the 1550 nm emission band fitting the principal optical communication band.

12.4.5 Modulators

There is extensive activity toward development of epitaxial fast electro-optical Si-based or Si-compatible modulators using various light-confinement ideas. The challenging target parameters are low energy consumption (1 fJ per bit) and high speed (1 Tb/s).

The general paradigm of electro-optical modulators is based on conversion of small changes in a material refractive index into strong changes in transmission of an optical component containing this material. Refractive index change may result from direct current-induced modulation of carrier density (carrier injection or depletion) through electron or hole plasma dispersion. This effect is similar to free electron gas dielectric function discussed in Chapter 4 in the context of metal reflection (see Figure 4.25), but with typical concentrations about 10^{18} cm^{-3}, i.e., four orders of the magnitude lower than in metals. High free carrier density results in small, <0.1%, change in refractive index n in the wide spectral range. Then, material heating from the current flow as well as the finite recombination time defining the time response will become issues. One way to get a refractive index modulation is to use the electric field effect on absorption (see Chapter 10) via Stark or Franz–Keldysh effects. Here, the main problem is to fit the desirable spectral range since refraction index changes only at the absorption edge, i.e., material band gap should fit the optical communication range(s). Ge quantum wells were suggested a decade ago for the 1550 nm range (Kuo et al. 2005). This approach seems to be free from the heating problem and has no limits in time response beyond the capacity-related restrictions of a real multilayer thin-film design. Nevertheless, a semiconductor structure, when external electric field is applied to and light shines thereon with the wavelength corresponding to interband absorption (otherwise electroabsorption effect is negligible), acts to a certain extent as a photodetector generating photocurrent, and therefore results in heating and power consumption.

To convert small refractive index change into strong transmission modulation, a number of optical solutions are used, each of them using light interference and/or tunneling phenomena, which are extremely sensitive to perfect resonance tuning (see also Section 10.3): a Fabry–Pérot interferometer with the cavity material refractive index control; a Mach–Zehnder interferometer with one "shoulder" containing material with refractive index control; a waveguide coupled to a microring or microdisk resonator whose resonant wavelength is tuned via material refractive index control; a Mach–Zehnder interferometer with one shoulder coupled to a microring or microdisk resonator whose resonant wavelength is tuned as was discussed above.

There is extensive activity in the field so that every attempt to give a comprehensive overview may appear obsolete at the time of publication. However, we have to highlight that all activities resemble a search for new approaches to get higher performance numbers (fast operation >100 Gb/s, low energy consumption about 1 fJ, multi-channel and multi-wavelength options, small size) provided that every approach should achieve CMOS compatibility. Among recent publications, the following should be highlighted: mode division multiplexing (MDM) in waveguides in addition to wavelength division/multiplexing (WDM) was suggested by Stern et al. (2015); Mach–Zehnder interferometer combined with a microring cavity to enhance phase shift induced modulation (Guha et al. 2010); thermally tuned microring resonator arrays for light modulation (Absil et al. 2015); sharp reduction in free carrier lifetimes from nano- to picosecond ranges in Si nano-waveguides (Turner-Foster et al. 2010).

12.5 QUANTUM OPTICS

The realm of quantum optics gives rise to the notions of quantum information, quantum teleportation, and quantum computing. This is an exciting and challenging field of modern science that promises unprecedented phenomena with unusual practical applications that had been never considered before. It can be viewed as quantum electrodynamics combined with atomic physics and merged with nanophotonics. At first glance, we do believe each of us adheres to quantum theory when considering light–matter interaction phenomena, by separating these into matter excitation and relaxation owing to energy exchange with the electromagnetic field. However, this is still a semi-classical consideration even if matter is considered to absorb and emit electromagnetic energy in portions termed photons. It is quantum electrodynamics which a priori treats electromagnetic radiation in terms of photons and describes electromagnetic field states based on the photon concept rather than wave vector, frequency, or intensity. The thorough discussion of quantum optics ideas, experimental proofs, and consequences is beyond the scope of this book and the interested reader is referred to the topical works by Kira and Koch (2011), Duarte (2014), and Yamamoto, et al. (2000). In what follows we briefly highlight a few non-trivial phenomena related to quantum optics.

12.5.1 Shot noise

When a coin is tossed a number of times, what will be the number of heads or tails? The higher the number of tosses, the closer are both probabilities to 0.5. When photons enter a detector input aperture, what will be the variance of the counts from the real photon numbers? The theory gives $n^{1/2}$ deviation for n photons entering the input aperture provided that arrival of every photon does not depend on the number of photons that arrived earlier. This result is a consequence of the *Poisson distribution*, a function $P(n)$ describing the probability to get n events in the unit time interval in the sequence of independent events with average rate of N events per unit time. It reads,

$$P(n) = \frac{N^n e^{-N}}{n!} \tag{12.4}$$

and is applicable for integer N and n.

The Poisson function is plotted in Figure 12.17. For higher N, the Poisson distribution merges with the normal (Gaussian) distribution with the mean and variance of N. The noise in signal detection arising from the Poisson statistic is referred to as the *shot noise*. It is inherent in photon detection, in electric current (because of charge discreteness), and in many other experimental processes. Since the noise level scales as the square root of the signal, the signal-to-noise ratio will be equal to $N^{1/2}$ as well. One can see that if shot noise dominates, the signal-to-noise ratio will be higher for higher numbers of photons detected. The shot noise is a common phenomenon in light detection and is considered as a confirmation of the discrete structure of electromagnetic radiation.

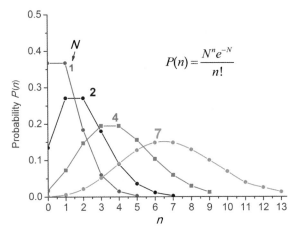

$$P(n) = \frac{N^n e^{-N}}{n!}$$

Figure 12.17 Poisson distribution function gives the probability of n events during a time interval in a case when the mean rate of events during the same time interval is N. It is valid for integer n only (dots); the lines are the guides for the eye. For higher N the Poisson distribution merges with the normal (Gaussian) distribution.

12.5.2 Photon antibunching

A single given quantum emitter (an atom, a molecule, a quantum dot, a defect in a semi-conductor or dielectric) features a *sub-Poisson* statistic. Photon emission events by a single emitter cannot occur more often than once per lifetime period. Therefore, an excited quantum emitter cannot emit two photons simultaneously. This phenomenon is termed *photon antibunching*. It has been experimentally proved for different single emitters. Figure 12.18 shows a representative example for a single CdSe/ZnS colloidal quantum dot luminescence.

A 488 nm Ar laser radiation was focused on the sample by an oil-immersion objective in an inverted confocal microscope. The emitted photons collected by the same objective were filtered from the scattered excitation light by a band pass filter (or dichroic mirror) and sent through a 50/50 beam splitter onto two (start and stop) single photon counting detectors

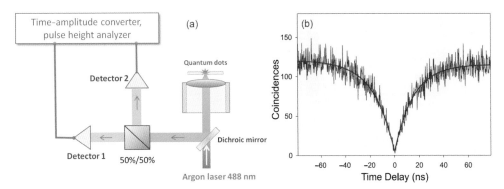

Figure 12.18 Photon antibunching experiment with a single CdSe/ZnS core–shell colloidal quantum dot. (a) experimental setup; (b) coincidence of the two detectors' counts versus delay time reported by Lounis et al. (2000). Reprinted from Lounis et al. (2000) with permission from Elsevier.

(avalanche photodiodes). A short pass filter was inserted before one of the photodiodes to suppress cross-talk between the two detectors. Single quantum dot emission was detected with a signal-to-background ratio greater than 500 (the maximum signal rate was 500 kHz per detector, while the background signal is less than 1 kHz). The signals from the detectors were sent to a time-to-amplitude converter (TAC) followed by a pulse-height analyzer to create a histogram of the delays between consecutive photons. To maximize the signal-to-noise ratio, the excitation beam was switched off by an acousto-optic modulator during the TAC dead time of approximately 7 µs. The histogram in Figure 12.18(b) was recorded with a 200 ns TAC time window and a bin width of 0.2 ns. The dip in coincidences around zero delay is an unambiguous signature of antibunching.

12.5.3 Single photon emitters and squeezed light

There is a challenge for modern quantum optics to design an emitter that could provide the desirable number of photons on demand. Then the so-called radiation *Fock states* (after the Russian physicist V. A. Fock) or *number-states* will be created. A "number-state" is a quantum state in which the number of photons, n, is a precisely fixed integer and the radiation is viewed as a quantum harmonic oscillator with the spectrum

$$E = \hbar\omega(n + \tfrac{1}{2}). \tag{12.5}$$

Number-states of light are very hard to obtain in practice except for the trivial vacuum-state corresponding to $n = 0$ over all possible modes. A laser generates monochromatic radiation that seems extremely regular in amplitude, but this is only because it contains so many photons that their number fluctuations become negligible on a macroscopic scale. Laser radiation is a superposition of many number-states with different n. Restricting fluctuations in the number of photons seems easiest for states with $n = 1$ (single-photon states). A microscopic emitter of light, e.g., an atom, a molecule, or a quantum dot, can serve as a single-photon

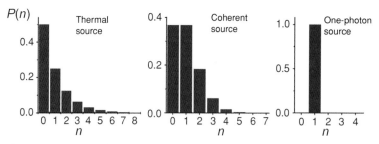

Figure 12.19 Probability distribution of the number of photons for three sources with an average photon number $\langle n \rangle = 1$. The thermal source obeys the Bose–Einstein statistic of black body radiation. The coherent light source presents a Poisson distribution (compare to Figure 12.17), narrower than that of thermal light, but still with a strong number fluctuation, called photon noise. An ideal squeezed-light source delivers a number-state with $n = 1$. A single-photon source can match this distribution by delivering single photons at regular time intervals. Adapted from Lounis and Orrit (2005). © IOP Publishing. Reproduced with permission. All rights reserved.

source if it is coupled to a resonant high Q-factor cavity, which can fulfill several functions. First, it can enhance the spontaneous emission rate and thereby the rate of photon generation (Purcell effect, local density of states effect; see Sections 5.3 and 5.7). Second, it emits radiation into a well-defined mode, thus enabling efficient light harvesting. Third, a high Q-factor cavity will make the emission spectrum narrower (see, e.g., Figure 5.12).

Consider fluctuations of radiation states with the average photon number $N = 1$ provided by different sources (Figure 12.19). For thermal radiation, the photon number distribution takes the Bose form,

$$P(n) = \frac{\langle N \rangle^n}{(1 + \langle N \rangle)^{n+1}}, \tag{12.6}$$

where $\langle N \rangle$ is the mean number of photons in the mode. For thermal radiation, the state with zero photons ($n = 0$) always has the largest probability of occupation. This distribution is far from that of a desired source of single photons, which should feature a sharp maximum at $n = 1$. The number of photons in a coherent state is a variable that fluctuates according to the Poisson distribution (12.4). This statistic is very different from those of thermal radiation. The maximum probability is to find N photons in the mode. The resulting noise is the familiar *shot noise*; it is an absolute minimum for the noise of a macroscopic laser. The probability distribution is still rather different from the desired one.

Narrower distribution can be achieved by means of the so-called *squeezed states* of light. In accordance with the complementarity principle, fluctuations of conjugate variables meet the Heisenberg uncertainty relations. Principally, the states become possible where fluctuations in one of the variables can be reduced (as compared with those of a coherent state) at the expense of increased fluctuations of the conjugate variable. This is referred to as a squeezed state. A squeezed state with reduced amplitude fluctuations (i.e., photon number fluctuations) will therefore exhibit enhanced phase noise. An ideal amplitude–squeezed-light source would deliver a regular stream of photons at regular time periods. The fluctuations in the numbers of photons emitted by a squeezed source are weaker than those of a coherent state (Figure 12.19, the right-hand panel). Single photon sources are important for applications in quantum computing, quantum cryptography, ultimate experiments in quantum science testing the basic and the questionable notions, and challenging experiments like quantum teleportation (Lodahl et al. 2015).

12.5.4 Strong light–matter coupling in a microcavity and in a photonic crystal

The Purcell effect introduced in Sections 5.3–5.5 deals with the probability of an excited quantum system (atom, molecule, quantum dot) emitting photons spontaneously. In microdisks, microspheres, and PC cavities very high Q-factors (up to 10^6) can be obtained (see Chapter 10). In these cases, an excited quantum system plus electromagnetic radiation should be considered as a single system characterized by means of quantum electrodynamics. It is possible to say that a quantum system is disturbed by photons it tends to emit, though this is also a simplified interpretation. Joint states of matter and radiation in high Q-cavities constitute the subject of *cavity quantum electrodynamics*. The notions of spontaneous emission rate and excited state lifetime vanishes; these terms correspond to the so-called perturbational approach in which matter states and field states are described independently. The conception of photon density of states effect on excited state lifetime cannot be used either. The reader is referred to special volumes for detail (Yamamoto et al. 2000; Gaponenko 2010, chapter 15).

In PCs with full omnidirectional band gap, no propagating mode exists within the gap. One can expect the photon DOS to go to zero, but this is inaccurate and incorrect treatment since the photon DOS function becomes discontinuous and cannot be defined within the gap interval. In this case the joint atom–field states (molecules and quantum dots are considered alike) develop, featuring a long-lived oscillating population of an excited state of an atom in a PC. The situation is referred to as a "frozen excited state." These states were intuitively foreseen in the 1970s by V. P. Bykov, a Russian physicist from the Lebedev Physical Institute at Moscow (Bykov 1972; 1993), and this idea has become the driving concept in the elaboration of PC trends in modern optics. The consistent quantum electro-dynamics-based consideration in terms of atom–field states developed later (Mogilevtsev et al. 2007; Mogilevtsev and Kilin 2007 and references therein; Gaponenko 2010, chapter 15).

12.6 METAMATERIALS

The notion of metamaterials was coined to identify the complicated, composite, structured materials that acquire unusual effective medium parameters (thus justifying the notion of materials) defined by the structure on a subwavelength scale whose properties have no natural analog. For the optical range, these materials should definitely have special substructure on a nanometer scale. Under certain conditions, spatially organized materials on a subwavelength scale feature negative refractive index, a property that cannot be observed for any natural material but that is not totally forbidden by the physical laws describing electromagnetic radiation.

12.6.1 Optics with magnetic materials

In the optical range, magnetic permeability of materials typically equals unity and for this reason the "μ" factor if present in formulas is not recognized at all. This is a reasonable consequence of the properties of all existing *natural* materials. However, composite, nano-structured materials may have $\mu \neq 1$. It is instructive to consider what will happen with basic formulas where μ enters but has not been analyzed because of the "$\mu = 1$" convention.

Whenever $\mu \neq 1$ holds, the notion of *impedance* $Z(\omega)$ as a property of a medium becomes essential. It reads

$$Z(\omega) = \sqrt{\frac{\mu_0 \mu(\omega)}{\varepsilon_0 \varepsilon(\omega)}} \equiv Z_0 \sqrt{\frac{\mu(\omega)}{\varepsilon(\omega)}} \tag{12.7}$$

BOX 12.3 INGENUOUS IDEAS IN NANOPHOTONICS: PHOTONIC CRYSTALS AND METAMATERIALS

Ingenuous ideas in science often come to brilliant minds well before they can be really accomplished. But the whole history of science proves that high technologies are born owing to high science. In 1967 the Russian physicist V. G. Veselago predicted the possibility of strange matter with a negative refractive index. These are now called "metamaterials," and owing to many talented scientists have become a mature field of modern optics. J. B. Pendry, N. I. Zheludev, and V. M. Shalaev have made outstanding contributions to the field.

V. G. Veselago

J. B. Pendry

N. I. Zheludev

V. M. Shalaev

In 1972 another Russian physicist, V. P. Bykov, advanced the idea of a periodic dielectric medium in which no propagating electromagnetic mode exists. He understood this medium should inhibit spontaneous decay of excited atoms and molecules.

In 1987, E. Yablonovich proposed using this effect in lasers. This unusual medium gained the name "photonic crystal" and has become one of the driving concepts in photonic circuitry and lasers.

and has a very important meaning in electrodynamics with $\mu \neq 1$, which is the typical case in radiophysics. In optics ($\mu = 1$), reflection and transmission of electromagnetic waves at an interface of two media is governed by the relative refractive index n. In radiophysics $\mu \neq 1$ results in transmission and reflection at an interface governed by impedances of the interfacing media (Figure 12.20). For example, the known formulas for light intensity reflection R and transmission T coefficients at normal incidence,

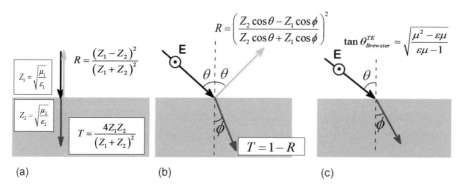

Figure 12.20 A few examples of electromagnetic wave reflection/transmission at an interface of two media with different magnetic permeabilities. (a) Normal incidence transmission and reflection are governed by impedances (mis)matching. (b) Modification of Fresnel formula in terms of impedances. (c) Existence of the "magnetic" Brewster angle for transverse electric wave under condition $\theta + \phi = \pi / 2$.

$$R = \frac{(n-1)^2}{(n+1)^2} \ , \quad T = \frac{4n}{(n+1)^2} \ , \quad n = \frac{n_2}{n_1} = \frac{\sqrt{\varepsilon_2}}{\sqrt{\varepsilon_1}} \tag{12.8}$$

are to be modified using

$$\frac{n_2}{n_1} = \frac{\sqrt{\varepsilon_2}}{\sqrt{\varepsilon_1}} \ \rightarrow \ \frac{Z_1}{Z_2} = \frac{\sqrt{\varepsilon_2 / \mu_2}}{\sqrt{\varepsilon_1 / \mu_1}} \tag{12.9}$$

to arrive at

$$R = \frac{(Z_1 - Z_2)^2}{(Z_1 + Z_2)^2} \ , \quad T = 1 - R = \frac{4Z_1 Z_2}{(Z_1 + Z_2)^2} . \tag{12.10}$$

Reflection at the interface vanishes when $Z_1 = Z_2$ and develops otherwise. Reflectionless propagation occurs under conditions of impedance matching rather than refractive indices matching at the interface. Therefore, reflection vanishes when $\varepsilon_1 / \mu_1 = \varepsilon_2 / \mu_2$. Considering a vacuum (or air, $\varepsilon_1 = \mu_1 = 1$) interface with a material with finite ε and μ, one can see matching occurs for every material with $\varepsilon = \mu$.

For oblique incidence, Figure 12.20(b) shows that for a transverse electric wave (*s*-polarization) the reflection angle equals the incidence angle as usual, and Snell's law $n_2 \sin \theta = n_1 \sin \phi$, $n_i = \sqrt{\varepsilon_i \mu_i}$ holds. The intensity reflection and transmission coefficients read

$$R = \left(\frac{Z_2 \cos \theta - Z_1 \cos \phi}{Z_2 \cos \theta + Z_1 \cos \phi} \right)^2, \quad T = 1 - R \tag{12.11}$$

differing from the familiar Fresnel formula by using impedances instead of refractive indices.

For a transverse magnetic wave (*p*-polarization) in the traditional "$\mu = 1$" optics, reflection vanishes ($R = 0$) at the *Brewster angle* defined by the relation

$$\tan\theta^{\text{TM}}_{\text{Brewster}} = \frac{n_2}{n_1} = \sqrt{\varepsilon_2 / \varepsilon_1} \qquad (12.12)$$

corresponding to the case $\theta + \phi = \pi/2$. Electric dipoles do not emit along their oscillation direction. What happens if $\mu_2 \neq 1$? In this case, Eq. (12.12) modifies

$$\tan\theta^{\text{TM}}_{\text{Brewster}} = \sqrt{\frac{\left(\varepsilon_2/\varepsilon_1\right)^2 - \left(\varepsilon_2/\varepsilon_1\right)\left(\mu_2/\mu_1\right)}{\left(\varepsilon_2/\varepsilon_1\right)\left(\mu_2/\mu_1\right) - 1}}. \qquad (12.13)$$

Furthermore, coming back to the transverse electric wave (*s*-polarization), we now arrive at the notion of the "magnetic" Brewster angle in the case $\theta + \phi = \pi/2$. It is defined by the relation

$$\tan\theta^{\text{TE}}_{\text{Brewster}} = \sqrt{\frac{\left(\mu_2/\mu_1\right)^2 - \left(\varepsilon_2/\varepsilon_1\right)\left(\mu_2/\mu_1\right)}{\left(\varepsilon_2/\varepsilon_1\right)\left(\mu_2/\mu_1\right) - 1}}. \qquad (12.14)$$

For the particular case of $\varepsilon_1 = \varepsilon_2$ it reduces to

$$\tan\theta^{\text{TE}}_{\text{Brewster}} = \sqrt{\mu_2/\mu_1} \equiv \frac{n_2}{n_1}, \qquad (12.15)$$

supporting the notation of the "magnetic" Brewster angle. Reflection vanishes because magnetic dipoles do not emit along their oscillation direction. These selected examples clearly show a wealth of effects and phenomena in electrodynamics for media with $\mu \neq 1$.

12.6.2 "Left-handed" materials

The whole multitude of material dielectric and magnetic properties can be placed in an ε–μ plane (Figure 12.21) as was suggested by V. G. Veselago (1968). Segment I (positive ε, positive μ) corresponds to traditional dielectric and magnetic materials. As we mentioned, in the optical range $\mu = 1$ typically holds. Furthermore, $\varepsilon > 1$ holds for dielectrics and semiconductors. This is indicated by a red line in the figure. Segment II corresponds to plasma (electrons in metal, ion plasma, etc.). The plasma area extends to $\varepsilon = 1$. This is shown by the gray square. These media are well identified, understood, and attainable both as natural and manmade materials. What about the remaining two segments forming a semiplane with negative magnetic permeability? Such materials are not identified in nature but their existence is not forbidden in principle. Segment IV, with positive dielectric permittivity but negative magnetic permeability, can be assigned, by analogy with segment II, to the hypothetical "magnetic" plasma. Segment III implies simultaneously negative dielectric permittivity and magnetic permeability. Such a combination has not been known to date, but since there is no basic restriction for such a combination, V. G. Veselago suggested examining what will happen if such materials are found or developed.

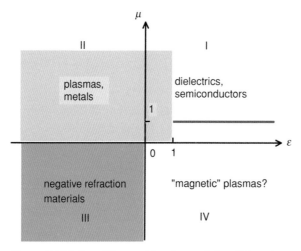

Figure 12.21 Possible combinations of the dielectric permittivity and the magnetic permeability of materials.

For $\varepsilon < 0, \mu < 0$ the refraction index becomes negative. In the relation

$$n = \pm\sqrt{\varepsilon\mu}, \qquad (12.16)$$

with typically positive values of permittivity and permeability, refractive index is positive and takes the meaning of the factor defining speed of wave propagation with respect to a vacuum. In the case of both negative ε and μ, formally in terms of purely mathematical treatment, the positive square root of the positive $\varepsilon\mu$ product seems reasonable as well. However, this is not the case. The surprising consequence of both negative permittivity and permeability is the inverse direction of the wave vector $\mathbf{k}$ with respect to the vectorial product of $\mathbf{E}$ and $\mathbf{H}$, $[\mathbf{E}\times\mathbf{H}]$. For a plane monochromatic wave, Maxwell equations reduce to the couple of equations (SI units)

$$[\mathbf{k}\times\mathbf{E}] = \omega\mu_0\mu\mathbf{H}, \quad [\mathbf{k}\times\mathbf{H}] = -\omega\varepsilon_0\varepsilon\mathbf{E}, \qquad (12.17)$$

whence for both positive ε and μ the three vectors $\mathbf{E}, \mathbf{H}, \mathbf{k}$ form the "right-hand" set of vectors, i.e., $\mathbf{k}$ direction coincides with the $[\mathbf{E}\times\mathbf{H}]$ defined direction. For both negative ε and μ, these three vectors form a "left-hand" set. The $\mathbf{k}$ vector direction is now opposite to that defined by $[\mathbf{E}\times\mathbf{H}]$. On the other hand, the Poynting vector, which defines the energy flux density transfer, reads

$$\mathbf{S} = [\mathbf{E}\times\mathbf{H}] \qquad (12.18)$$

and appears to have the opposite direction with respect to **k**. Therefore, when an electro-magnetic wave propagates in a "left-handed" material, the "left-hand" orientation of the **E, H, k** vector set gives rise to counterpropagating phase and group velocities.

To account for the opposite directions of phase and group velocity as well as "left-handness" of a material with negative ε and μ, the minus sign in the square root of Eq. (12.16) should be chosen. Therefore, such materials are often referred to as "negative refraction materials." V. G. Veselago proposed introducing "handedness" p as a material property with $p = +1$ for "right-handed" and $p = -1$ for "left-handed" ones. With this notation, the generalized Snell's law takes the form

$$\frac{\sin\theta}{\sin\phi} = \frac{n_2}{n_1} = \frac{p_2}{p_1}\left|\sqrt{\frac{\varepsilon_2\mu_2}{\varepsilon_1\mu_1}}\right| \tag{12.19}$$

with notations for angles θ, ϕ shown in Figure 12.20(b). The reflection angle remains equal to the incidence angle independently of material "handedness." One can see that when light is coming from a right-handed material to a left-handed one, the refraction angle changes sign as compared to traditional optics.

Consider refraction at the border of a vacuum (or air), $\varepsilon = \mu = 1$ and a left-handed medium (Figure 12.22(a)) with $\varepsilon = \mu = -1$. In this case, reflected intensity is zero because impedances are matched and only the refracted wave presents. One can see that the **E** vector, **H** vector, and wave vector change symmetrically with respect to the interface bor-der. Refraction angle equals the angle of incidence but has the opposite sign. The phase velocity direction defined by the **k** vector has the opposite direction in the negative refrac-tion medium with respect to energy flow defined by the Poynting vector. Figure 12.22(b) shows the results of numerical modeling for this case for a Gaussian beam. Notably, a plane-parallel slab of a negative refraction material can serve as a lens at normal incidence (Figure 12.22(c)). Many further instructive examples of unusual properties of negative refraction materials (not necessarily from optics) can be found in special papers and books (Veselago and Narimanov 2006; Cai and Shalaev 2010; Zheludev and Kivshar 2012).

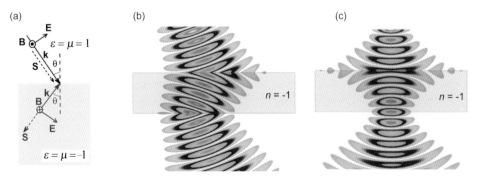

Figure 12.22 Propagation of electromagnetic waves impinging a slab of negative refractive index material from air. (a) Negative refraction at a matched interface; (b) calculated Gaussian beam propagation for oblique incidence; (c) calculated Gaussian beam propagation at normal incidence with focusing effect. Courtesy of R. Ziolkowski.

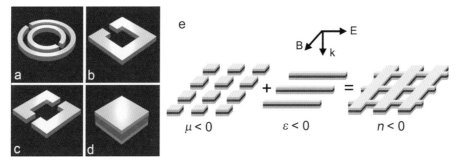

Figure 12.23 Examples of elementary building units to get magnetic permeability in a regular array (a–d) and design of a negative refraction material by means of combined arrays with negative permeability and negative permittivity. Reprinted from Busch et al. (2007), with permission from Elsevier.

There is no natural material that features negative magnetic permeability in the optical range. J. Pendry (2000) proposed a design of a metamaterial with negative magnetic permeability. At a subwavelength scale, such a material should resemble a broken conductive ring (single or concentric coupled) or even a couple of plane metal plates separated by a dielectric (Figure 12.23(a–d)). Such elements represent an LC circuit, the inductance coming from a ring and capacitance coming from the air spacing. Negative permeability develops in a regular array of such LC circuits. When such a material is combined with another periodic structure (Figure 12.23(e)), the overall response of the whole system features negative refraction. This idea has been followed by many research groups, mainly in microwaves, where the length scale allows for complicated subwavelength structures to be developed. For the optical region, fine techniques on the nanometer scale using high-resolution lithography, etching, vacuum deposition, etc. are necessary. Such an approach has been performed by a number of groups.

12.7 BIOINSPIRED NANOPHOTONICS

The wildlife of the world offers many prototype solutions suitable for various photonic components. There are several trends in bioinspired nanophotonics:

1. mimicking periodic biostructures (photonic crystals) to develop desirable optical properties and templating with periodic biostructures (photonic crystals);

2. spatial arrangement of nanostructures using biocomponents for chemical linking;

3. using biomaterials as photonic components, especially for bio-applications.

In what follows a few representative examples are provided for these approaches.

12.7.1 Biophotonic crystals as design prototypes and templates

There are numerous cases of periodic structures in nature enabling iridescent colors owing to interference in periodic structures. Male peacocks' feathers, wings of many butterflies, and the scales of several fish represent colors from interference in multilayer structures. Eyes of many insects comprise two-dimensional periodicity. Many beetles have shining colors from three-dimensional periodicity of scales. These cases have become common knowledge and were identified, described in numerous publications, and summarized in reviews (Srinivasarao 1999; Parker 2000; Starkey and Vukusic 2013). Among the biophotonic crystal structures, a few particular cases are noteworthy.

Vukusic and Hooper (2005) described directionally controlled fluorescence emission in butterflies, which is an important approach to improving light extraction from LEDs (Zhmakin 2011; see Chapter 8 for representative examples). Another example is the dynamically tuned periodic two-dimensional structure of chameleons' skin, discussed in detail in Section 4.3 (Figure 4.23). Another interesting example is multiple layers consisting of aligned fibers present in the human cornea. Here, periodic structuring enables both high optical transmission and openness for chemical exchange processes, e.g., tear injection on blinking.

Mimicking of a moth eye was among the earlier cases of photonic crystal bioinspired design. It was described in detail in Chapter 11 (Section 11.3) as a prototype to efficient broad-band antireflective coating.

Multilayer structures present in certain tropical plant seeds have inspired designers to manufacture stretchable iridescent fibers. Another example is a synthetic bioinspired elastic textile. This material (Morphotex®) consists of alternating polyester/nylon layers featuring colors without pigment. Since absorption is not present, photoinduced bleaching is avoided to give highly durable colors (Kolle et al. 2013).

Morpho butterflies are known to exhibit efficient reflectance visible at a hundred meters, with blue color independent of angle of observation. Samsung researchers and university colleagues have managed to reproduce a similar angle-independent high reflectance using specially tailored layer-by-layer deposition of dielectric films (Chung et al. 2012).

Mimicking certain natural nanostructures can help to avoid overheating in hot climates (Smith et al. 2016), e.g., desert silver ants owing to their nanostructures have surfaces that reflect light and thus they can look for food in the hot daytime to avoid nighttime predators. In hot climates, many plants evolved in a way that allows their leaves to efficiently scatter IR radiation to prevent excessive heating of the water contained therein.

Biotemplating has been suggested to develop morphology-controllable materials with structural specificity that either cannot be obtained otherwise or that will need unreasonable technological efforts (Lu et al. 2016). This can be implemented either by direct replication (using a biosystem to make replicas with chemical reactions or physical processes) or by exploiting biophotonic crystal matrices as scaffolds for spatial arrangement of nanocomponents as dots or rods. Sol-gel techniques are actively involved in replication procedures. Stimuli-responsive structures can be used for sensing applications – e.g., the humidity-induced color change observed in *Dynastes hercules* beetles has been outlined as a possible route to humidity sensors; pH sensing with biotemplates is considered by many authors as well.

12.7.2 Spatial arrangement of nanostructures using biocomponents for chemical linking

Chemical linking of specially developed semiconductor colloidal quantum dots and metal nanoparticles can be used to get their spatial arrangement based on DNA fragment chemical recognition (specific binding). This technique is often referred to as the puzzle approach or DNA origami. The general idea of this approach is sketched in Figure 12.24.

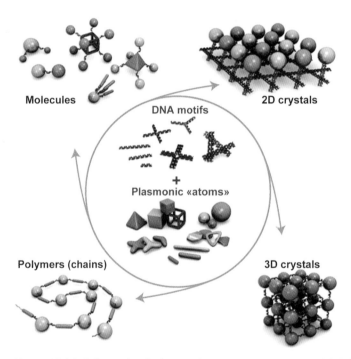

Figure 12.24 Schematic of plasmonic nanostructures assembled from libraries of plasmonic atoms with various DNA motifs. A vast library of plasmonic atoms can be synthesized using wet-chemistry approaches; various DNA motifs can be created using DNA nanotechnology; the plasmonic atoms and DNA can then be used to rationally design and synthesize a range of plasmonic nanostructures. Reprinted by permission from Macmillan Publishers Ltd.; Tan et al. (2011).

12.7.3 Using biomaterials as photonic components

Many biomolecules fluoresce and a few of them have quantum yield, enabling optical gain at pulse excitation. Additionally, biopolymers with different refractive indices in a thin-film form can serve as building blocks for cavity mirrors. Thus, fully biocompatible and human-safe microlasers with optical pumping can be developed. An example is given in Figure 12.25. For the gain medium, flavin mononucleotide (FMN), a biomolecule produced from vitamin B2, in glycerol-mixed microspheres was used as a gain medium. Vitamin microspheres with diameters of 10–40 μm were formed by spraying *in situ* and encapsulated in patterned super-hydrophobic polymer films. The spheres support lasing at optical pump energies as low as 15 nJ per pulse with dielectric mirrors. FMN serves as coenzyme in a series of oxidation–reduction catalysts and is found in many types of human tissue, including heart, liver, and kidney tissue. The distance between the mirrors was adjusted to ≈23 μm using microbeads. After optical gain was established and lasing

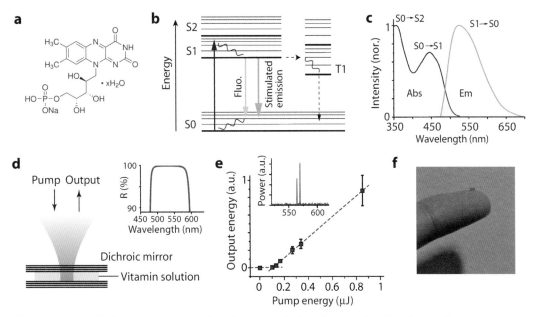

Figure 12.25 All-biomaterial laser using vitamin and biopolymers: (a) The chemical structure of flavin mononucleotide, riboflavin 5′-monophosphate sodium salt hydrate (also known as FMN-Na). (b) Energy-level diagram transitions corresponding to optical excitation (S0 → S1), spontaneous fluorescence emission and stimulated emission (S1 → S0). (c) Measured absorption and emission spectra. (d) Schematic of the vitamin solution laser. Inset, the measured reflectivity profile of the cavity mirror. (e) Laser output energy as a function of pump energy (per pulse); the inset shows a normalized laser spectrum at a pump energy of 1 μJ; (f) A miniature biomaterial gain chip on a fingertip. Reprinted from Nizamoglu et al. (2013) with permission from John Wiley and Sons. © 2013 WILEY-VCH Verlag GmbH & Co. KGaA, Weinheim.

obtained with dielectric mirrors, a fully biocompatible version was designed using aqueous microdroplets with whispering gallery modes on a hydrophobic biopolymer film. Such a laser can serve as a basic component for *in vivo* sensing when combined with biosensor molecules with a biopolymer fiber to deliver optical pump and to transport the optical signal from a biosensing molecule to a detector.

Conclusion

The brief overview offered in this chapter on a variety of trends in applied research related to nanophotonics clearly shows that nanophotonics is an open and extensively expanding field that promises many new technological solutions, including colloidal platform, silicon-based photonic components, biocompatible devices, various sensing systems, and integration of dielectric, semiconductor, and metal nanostructures in a single device for optimal performance. Nanophotonics, when being biocompatible or even sometimes bio-inspired, promises *in vivo* interfacing of optoelectronic components (lasers, fluorophores) with the human body, flexible and cheap light-emitting devices, and higher integration of optical components with electronic counterparts.

Further reading

Berini, P. (2014). Surface plasmon photodetectors and their applications. *Laser Photonics Rev*, **8**, 197–220.

Cornet, C., Léger, Y., and Robert, C. (2016). *Integrated Lasers on Silicon*. Elsevier.

Duan, G. H., Jany, C., Le Liepvre, A., et al. (2014). Hybrid III–V on silicon lasers for photonic integrated circuits on silicon. *IEEE J Select Topics Quantum Electronics*, **20**, 158–170.

Duarte, F. J. (2014). *Quantum Optics for Engineers*. CRC.

Enoch, S., and Bonod, N. (eds.) (2012). *Plasmonics: From Basics to Advanced Topics*, vol. **167**. Springer.

Gaponenko, S. V., Gaiduk, A. A., Kulakovich, O. S., et al. (2001). Raman scattering enhancement using crystallographic surface of a colloidal crystal. *JETP Lett*, **74**, 309–313.

Gerasimos, K., and Sargent E. H. (eds.) (2013). *Colloidal Quantum Dot Optoelectronics and Photovoltaics*. Cambridge University Press.

Khriachtchev, L. (2009). *Silicon Nanophotonics: Basic Principles, Current Status and Perspectives*. Pan Stanford Publishing.

Kira, M., and Koch, S. W. (2011). *Semiconductor Quantum Optics*. Cambridge University Press.

Kneipp, J., Kneipp, H., and Kneipp, K. (2008). SERS: a single-molecule and nanoscale tool for bio-analytics. *Chem Soc Rev*, **37**, 1052–1060.

Kovalenko, M. V., Manna, L., Cabot, A., et al. (2015). Prospects of nanoscience with nanocrystals. *ACS Nano*, **9**, 1012–1057.

Le Ru, E., and Etchegoin, P. (2008). *Principles of Surface-Enhanced Raman Spectroscopy and Related Plasmonic Effects*. Elsevier.

Li, Z. Y. (2015). Optics and photonics at nanoscale: principles and perspectives. *Europhysics Lett*, **110**, 14001.

Liao, H., Nehl, C. L., and Hafner, J. H. (2006). Biomedical applications of plasmon resonant metal nanoparticles. *Nanomedicine*, **1**, 201–208.

Liu, A. Y., Zhang, C., Norman, J., et al. (2014). High performance continuous wave 1.3 μm quantum dot lasers on silicon. *Appl Phys Lett*, **104**, 041104.

Lu, T., Peng, W., Zhu, S., and Zhang, D. (2016). Bio-inspired fabrication of stimuli-responsive photonic crystals with hierarchical structures and their applications. *Nanotechnology*, **27**, 122001.

Mayer, K. M., and Hafner, J. H. (2011). Localized surface plasmon resonance sensors. *Chem Rev*, **111**, 3828–3857.

Miller, D. A. B. (2017). Attojoule optoelectronics for low-energy information processing and communications: a tutorial review. *J Lightwave Technol*, **35**, 346–396.

Priolo, F., Gregorkiewicz, T., Galli, M., and Krauss, T. F. (2014). Silicon nanostructures for photonics and photovoltaics. *Nature Nanotechn*, **9**, 19–32.

Resch-Genger, U., Grabolle, M., Cavaliere-Jaricot, S., Nitschke, R., and Nann, T. (2008). Quantum dots versus organic dyes as fluorescent labels. *Nat Methods*, **5**, 763–775.

Sarychev, A. K., and Shalaev, V. M. (2007). *Electrodynamics of Metamaterials*. World Scientific.

Shamirian, A., Ghai, A., and Snee, P. T. (2015). QD-based FRET probes at a glance. *Sensors*, **15**, 13028–13051.

Starkey, T., and Vukusic, P. (2013). Light manipulation principles in biological photonic systems. *Nanophotonics*, **2**, 289–307.

Veselago, V. G. (2009). Energy, linear momentum and mass transfer by an electromagnetic wave in a negative-refraction medium. *Physics-Uspekhi*, **52**, 649–654.

Yamamoto, Y., Tassone, F., and Cao, H. (2000). *Semiconductor Cavity Quantum Electrodynamics*. Springer.

Zenkevich, E., and von Borczyskowski, C. (eds.) (2016). *Self-Assembled Organic–Inorganic Nanostructures: Optics and Dynamics*. Pan Stanford Publishing.

References

Absil, P. P., Verheyen, P., De Heyn, P., et al. (2015). Silicon photonics integrated circuits: a manufacturing platform for high density, low power optical I/O's. *Optics Expr*, **23**, 9369–9378.

Achtstein, A. W., Schliwa, A., Prudnikau, A., et al. (2012). Electronic structure and exciton–phonon interaction in two-dimensional colloidal CdSe nanosheets. *Nano Lett*, **12**, 3151–3157.

Achtstein, A. W., Prudnikau, A. V., Ermolenko, M. V., et al. (2014). Electroabsorption by 0D, 1D, and 2D nanocrystals: a comparative study of CdSe colloidal quantum dots, nanorods, and nanoplatelets. *ACS Nano*, **8**, 7678–7686.

Artemyev, M. V., Bibik, A. I., Gurinovich, L. I., Gaponenko, S. V., and Woggon, U. (1999). Evolution from individual to collective electron states in a dense quantum dot ensemble. *Phys Rev B*, **60**, 1504.

Bae, W. K., Brovelli, S., and Klimov, V. I. (2013). Spectroscopic insights into the performance of quantum dot light-emitting diodes. *MRS Bull*, **38**, 721–730.

Borrelli, N. F., Hall, D. W., Holland, H. J., and Smith, D. W. (1987). Quantum confinement effects of semiconducting microcrystallites in glass. *J Appl Phys*, **61**, 5399–5409.

Bruchez, M., Moronne, M., Gin, P., Weiss, S., and Alivisatos, A. P. (1998). Semiconductor nanocrystals as fluorescent biological labels. *Science*, **281**, 2013–2016.

Brus, L. E. (1983). A simple model for the ionization potential, electron affinity, and aqueous redox potentials of small semiconductor crystallites. *J Chem Physics*, **79**, 5566–5571.

Brus, L. E. (1984). Electron–electron and electron–hole interactions in small semiconductor crystallites: the size dependence of the lowest excited electronic state. *J Chem Phys*, **80**, 4403–4409.

Busch, K., von Freymann, G., Linden, S., et al. (2007). Periodic nanostructures for photonics. *Phys Rep*, **444**, 101–202.

Bykov, V. P. (1972). Spontaneous emission in a periodic structure. *Soviet Physics-JETP*, **35** 269–273.

Bykov, V. P. (1993). *Radiation of Atoms in a Resonant Environment*. World Scientific.

Cai, W., and Shalaev, V. M. (2010). *Optical Metamaterials*, vol. **10**. Springer.

Chan, W. C. W., and Nie, S. (1998). Quantum dot bioconjugates for ultrasensitive nonisotopic detection. *Science*, **281**, 2016–2018.

Chen, M., Shao, L., Kershaw, S. V., et al. (2014). Photocurrent enhancement of HgTe quantum dot photodiodes by plasmonic gold nanorod structures. *ACS Nano*, **8**, 8208–8216.

Chung, K., Yu, S., Heo, C.-J., et al. (2012). Flexible, angle-independent, structural color reflectors inspired by *Morpho* butterfly wings. *Adv Mater*, **24**, 2375–2379.

Colvin, V. L., Schlamp, M. C., and Alivisatos, A. P. (1994). Light-emitting diodes made from cadmium selenide nanocrystals and a semiconducting polymer. *Nature*, **357**, 354–357.

Dang, C., Lee, J., Breen, C., et al. (2012). Red, green and blue lasing enabled by single-exciton gain in colloidal quantum dot films. *Nat Nanotechnol*, **7**(5), 335–339.

Demir, H. V., Nizamoglu, S., Erdem, T., et al. (2011). Quantum dot integrated LEDs using photonic and excitonic color conversion. *Nano Today*, **6**, 632–647.

Deng, Z., Jeong, K. S., and Guyot-Sionnes, P. (2014). Colloidal quantum dots intraband photodetectors. *ACS Nano*, **8**, 11707–11714.

Dennis, A. M., Rhee, W. J., Sotto, D., Dublin, S. N., and Bao, G. (2012). Quantum dot-fluorescent protein FRET probes for sensing intracellular pH. *ACS Nano*, **6**, 2917–2924.

Docampo, P., and Bein, T. (2016). A long-term view on perovskite optoelectronics. *Acc Chem Res*, **49**, 339–346.

Duarte, F. J. (2014). *Quantum Optics for Engineers*. CRC.

Efros, A. L., and Efros, A. L. (1982). Interband absorption of light in a semiconductor sphere. *Soviet Physics Semiconductors-USSR*, **16**, 772–775.

Ekimov, A. I., and Onushchenko, A. A. (1981). Quantum size effect in three-dimensional microscopic semiconductor crystals. *JETP Lett*, **34**, 345–349.

Ekimov, A. I., and Onushchenko, A. A. (1984). Size quantization of the electron energy spectrum in a microscopic semiconductor crystal. *JETP Lett*, **40**, 1136–1139.

Ellingson, R. J., Beard, M. C., Johnson, J. C., et al. (2005). Highly efficient multiple exciton generation in colloidal PbSe and PbS quantum dots. *Nano Lett*, **5**, 865–871.

Erdem, T., and Demir, H. V. (2013). Color science of nanocrystal quantum dots for lighting and displays. *Nanophotonics*, **2**, 57–81.

Gaponenko, N. V. (2001). Sol-gel derived films in mesoporous matrices: porous silicon, anodic alumina and artificial opals. *Synth Met*, **124**, 125–130.

Gaponenko, S. V. (1998). *Optical Properties of Semiconductor Nanocrystals*. Cambridge University Press.

Gaponenko, S. V. (2010). *Introduction to Nanophotonics*. Cambridge University Press.

Gaponenko, S. V., and Guzatov, D. V. (2009). Possible rationale for ultimate enhancement factor in single molecule Raman spectroscopy. *Chem Phys Lett*, **477**, 411–414.

Gaponenko, S. V., Germanenko, I. N., Petrov, E. P., et al. (1994). Time-resolved spectroscopy of visibly emitting porous silicon. *Appl Phys Lett*, **64**, 85–87.

Gaponenko, S. V., Kapitonov, A. M., Bogomolov, V. N., et al. (1998). Electrons and photons in mesoscopic structures: quantum dots in a photonic crystal. *JETP Lett*, **68**, 142–147.

Goldman, E. R., Medintz, I. L., Whitley, J. L., et al. (2005). A hybrid quantum dot–antibody fragment fluorescence resonance energy transfer-based TNT sensor (2005). *J Amer Chem Soc*, **127**, 6744–6751.

Gonzalez-Carrero, S., Galian, R. E., and Pérez-Prieto, J. (2016). Organic–inorganic and all-inorganic lead halide nanoparticles [Invited]. *Opt Expr*, **24**, A285–A301.

Gramotnev, D. K., and Bozhevolnyi, S. I. (2010). Plasmonics beyond the diffraction limit. *Nat Photonics*, **4**, 83–91.

Guha, B., Kyotoku, B. B., and Lipson, M. (2010). CMOS-compatible athermal silicon microring resonators. *Opt Express*, **18**, 3487–3493.

Guzelturk, B., Martinez, P. L. H., Zhang, Q., et al. (2014a). Excitonics of semiconductor quantum dots and wires for lighting and displays. *Laser Photonics Rev*, **8**, 73–93.

Guzelturk, B., Kelestemur, Y., Olutas, M., Delikanli, S., and Demir, H. V. (2014b). Amplified spontaneous emission and lasing in colloidal nanoplatelets. *ACS Nano*, **8**, 6599–6605.

He, L., Özdemir, S. K., Zhu, J., Kim, W., and Yang, L. (2011). Detecting single viruses and nanoparticles using whispering gallery microlasers. *Nat Nanotechnol*, **6**, 428–432.

Jain, J. R., Hryciw, A., Baer, T. M., et al. (2012). A micromachining-based technology for enhancing germanium light emission via tensile strain. *Nat Photonics*, **6**, 398–405.

Jain, P. K., and El-Sayed, M. A. (2007). Surface plasmon resonance sensitivity of metal nanostructures: physical basis and universal scaling in metal nanoshells. *J Phys Chem C*, **111**, 17451–17454.

Kamat, P. V. (2008). Quantum dot solar cells: semiconductor nanocrystals as light harvesters. *J Phys Chem C*, **112**, 18737–18753.

Kira, M., and Koch, S. W. (2011). *Semiconductor Quantum Optics*. Cambridge University Press.

Klimov, V. I., Ivanov, S. A., Nanda, J., et al. (2007). Single-exciton optical gain in semiconductor nanocrystals. *Nature*, **447**, 441–446.

Klyachkovskaya, E., Strekal, N., Motevich, I., et al. (2011). Enhanced Raman scattering of ultramarine on Au-coated Ge/Si-nanostructures. *Plasmonics*, **6**, 413–418.

Kneipp, K., Kneipp, H., and Kneipp, J. (2006). Surface-enhanced Raman scattering in local optical fields of silver and gold nanoaggregates from single-molecule Raman spectroscopy to ultrasensitive probing in live cells. *Acc Chem Res*, **39**, 443–450.

Knight, M. W., Sobhani, H., Nordlander, P., and Halas, N. J. (2011). Photodetection with active optical antennas. *Science*, **332**, 702–704.

Kolle, M., Lethbridge, A., Kreysing, M., et al. (2013). Bio-inspired band-gap tunable elastic optical multilayer fibers. *Adv Mater*, **25**, 2239–2245.

Konstantatos, G., and Sargent, E. H. (2009). Solution-processed quantum dot photodetectors. *Proceedings IEEE*, **97**, 1666–1683.

Kovalenko, M. V., Manna, L., Cabot, A., et al. (2015). Prospects of nanoscience with nanocrystals. *ACS Nano*, **9**, 1012–1057.

Kulakovich, O. S., Shabunya-Klyachkovskaya, E. V., Matsukovich, A. S., et al. (2016). Nanoplasmonic Raman detection of bromate in water. *Opt Expr*, **24**(2), A174–A179.

Kuo, Y. H., Lee, Y. K., Ge, Y., et al. (2005). Strong quantum-confined Stark effect in germanium quantum-well structures on silicon. *Nature*, **437**, 1334–1336.

Lee, J. S., Kovalenko, M. V., Huang, J., Chung, D. S., and Talapin, D. V. (2011). Band-like transport, high electron mobility and high photoconductivity in all-inorganic nanocrystal arrays. *Nat Nanotechnol*, **6**(6), 348–353.

Lesnyak, V., Gaponik, N., and Eychmüller, A. (2013). Colloidal semiconductor nanocrystals: the aqueous approach. *Chem Soc Rev*, **42**, 2905–2929.

Liao, H., Nehl, C. L., and Hafner, J. H. (2006). Biomedical applications of plasmon resonant metal nanoparticles. *Nanomedicine*, **1**, 201–208.

Liu, A. Y., Zhang, C., Norman, J., et al. (2014). High performance continuous wave 1.3 μm quantum dot lasers on silicon. *Appl Phys Lett*, **104**, 041104.

Lodahl, P., Mahmoodian, R., and Stobbe, S. (2015). Interfacing single photons and single quantum dots with photonic nanostructures. *Rev Mod Phys*, **87**, 347–400.

Lounis, B., and Orrit, M. (2005). Single-photon sources. *Rep Progr Physics*, **68**, 1129–1179.

Lounis, B., Bechtel, H. A., Gerion, D., Alivisatos, P., and Moerner, W. E. (2000). Photon antibunching in single CdSe/ZnS quantum dot fluorescence. *Chem Phys Lett*, **329**, 399–404.

Lu, T., Peng, W., Zhu, S., and Zhang, D. (2016). Bio-inspired fabrication of stimuli-responsive photonic crystals with hierarchical structures and their applications. *Nanotechnology*, **27**, 122001.

Mahmoud, K. H., and Zourob, M. (2013). Fe_3O_4/Au nanoparticles/lignin modified microspheres as effectual surface enhanced Raman scattering (SERS) substrates for highly selective and sensitive detection of 2,4,6-trinitrotoluene (TNT). *Analyst*, **138**, 2712–2719.

Mayer, K. M., and Hafner, J. H. (2011). Localized surface plasmon resonance sensors, *Chem Rev*, **111**, 3828–3857.

Medintz, I. L., Uyeda, H. T., Goldman, E. R., and Mattoussi, H. (2005). Quantum dot bioconjugates for imaging, labelling and sensing. *Nat Mater*, **4**, 435–446.

Miller, M. M., and Lazarides, A. A. (2005). Sensitivity of metal nanoparticle surface plasmon resonance to the dielectric environment, *J Phys Chem B*, **109**, 21556–21565.

Mogilevtsev, D. S., and Kilin, S. Ya. (2007). *Quantum Optics Methods of Structured Reservoirs*. Belorusskaya Nauka. In Russian.

Mogilevtsev, D., Moreira, F., Cavalcanti, S. B., and Kilin, S. (2007). Field–emitter bound states in structured thermal reservoirs. *Phys Rev A*, **75**, 043802.

Murray, C. B., Kagan, C. R., and Bawendi, M. G. (1995). Self-organization of CdSe nanocrystallites into three-dimensional quantum dot superlattices. *Science*, **270**, 1335–1338.

Nizamoglu, S., Gather, M. C., and Yun, S. H. (2013). All-biomaterial laser using vitamin and biopolymers. *Adv Mater*, **25**, 5943–5947.

Nudelman, S. (1962). The detectivity of infrared photodetectors, *Appl Opt*, **1**, 627–636.

Page, L. E., Zhang, X., Jawaid, A. M., and Snee, P. T. (2011). Detection of toxic mercury ions using a ratiometric CdSe/ZnS nanocrystal sensor. *Chem Commun*, **47**, 7773–7775.

Palui, G., Aldeek, F., Wang, W., and Mattoussi, H. (2015). Strategies for interfacing inorganic nanocrystals with biological systems based on polymer-coating. *Chem Soc Rev*, **44**, 193–227.

Panarin, A. Yu., Khodasevich, I. A., Gladkova, O. L., and Terekhov, S. N. (2014). Determination of antimony by surface-enhanced Raman spectroscopy. *Appl Spectr*, **68**, 297–306.

Parker, A. R. (2000). 515 million years of structural color. *J Optics A*, **2**, R15–R28.

Pavesi, L., Gaponenko, S., and Dal Negro, L. (eds.) (2012). *Towards the First Silicon Laser*. Springer Science & Business Media.

Pendry, J. B. (2000). Negative refraction makes a perfect lens. *Phys Rev Lett*, **85**, 3966–3969.

Petrov, E. P., Bogomolov, V. N., Kalosha, I. I., and Gaponenko, S. V. (1998). Spontaneous emission of organic molecules in a photonic crystal. *Phys Rev Lett*, **81**, 77–80.

Pompa, P. P., Martiradonna, L., Della Torre, A., et al. (2006). Metal-enhanced fluorescence of colloidal nanocrystals with nanoscale control. *Nat Nanotechnol*, **1**(2), 126–130.

Reckmeier, C. J., Schneider, J., Susha, A. S., and Rogach, A. L. (2016). Luminescent colloidal carbon dots: optical properties and effects of doping [Invited]. *Opt Express*, **24**, A312–A340.

Resch-Genger, U., Grabolle, M., Cavaliere-Jaricot, S., Nitschke, R., and Nann, T. (2008). Quantum dots versus organic dyes as fluorescent labels. *Nat Methods*, **5**, 763–775.

Roelkens, G., Abassi, A., Cardile, P., et al. (2015). III–V-on-silicon photonic devices for optical communication and sensing. *Photonics*, **2**, 969–1004.

Rumyantseva, A., Kostcheev, S., Adam, P. M., et al. (2013). Nonresonant surface-enhanced Raman scattering of ZnO quantum dots with Au and Ag nanoparticles. *ACS Nano*, **7**, 3420–3426.

Sargent, E. H. (2012). Colloidal quantum dot solar cells. *Nat Photonics*, **6**, 133–135.

Schaller, R. D., and Klimov, V. I. (2004). High efficiency carrier multiplication in PbSe nanocrystals: implications for solar energy conversion. *Phys Rev Lett*, **92**, 186601.

Shabunya-Klyachkovskaya, E., Kulakovich, O., Gaponenko, S., Vaschenko, S., and Guzatov, D. (2016). Surface enhanced Raman spectroscopy application for art materials identification. *Eur J Sci Theol*, **12**, 211–220.

Shirasaki, Y., Supran, G. J., Bawendi, M. G., and Bulović, V. (2013). Emergence of colloidal quantum-dot light-emitting technologies. *Nat Photonics*, **7**, 13–23.

Smith, G., Gentle, A., Arnold, M., and Cortie, M. (2016). Nanophotonics-enabled smart windows, buildings and wearables. *Nanophotonics*, **5**, 55–73.

Srinivasarao, M. (1999). Nano-optics in the biological world: beetles, butterflies, birds, and moths *Chem Rev*, **99**, 1935–1961.

Starkey, T., and Vukusic, P. (2013). Light manipulation principles in biological photonic systems. *Nanophotonics*, **2**, 289–307.

Stern, B., Zhu, X., Chen, C. P., et al. (2015). On-chip mode-division multiplexing switch. *Optica*, **2**, 530–535.

Su, L., Zhang, X., Zhang, Y., and Rogach, A. L. (2016). Recent progress in quantum dot based white light-emitting devices. *Top Curr Chem*, **374**, 1–25.

Sun, Y., and Xia, Y. (2002). Increased sensitivity of surface plasmon resonance of gold nanoshells compared to that of gold solid colloids in response to environmental changes. *Anal Chem*, **74** 5297–5305.

Talapin, D. V., Lee, J. S., Kovalenko, M. V., and Shevchenko, E. V. (2009). Prospects of colloidal nanocrystals for electronic and optoelectronic applications. *Chem Rev*, **110**, 389–458.

Tam, F., Moran, C., and Halas, N. (2004). Geometrical parameters controlling sensitivity of nanoshell plasmon resonances to changes in dielectric environment, *J Phys Chem B*, **108**, 17290–17294.

Tan, S. J., Campolongo, M. J., Luo, D., and Cheng, W. (2011). Building plasmonic nanostructures with DNA. *Nat Nanotechnol*, **6**, 268–276.

Tang, Y., Yang, Q., Wu, T., et al. (2014). Fluorescence enhancement of cadmium selenide quantum dots assembled on silver nanoparticles and its application for glucose detection. *Langmuir*, **30**, 6324–6330.

Törmä, P., and Barnes, W. L. (2015). Strong coupling between surface plasmon polaritons and emitters: a review. *Rep Prog Phys*, **78**, 013901.

Turner-Foster, A. C., Foster, M. A., Levy, J. S., et al. (2010). Ultrashort free-carrier lifetime in low-loss silicon nanowaveguides. *Opt Express*, **18**, 3582–3591.

Veselago, V. G. (1968). The electrodynamics of substances with simultaneously negative values of ε and μ. *Soviet Physics Uspekhi*, **10**, 509–514.

Veselago, V. G., and Narimanov, E. E. (2006). The left hand of brightness: past, present and future of negative index materials. *Nat Mater*, **5**, 759–766.

Vogel, N., Weiss, C. K., and Landfester, K. (2012). From soft to hard: the generation of functional and complex colloidal monolayers for nanolithography. *Soft Matter*, **8**, 4044–4061.

Vollmer, F., and Arnold, S. (2008). Whispering-gallery-mode biosensing: label-free detection down to single molecules. *Nat Methods*, **5**, 591–596.

Vukusic, P., and Hooper, I. (2005). Directionally controlled fluorescence emission in butterflies. *Science*, **310**(5751), 1151.

Wang, Y., Li, X., Song, J., et al. (2015). All-inorganic colloidal perovskite quantum dots: a new class of lasing materials with favorable characteristics. *Adv Mater*, **27**, 7101–7108.

Werner, J., Oehme, M., Schmid, M., et al. (2011). Germanium–tin p–i–n photodetectors integrated on silicon grown by molecular beam epitaxy. *Appl Phys Lett*, **98**, 061108.

Woggon, U. (1997). *Optical Properties of Semiconductor Quantum Dots*. Springer.

Wood, V., and Bulović, V. (2010). Colloidal quantum dot light-emitting devices. *Nano Rev*, **1**, 5202–5210.

Yakunin, S., Protesescu, L., Krieg, F., et al. (2015). Low-threshold amplified spontaneous emission and lasing from colloidal nanocrystals of caesium lead halide perovskites. *Nat Commun*, **6**, 8056–8060.

Yamamoto, Y., Tassone, F., and Cao, H. (2000). *Semiconductor Cavity Quantum Electrodynamics*. Springer.

Zenkevich, E. I., Gaponenko, S. V., Sagun, E. I., and Borczyskowski, C. V. (2013). Bioconjugates based on semiconductor quantum dots and porphyrin ligands: properties, exciton relaxation pathways and singlet oxygen generation efficiency for photodynamic therapy applications. *Rev Nanosci Nanotechnol*, **2**, 184–207.

Zheludev, N. I., and Kivshar, Y. S. (2012). From metamaterials to metadevices. *Nat Mater*, **11**, 917–924.

Zhmakin, A. I. (2011). Enhancement of light extraction from light emitting diodes. *Phys Rep*, **498**, 189–241.

Zhu, J., Zhang, F., Li, J., and Zhao, J. (2013). Optimization of the refractive index plasmonic sensing of gold nanorods by non-uniform silver coating, *Sens Actuators, B: Chem*, **183**, 143–150.

Zhukovsky, S. V., Ozel, T., Mutlugun, E., et al. (2014). Hyperbolic metamaterials based on quantum-dot plasmon-resonator nanocomposites. *Opt Express*, **22**(15), 18290–18298.

Zimin, L. G., Gaponenko, S. V., Lebed, V. Y., Malinovskii, I. E., and Germanenko, I. N. (1990). Nonlinear optical absorption of CuCl and CdS_xSe_{1-x} microcrystallites under quantum confinement. *J Luminescence*, **46**, 101–107.

INDEX

SESAM, 195
Shockley–Queisser limit, 365
Si
    band structure, 33
Si electro-optical modulator, 353
SiC
    parameters, 38
silicon photonics, 346
silicon-on-insulator (SOI), 346
skin effect, 124
slow light, 357
Snell's law, 100
solar radiation spectrum, 150
solid solution
    GaN, InN, AlN, 48
spectral hole burning, 85
spherical coordinates, 22
spontaneous transitions, 147
standard luminosity function, 230
stimulated transitions, 147
Stokes shift, 234
stop band, 105
superlattice, 19, 285

thermal radiation, 148
threshold current density, 281
TNT (trinitrotoluene) detector, 396, 397
transmission coefficient, 131
transparency current, 280
tumor markers, 393
tunneling, 19
    in optics, 126
    resonant, 285
    resonant in optics, 21, 105

Urbach–Martienssen rule, 46

valence band, 32
VCSEL, 202
velocity, 92
    phase, 93
vision, 136

wall-plug efficiency (WPE), 248
wave equation, 96
wave function, 11
wave number, 92
wave vector, 92
waveguide, 101
wavelength, 92
wavelength in a medium, 97
whispering gallery modes, 128
Wigner spin, 219
wurtzite, 32

Yagi–Uda antenna, 135

zinc blende, 32
ZnO
    parameters, 38
ZnO nanocrystals
    quantum size effect, 80
ZnS
    parameters, 38
ZnSe
    absorption spectrum, 45
    parameters, 38
ZnSe nanocrystals
    quantum size effect, 80
ZnSe quantum dot
    luminescence, 84
ZnTe
    parameters, 38